MODERN POWER STATION PRACTICE

Third Edition

(in 12 volumes)

Incorporating Modern Power System Practice

Main Editorial Panel

D. J. Littler, BSc, PhD, ARCS, CPhys, FInstP, CEng, FIEE (Chairman)

Professor E. J. Davies, DSc, PhD, CEng, FIEE

H. E. Johnson

F. Kirkby, BSc, CEng, MIMechE, AMIEE

P. B. Myerscough, CEng, FIMechE, FINucE

W. Wright, MSc, ARCST, CEng, FIEE, FIMechE, FInstE, FBIM

Volume Advisory Editor

R. M. Clapp, BSc (Eng), PhD, ACGI

Authors

N. U. Akturk, BSc, MIMechE

R. E. Allan, BSc (Hons), CEng, MIMechE

A. A. Barrett, CEng, MIMechE

W. J. D. Brooks, CEng, MIMechE

J. R. P. Cooper, HND

C. P. Harris, CEng, MIMechE

G. Willis, CEng, MIMechE

Series Production

Managing Editor	P. M. Reynolds
Production Editor	H. E. Johnson
Resources and Co-ordination	T. A. Dolling J. R. Jackson

MODERN POWER STATION PRACTICE

Third Edition

Incorporating Modern Power System Practice

British Electricity International, London

Volume B

Boilers and Ancillary Plant

PERGAMON PRESS

Member of Maxwell Macmillan Pergamon Publishing Corporation

OXFORD · NEW YORK · BEIJING · FRANKFURT
SÃO PAULO · SYDNEY · TOKYO · TORONTO

U.K.	Pergamon Press plc., Headington Hill Hall, Oxford OX3 OBW, England
U.S.A.	Pergamon Press, Inc., Maxwell House, Fairview Park, Elmsford, New York 10523, U.S.A.
PEOPLE'S REPUBLIC OF CHINA	Pergamon Press, Room 4037, Qianmen Hotel, Beijing, People's Republic of China
GERMANY	Pergamon Press GmbH, Hammerweg 6, D-6242 Kronberg, Germany
BRAZIL	Pergamon Editora Ltda, Rua Eça de Queiros, 346, CEP 04011, Paraiso, São Paulo, Brazil
AUSTRALIA	Pergamon Press Australia Pty Ltd., P.O. Box 544, Potts Point, N.S.W. 2011, Australia
JAPAN	Pergamon Press, 5th Floor, Matsuoka Central Building, 1-7-1 Nishishinjuku, Shinjuku-ku, Tokyo 160, Japan
CANADA	Pergamon Press Canada Ltd., Suite No. 271, 253 College Street, Toronto, Ontario, Canada M5T 1R5

First edition 1963

Second edition 1971

Third edition 1991

Library of Congress Cataloging in Publication Data

Modern power station practice: incorporating modern power system practice/British Electricity International.—
3rd ed. p. cm.
Includes index.
1. Electric power-plants. I. British Electricity International.
TK1191.M49 1990
62.31'21 — dc20 90-43748

British Library Cataloguing in Publication Data

British Electricity International
Modern power station practice.—3rd. ed.
1. Electric power-plants. Design and construction
I. Title II. Central Electricity Generating Board
621.3121.

ISBN 0-08-040510-X (12 Volume Set)
ISBN 0-08-040512-6 (Volume B)

Printed in the Republic of Singapore by Singapore National Printers Ltd

Contents

Colour Plates

(between pp 66 and 67)

Foreword

G. A. W. Blackman, CBE, FEng
Chairman, Central Electricity Generating Board
and Chairman, British Electricity International Ltd

For over thirty years, since its formation in 1958, the Central Electricity Generating Board (CEGB) has been at the forefront of technological advances in the design, construction, operation, and maintenance of power plant and transmission systems. During this time capacity increased almost fivefold, involving the introduction of thermal and nuclear generating units of 500 MW and 660 MW, to supply one of the largest integrated power systems in the world. In fulfilling its statutory responsibility to ensure continuity of a safe and economic supply of electricity, the CEGB built up a powerful engineering and scientific capability, and accumulated a wealth of experience in the operation and maintenance of power plant and systems. With the privatisation of the CEGB this experience and capability is being carried forward by its four successor companies — National Power, PowerGen, Nuclear Electric and National Grid.

At the heart of the CEGB's success has been an awareness of the need to sustain and improve the skills and knowledge of its engineering and technical staff. This was achieved through formal and on-job training, aided by a series of textbooks covering the theory and practice for the whole range of technology to be found on a modern power station. A second edition of the series, known as Modern Power Station Practice, was produced in the early 1970s, and it was sold throughout the world to provide electricity undertakings, engineers and students with an account of the CEGB's practices and hard-won experience. The edition had substantial worldwide sales and achieved recognition as the authoritative reference work on power generation.

A completely revised and enlarged (third) edition has now been produced which updates the relevant information in the earlier edition together with a comprehensive account of the solutions to the many engineering and environmental challenges encountered, and which puts on record the achievements of the CEGB during its lifetime as one of the world's leading public electricity utilities.

In producing this third edition, the opportunity has been taken to restructure the information in the original eight volumes to provide a more logical and detailed exposition of the technical content. The series has also been extended to include three new volumes on 'Station Commissioning', 'EHV Transmission' and 'System Operation'. Each of the eleven subject volumes had an Advisory Editor for the technical validation of the many contributions by individual authors, all of whom are recognised as authorities in their particular field of technology.

All subject volumes carry their own index and a twelfth volume provides a consolidated index for the series overall. Particular attention has been paid to the production of draft material, with text refined through a number of technical and language editorial stages and complemented by a large number of high quality illustrations. The result is a high standard of presentation designed to appeal to a wide international readership.

It is with much pleasure therefore that I introduce this new series, which has been attributed to British Electricity International on behalf of the CEGB and its successor companies. I have been closely associated with its production and have no doubt that it will be invaluable to engineers worldwide who are engaged in the design, construction, commissioning, operation and maintenance of modern power stations and systems.

March 1990

Preface

THIS THIRD EDITION reflects the great experience obtained in the design, manufacture and operation of large coal- and oil-fired boilers in the last 20 years. Introduction of the 500 MW boilers was not without its problems and this provided an impetus to the analysis of boiler plant in all its aspects — combustion, furnace heat transfer, the design of superheaters and reheaters operating in the creep range, boiler tube corrosion and its interaction with creep life, the control of slagging and fouling and the maintenance of boiler efficiency.

A series of trials on the first unit of each boiler family was carried out and the analysis of this plant-derived data was consolidated into computer programs, which now allow boiler performance to be predicted with great accuracy. In addition, operating temperatures within tube banks and on individual header stubs can be calculated to allow residual life estimates on operating plant to be made. This process of analysis and consolidation has been applied across the whole boiler plant with its numerous auxiliary systems, allowing plant selection to be made with confidence.

While 15–20 years ago the main environmental consideration in the design and operation of boiler plant was the control of flue gas particulate emissions and, in the case of oil-fired boilers, the avoidance of acid smut production and subsequent emission, the last decade has seen pressures to reduce emission of the acid gases, NOx and SO_2. The former has been addressed by the modification of combustion to achieve a reducing condition in the core of the flame and the latter by the installation of major absorption plants between the boiler and the chimney. These developments, together with the advances that have occurred in the design and control of electrostatic precipitators, are fully reviewed.

This volume is, then, a review of modern practice in the design and operation of large boiler plant covering all aspects of furnace superheater and reheater design, the characteristics and selection of major auxiliaries and the control of gaseous and particulate emissions.

R. M. CLAPP
Advisory Editor — Volume B

Contents of All Volumes

Volume F — Control and Instrumentation

Volume G — Station Operation and Maintenance

Volume H — Station Commissioning

Volume J — Nuclear Power Generation

Volume K — EHV Transmission

CHAPTER 1

Furnace design, gas side characteristics and combustion equipment

Contents

1 Fundamentals of boiler design and factors affecting selection

Boiler design has been an evolutionary process and the modern power station boiler represents the accumulation of operating experience over some 40 years and parallel development of analytical techniques to achieve greater accuracy in the prediction of operating performance, particularly operating temperatures in the high temperature regions of the boiler. This process has allowed a steady increase in size and steam conditions to take place. The advance over a 25 year period is illustrated in Fig 1.1.

The CEGB has 660 MWe boilers operating at 165 bar outlet pressure with superheater and reheater outlet temperatures of 568°C and designs are now being offered of 900 MWe capacity with steam conditions

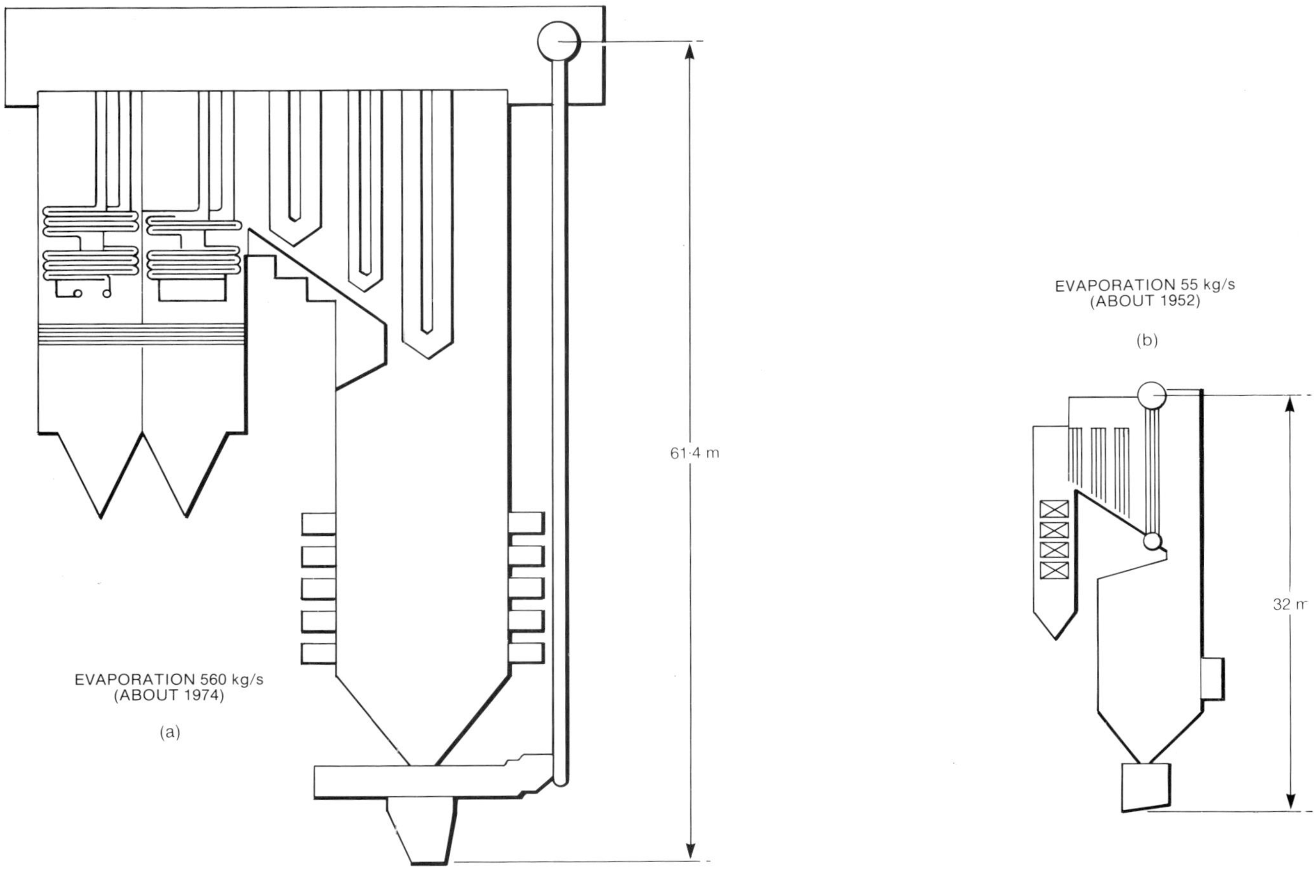

FIG. 1.1 Comparison of fossil-fired boilers over 25 years

of 182 bar/568°C. The superheater outlet pressure of these latter boilers represents the limit of evolution of the drum type boiler and further advance in pressure requires the use of a once-through instead of a recirculating steam/water system for furnace cooling. Studies have shown that a significant increase in pressure is required to gain sufficient improvement in the thermodynamic cycle to offset the additional pumping costs and the higher water treatment costs inherent in a once-through boiler. With the prospect of higher fuel costs in the future, however, there will be mounting pressure to consider more advanced steam conditions, the additional capital cost of the boiler and turbine plant is likely to be offset by the fuel cost savings. Introduction of plants with more advanced steam conditions than currently practised can result in some loss of availability, particularly in the early years of operation, and this has to be taken into account in the overall economic evaluation.

The factors that lead to a satisfactory and economic design of boiler are discussed in this and the following chapters. High boiler efficiency requires high combustion efficiency, minimum excess air and a low exhaust temperature to the chimney. High combustion efficiency results from correct sizing of the combustion chamber and the matching of fuel and air to each burner, together with maintenance of optimum particle size distribution in coal-fired boilers and droplet size in oil-fired boilers. Mismatch of fuel and air can often be overcome by greater excess air but this is at the expense of overall boiler efficiency. The exhaust temperature to the chimney is dictated by considerations of low temperature corrosion and fouling of air heaters. Within these constraints, the boiler has to be tolerant of changes in fuel characteristics, capable of operating with some equipment out of service for routine maintenance and able to respond to the requirements of the electrical system to which its associated turbine-generator is connected.

While some redundancy can be provided in fuel preparation plant, on load cleaning equipment and draught plant, the pressure circuit of a boiler is indivisible and a single failure leads inevitably to a unit outage. In recent years, close attention has been given to quality assurance in the manufacture of pressure parts and this, together with the greater accuracy in prediction of operating temperatures and adequate allowance for corrosion in pressure part design has noticeably increased boiler availability. The security of the pressure circuit and refinement of routine maintenance of equipment subject to high rates of wear has allowed periods of operation without major outages for maintenance to be steadily lengthened and CEGB practice is now to take units off for routine maintenance at 3 yearly intervals.

The modern boiler has then been developed to provide a high level of availability and generates steam at conditions giving very high station thermal efficiency. It has to handle fuel which is of variable quality, with characteristics which, if not adequately considered at the design stage, can lead to loss of availability and indeed to major outages. The nature of the fuels and the methods of preparing them for the combustion process are considered in the following sections.

2 The nature of fuel and its influence on design

2.1 Boiler slagging and fouling

Coals are formed by the compression of plant material over millions of years. During the process of coal formation incombustible mineral matter, chiefly clays in British coals, become intimately mixed with the combustible organic material such that they cannot be easily separated. It is this mineral material in coal which makes up the ash formed by coal combustion. The ash content of British coals varies widely (5–34%), as may the chemical composition of the ash.

In a *dry ash* coal-fired boiler between 15% and 30% of the ash will be removed at the furnace ash hopper, with the balance being removed at the electrostatic precipitators. A small proportion of ash may however be deposited on the heat transfer surfaces. Ash is deposited in two main ways:

- Slagging; where fused or resolidified molten material forms on surfaces exposed to high radiant heat fluxes and gas temperatures.
- Fouling; where sintered or cemented deposits form on convection surfaces.

Slagging and fouling considerations have a major effect on the design and operation of a pulverised fuel (PF) fired boiler. Some of the major considerations are:

- Deposits on heat transfer surfaces lead to a reduction in heat flow to the waterside, for which an allowance must be made at the boiler design stage.
- The formation of molten slags is increased by high furnace gas temperatures and heat fluxes. To reduce the probability of slagging; a PF boiler furnace must be made much larger than the furnace of an equivalently rated oil (ash content <0.1%) fired unit.
- Slagging and fouling, by decreasing the heat transfer at a particular location, can result in an increase in gas temperatures to subsequent tube banks which can increase corrosion rates.
- Corrosion is in many areas associated with bonded ash deposits.

The tendency of a particular coal ash to produce slagging and fouling problems can to some extent be predicted from the chemical composition and physical properties of its ash. The range of ash components found in British bituminous coals is shown in Table 1.1.

TABLE 1.1
Chemical composition of ashes prepared from Yorkshire and Midlands coals

CONSTITUENT % W/W	MEAN	MINIMUM	MAXIMUM
SiO_2	48.0	30.3	61.1
Al_2O_3	24.1	18.0	29.3
Fe_2O_3	11.7	2.9	26.9
CaO	3.7	1.1	11.9
MgO	1.4	0.8	2.4
K_2O	2.7	1.2	3.7
Na_2O	1.2	0.2	2.6
TiO_2	1.0	0.7	1.2

The majority of the mineral matter is in the form of clays, muscovite-illites (potassium alumino-silicates), kaolinites (alumino-silicates) and quartz (SiO_2).

The temperature at which ash fuses is widely used as an indicator of the slagging propensity of a coal. The British standard method uses an 8–13 mm high trigonometric pyramid of ash with an equilateral base, side 0.3–0.5 of the height. The pyramid is heated at 5°C/min in either an oxidising (air or CO_2) or a reducing atmosphere (50% CO + 50% H_2), and three temperatures observed:

Initial deformation temperature (IDT)	;	rounding of pyramid tip
Hemisphere temperature (HT)	;	height of ash = 0.5 × width
Flow temperature (FT)	;	height = 0.3 × base

The IDT provides a guide to the temperature above which the ash will tend to stick as opposed to accumulating as a non-fused *dry product*. The test is conducted in an oxidising and a reducing environment as the gas chemistry around the ash has a profound effect on its slagging propensity. In a reducing environment the ash fusion temperatures all decrease. In a reducing atmosphere the ferrous (Fe^{2+}) iron content of the ash will increase at the expense of the more highly oxidised ferric (Fe^{3+}) iron. The ferrous ions act as a flux, reducing the IDT by over 100°C for a typical British coal. Ash fusion temperatures should only be used as a guide, not only because of the sensitivity of the ash to gas chemistry, but also because of the relatively low accuracy to which the measurements may be made (the ASTM measurement of IDT allows a 55°C error margin). A slagging index based on IDT and HT is:

$$F_s = \frac{4\ IDT + HT}{5} \tag{1.1}$$

This index takes into account the tendency for ash to stick (IDT) and the temperature rise required to produce significant flow and hence the sintering propensity. Table 1.2 summarises some proposed guidelines for F_s.

TABLE 1.2
Variation of slagging propensity with slagging index

SLAGGING INDEX °C	BOILER SLAGGING
1230 — 1340	Medium
1050 — 1230	High
$<$ 1050	Severe

The compressive strength of laboratory-prepared sintered ash has been used successfully to predict the slagging propensity of coals. There is however no recognised standard method of testing the strength of sintered ash. Within the CEGB there is a system of describing ash samples removed from furnaces, using deposition probes, which provide a guide to the slagging potential of the fuel fired. Table 1.3 shows the system used to describe deposited samples.

Other indices of slagging propensity are based on the analysis of the coal ash. For ashes with less than 2.5% Na_2O and less than 7.5% CaO (most British coals) the proportions of iron and silica can both be used as a guide to slagging.

Table 1.4 provides some guidelines on the effect of iron and silica content on slagging propensity.

The principal constituents of sulphate free coal ash can be divided into acidic (SiO_2, Al_2O_3, TiO_2) and basic components (Fe_2O_3, CaO, MgO, K_2O, Na_2O). For bituminous ashes, increases in the iron, calcium and sodium content all increase slagging propensity, the ratio of basic to acidic components is therefore used as a slagging index.

$$R_{b/a} = \frac{Fe_2O_3 + CaO + MgO + K_2O + Na_2O}{SiO_2 + Al_2O_3 + TiO_2} \tag{1.2}$$

Most bituminous coals produce an $R_{b/a}$ of 0.2–0.4, a maximum value of 0.5 is generally recommended

TABLE 1.3
Ash type-nomenclature

TYPE NUMBER	DESCRIPTION	ULTIMATE COMPRESSIVE STRENGTH kN/m^2
0	Dust	0
1	Mostly dust, some very light sintering	$<$ 50
2	Mostly lightly sintered ash, some dust	50 — 100
3	Coherent light sintering	100 — 150
4	Light/medium sintering	150 — 400
5	Coherent medium sintering	400 — 750
6	Medium/strong sintering	750 — 1000
7	Strongly sintered ash	1000 — 1500
8	Strong sintered ash, some fusion	1500 — 2500
9	Mostly fused ash	2500 — $>$3000
10	Hard slag	$>>$3000

TABLE 1.4
Effect of iron and silica on slagging propensity

CLASSIFICATION OF COAL ASH	NON-SLAGGING	SOME SLAGGING	HIGH SLAGGING
Iron oxide in ash, weight percent Fe_2O_3	3 — 8	8 — 15	15 — 23
Silica ratio, $\frac{SiO_2 \times 100}{SiO_2 + Fe_2O_3 + CaO + MgO}$	72 — 80	65 — 72	50 — 65

for PF boilers. An alternative index uses the base/acid ratio multiplied by the weight percent of sulphur in dry coal.

$$R' = R_{b/a} S\% \qquad (1.3)$$

The basis of this index is that part of the sulphur is present as pyrite (FeS) whose residues on heating can act as fluxing agents. For bituminous coals a value of R′ of 0.65 has been suggested as a maximum in order to avoid severe slagging.

2.1.1 Furnace

Many continental and US power stations are supplied from a single uniform coal source. In such a situation it is possible to design a furnace in which up to 85% of the coal ash is deposited as running slag on the furnace walls which may be removed at the ash hopper. The advantage of such a *slag-tap* design is that convection surface fouling may be reduced along with a decrease in precipitator size. Trials in the UK in the 1950s established that the variable nature of British coal deliveries prevented the efficient operation of such furnaces, as uniform slagging properties could not be maintained. As a consequence all CEGB coal-fired units are based on a dry ash design in which all deposited ash should not flow and sinter, and should be easily dislodged by sootblowing. In order to maintain dry ash conditions it is important that ash impacting on convective heat transfer surfaces should be at a temperature below the IDT. This can be achieved by a short residence time to promote rapid cooling and an oxidising atmosphere. These requirements conflict with the need for complete combustion and the wish to reduce the excess air to minimise the waste heat exhausted from the boiler. In a PF boiler reducing conditions are indicated by the carbon monoxide content rising dramatically (see Fig 1.2), this corresponds to an excess air value of around 16%. It is desirable therefore to operate PF boilers with an O_2 in flue gas of around 3.5%.

If the furnace is operated in such a fashion that slagging occurs the major risks are from:

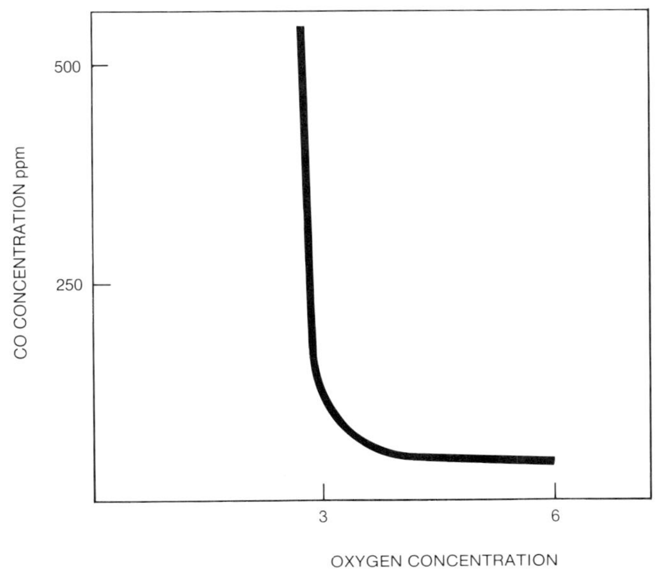

FIG. 1.2 Relationship between oxygen and carbon monoxide in PF-fired boilers

- Falling slag which may damage the hopper slope. If the resulting fragments are large enough an ash bridge may occur, in which a large mass of slag wedges in the hopper throat. If an ash bridge is left in-situ a total blockage of the ash hopper may result, which can prove extremely difficult to remove.
- Around the burners of a wall-fired unit heat fluxes of around 500 kW/m^2 can occur, whilst the inner surface of the burner quarls is cooled by the burner secondary/tertiary air. As a consequence running slag may form in the high flux zone above a burner, flow down to the quarl and solidify. Over a period of time burner eyebrows may form which may distort the burner air flow and may even block a burner which is out of service. Eyebrow formation may be promoted by local reducing conditions which may be brought about by aerodynamic recirculation of unburnt PF or the use of a large oil burn to maintain load in the event of a mill trip. These effects can be prevented by adequate burner/quarl design and the provision of sufficient secondary air. Careful selection of quarl material, e.g., the use of silicon

carbide with zirconium oxide and alumina will also reduce the tendency of slag to adhere.

2.1.2 Radiant superheaters and reheaters

Platen superheaters and reheaters are exposed to high radiant heat fluxes, especially at their tip sections. If ash above its IDT impacts on the platen it will be retained, lower temperature ash may also be retained by geometrical features such as displaced tubes. In either case the increased thermal impedance leads to increasing ash surface temperatures. As the ash surface temperature rises sintering of the deposits and eventually running slag may develop. If routine sootblowing proves unable to dislodge the ash, large deposits may form (see Fig 1.3). If large masses of slag are allowed to form they may ultimately bridge from platen to platen if their spacing is inadequate (<600 mm). The deposits can cause large tube-to-tube distortions and increased draught losses. A further problem of high level deposits is the risk of falling slag. The removal of fused platen deposits can involve off-load cleaning possibly using water hoses or manual methods.

However, even when large accumulations of slag form about their tip section, the heat transfer to platen superheaters is not always impaired. A platen superheater would receive heat by a number of modes:

- By direct radiation from the luminous furnace gases.
- By non-luminous radiation from bounding cavities.
- By inter-platen radiation.
- By convection.

The amount of heat transferred by convection is only typically 10% – 15% of the overall platen heat pick-up. If the tip section were heavily slagged, the proportion of heat picked up by direct radiation from the furnace would decrease and the amount of heat re-radiated back into the furnace from the platen would increase. This would slightly increase the furnace exit gas temperature and the gas temperature within the gas cavities bounding the platen tubes. Cavity radiation would then increase, shared between the cleaner vertical tubes of the platen, the bounding water walls, and in severe cases the superheater/reheater surface following. Thus, despite a significant increase in the slag build-up in an area of concern, it would not be immediately evident from a reduction in the steamside temperature pick-up.

Therefore, to avoid problems which may result from such an occurrence future boiler designs, and indeed the refurbishments carried out on operating plant, will include better tube alignment measures such as partial membraning of the tip sections of platens.

The platen metal temperature is an important factor in the bonding of deposits, metal temperatures should be held below 625°C in order to reduce the bond strength. In assessing probability of platen deposition by a coal the sinter strength of the ash is a better guide than IDT as it indicates the strength of the bonded material and its ability to resist sootblowing.

Before

During

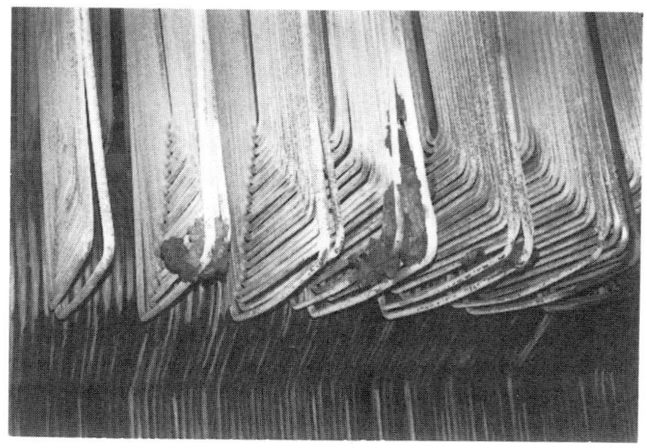

After

FIG. 1.3 Typical off-load water cleaning of platen superheater elements (see also colour photograph between pp 66 and 67)

2.1.3 Convection heating surfaces

Flue gas flowing into the superheaters and reheaters contains sodium and potassium volatilised from the coal and sulphur oxides formed from pyrites and organically bound sulphur. These components can react to form alkali sulphates which may condense forming an adhesive layer on the tubes. Fly ash may then be captured by the sulphate forming a porous surface sintered deposit, under which corrosion may occur. This method of formation results in a layered deposit with a whitish layer containing 50–70% sulphates next to the tube, and a brown outer layer consisting of chemically unaffected fly ash cemented by 20–40% pyrosulphates. Residual fuel oil-fired boilers burn oils whose ashes tend to contain vanadium, sodium and sulphur compounds. These can produce large deposits in the higher temperature convection surfaces. These deposits can increase corrosion. Accordingly oil-fired boilers are designed with a final steam temperature of <540°C in order to limit metal temperatures, and hence corrosion.

In fin-tubed economisers bonded dust deposits can build up. In the case of high calcium coals (>7% CaO in ash) insoluble calcium salts can form strong bonds which are difficult to remove by normal on-load cleaning. Enhanced sootblowing is an essential measure which can alleviate the problem from a boiler prone to these deposits.

While it is possible to design a boiler to burn any given coal or a relatively limited range of coals, it is not economically feasible to design it to burn *all* available coals. The range of coals can be broadly categorised as bituminous, anthracitic, lignitic, etc. The specific design is, therefore, tailored to meet the constraints imposed by the chosen range of coals.

Clearly, the fouling propensity is the major influence in the determination of the heating surface disposition and configuration.

The transverse pitch (element to element centres) of superheater and reheater banks in gas temperature regions above the IDT of the worst coal requires to be sufficient to reduce the risk of ash blockage, as discussed in the previous section. Convective superheater and reheater elements downstream of this point can be placed on a smaller pitch dependent mainly upon effective penetration of sootblowers and preventing high gas velocities to limit tube erosion.

The design of convective heating surface will also depend upon the desired steam/gas-side heat transfer, and acceptable gas-side pressure drop (i.e., draught loss), being determined at the design stage by geometrical factors such as tube diameter, tube spacing, number of tubes wide and deep.

From a long record of experience, given sets of conditions for each fuel to be burned have been effectively established for economic practice.

The design of convective heating surface for oil-fired boilers does not have to be limited to the low velocities required to avoid erosion damage in coal-fired plant as the quantity of ash in oil fuel is much smaller. However, residual fuel oils containing certain quantities of sodium and vanadium compounds in their ashes, may result in an increase in tube fouling and serious corrosion of the tube metal.

2.2 Fuel ash corrosion

Fuel ash corrosion, also referred to as fireside corrosion or high temperature corrosion, is basically the oxidation of steels affected to a greater or lesser extent by gaseous, molten or indeed solid components of the products of combustion of fossil fuels, i.e., coal and oil. This type of corrosion is invariably associated with deposits formed on boiler tubes. These deposits consist of species resulting from the combustion of the fuel together with the corrosion products of the tube material. The composition of the deposits depends on the type and nature of the fuel and the conditions of combustion. In many instances chemical elements which are present in relatively small quantities, in some cases only in trace proportions, can concentrate on the tube surfaces and constitute the majority of the corrosive deposits.

During combustion the major constituents of fossil fuels, namely carbon and hydrogen, are converted to carbon dioxide and water vapour. Therefore the flue gases mainly comprise carbon dioxide, water vapour, nitrogen (from the combustion air) and up to about 4% oxygen (since excess air is used to ensure complete combustion).

Both coal and oil contain organic and mineral impurities. Most of the impurity in coal is in the form of alumino-silicates which leads to the formation of ash. Some of this ash, formed during combustion, falls to the bottom, i.e., hopper, of the furnace but most of it is carried by the combustion gases. Residual fuel oil contains very much smaller quantities of ash producing alumino-silicates, and all of this ash is carried by the combustion gases. It is the ash carried by combustion gases that forms the deposits on boiler tubes.

The main elements in coal which influence corrosion are sulphur, sodium, potassium and chlorine. Some of the sulphur in coal is organically bound as part of the coal matter, and some is present as iron pyrites thoroughly mixed with the coal. Sodium and potassium are present mainly in the alumino-silicate minerals. Chlorine, in British coals, is present almost entirely as ions weakly-bonded to the coal substance.

CEGB experience shows that chlorine content of the coal burnt in a boiler is a good indication of its corrosive potential. During the initial stages of the combustion of pulverised fuel particles the chlorine is released as hydrogen chloride gas, the hydrogen chloride concentration being directly proportional to the chlorine content of coal and the amount of excess air. Its role in the corrosion process is complex and possibly indirect. Recent work has shown that chlorine promotes the release of both sodium and potassium, and as there is always sufficient sulphur present, the released sodium and potassium compounds are converted to fusible sulphates. It is these fusible sulphates in the tube deposits that cause corrosion.

The alkali metal release is also influenced by the ash content. The amount of ash affects the partition of alkali metals between the ash melt and the vapour phase with the result that the higher the ash content of a coal the greater the alkali metal sulphate rendered innocuous.

The main constituents of residual fuel oil which affect corrosion are sulphur, vanadium and sodium. Although vanadium is present in similar quantities in both coal and oil, because oil has very much less ash producing alumino-silicates, vanadium constitutes a much larger fraction of the ash in oil-fired boilers. During the combustion of oil, sodium sulphates are formed in a similar manner to coal combustion. These sulphates combine with vanadium oxide, also produced during the combustion process, to form mixed sulphate/ vanadates. The vanadate-containing deposits are particularly aggressive when molten.

Both coal and oil-fired boiler deposits melt at temperatures prevailing under normal boiler operating conditions. As sulphates and sulphate/vanadates play the most important role in the corrosion mechanisms, the lower their melting temperature the larger are the areas of a boiler affected by corrosion.

One of the main reasons why the CEGB oil-fired plant is designed to operate at lower final steam temperatures, hence lower metal temperatures, is the known lower melting point of the vanadate component of the tube deposits. This coupled with the fact that oil-fired deposits attack the higher strength austenitic tube materials, which have to be used for higher temperature plant, much more severely than ferritic tube materials, restricts the maximum metal temperatures to about 590°C, which in turn restricts the final steam temperatures to about 540°C for economic design.

Furnace wall and superheater/reheater corrosion are influenced by fuel impurities in different ways, although in both cases the relatively stable and protective oxide scale that can be formed under ideal conditions by the oxidation of the metal surfaces by the oxygen and carbon dioxide present in the combustion gases is undermined or even prevented from forming in the first instance.

The protectiveness of the scale depends upon the rate of transport of the elemental species through it, its mechanical properties, and its adhesion to the metal surface.

Alloys with high chromium content produce thin and highly protective chromia scales. The adhesion of these scales to the metal substrate may be improved by additions of trace quantities of rare earth elements such as yttrium and cerium. If the strength of this bond is insufficient the differential thermal expansion of the scale and the metal can act as the driving force for scale separation. The propensity to void formation at or near the interface also influences the bond strength.

While there is some evidence to suggest that high chromium, i.e., greater than about 20%, ferritic alloys may be more corrosion resistant that those alloys with an austenitic lattice structure, the austenitic alloys are preferred tube materials due to their high creep strength at elevated temperatures and better general mechanical integrity; also they do not suffer from service embrittlement as do the high chromium ferritics.

Although the development of alloys with good oxidation resistance at high temperatures has been successful, this development has not eliminated the corrosion problems caused by sulphates and vanadates present in the ash deposits, which can gain direct access to the metal surface by undermining the otherwise protective scales, or departure from ideal conditions preventing the formation of protective scales.

2.2.1 Furnace wall corrosion

The surface metal temperature of furnace wall evaporator tubes in CEGB boilers operating at 165 bar, the majority of the generation capacity, is about 450°C. At this temperature a protective oxide scale is formed when the flue gas adjacent to the wall contains excess oxygen (see Fig 1.4).

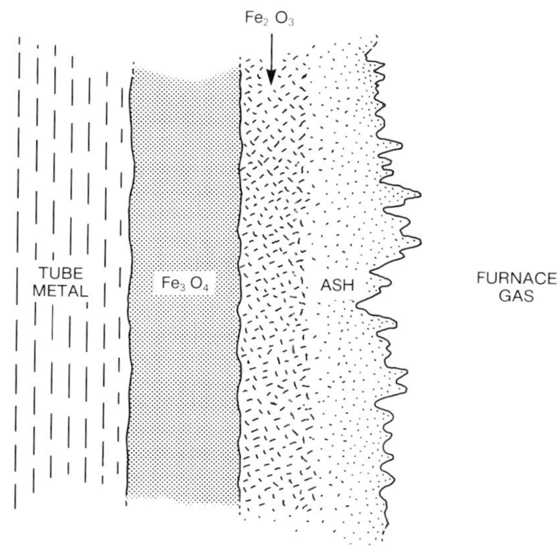

Fig. 1.4 Protective oxide scale formed in oxidising conditions

However, combustion is a dynamic process subject to considerable variations with coal-fired burners, and can lead to the occurrence of reducing conditions, i.e., high carbon monoxide and low oxygen concentrations, in the vicinity of these scales. Reducing conditions and other factors which promote aggressive corrosion can be created by:

- Unsatisfactory fuel and/or air distribution.
- Faulty burners.
- Insufficient combustion air.
- High chlorine concentration in the coal.
- The impingement of partially-burnt fuel particles onto the tubing, etc.

Complex gaseous interaction between sulphur, chlorine, carbon and oxygen in the presence of locally-high heat fluxes yield a range of non-protective corrosion scales. Slagging of furnace walls is also detrimental because it prevents the access of oxygen to the scale, thus encouraging the reducing conditions. The result of the abnormalities and the complex reactions is the formation of iron sulphide which renders the scales non-protective and causes mechanical weakening due to its greater volume (see Fig 1.5).

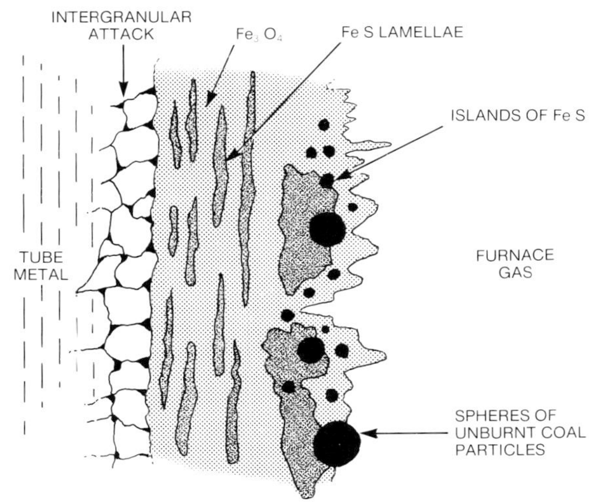

FIG. 1.5 Non-protective fast-growing scales caused by reducing conditions

The most serious reducing/sulphidising conditions, and therefore corrosion, occurs when incompletely-combusted coal particles impinge on and stick to furnace wall tubes. The smouldering of these increases the carbon monoxide release local to the tube surface. The high carbon monoxide concentration assists sulphidation and the presence of hydrogen chloride in the local environment also enhances corrosion. The resultant corrosive scales are a mixture of sulphide and oxide with ferrous chloride as a thin layer adjacent to the metal.

Regarding the role of chlorine, previously it had been postulated that the hydrogen chloride diffused rapidly through the scale to the metal surface where volatile iron chloride was formed. This subsequently diffused back to the scale/gas interface and decomposed to give Fe_3O_4, FeS and HCl. However, it has since been established that at high sulphidation potentials insufficient volatile transport takes place to account for the metal loss rates observed.

An alternative proposal is that in the presence of reducing/sulphidising conditions, the volatile hydrogen chloride enhances the porosity of the sulphide-containing scale. This weakens the scale leading to its cracking and spalling when much thinner than would be the case in the absence of hydrogen chloride. It also manifests itself in the intergranular attack on the metal surface. Hence the detailed mechanism is still not fully known, and work is continuing to ascertain this mechanism.

It is clear from the foregoing commentary that furnace wall corrosion is confined to localised regions of coal-fired boiler furnaces, mainly side walls, near burners, and less commonly rear walls of frontwall-fired boilers as shown in Fig 1.6. Figure 1.7 shows the effect of fuel ash corrosion on a coal-fired boiler furnace wall tube.

Under one or more of the adverse conditions discussed here, severe corrosion, such as that shown in Fig 1.8, has occurred leading to failures such as can be seen in Fig 1.9.

The corrosion of evaporator (furnace wall) tubes is not a problem in CEGB oil-fired boilers despite the low excess oxygen combustion regime employed to reduce the acid smut problem. The combustion of fuel oil is much more rapid and efficient, and the oil flow more controllable. This explains why corrosion through sulphide formation does not occur, although local reducing conditions may still be present. A further reason is that there is considerably less chlorine in residual fuel oil than in British coals, and chloride attack on metal does not take place.

Furnace wall corrosion is also uncommon in many large coal-fired boilers where the combustion conditions are well controlled, where flame and pulverised coal particle impingement is minimal, where low chlorine coals are burnt, and where the geometry and aerodynamics of the furnace chamber make it more difficult for reducing conditions to occur in the vicinity of the furnace walls.

2.2.2 Superheater and reheater corrosion

Coal ash corrosion of austenitic superheater and reheater tubes manifests itself by either two distinct wastage flats on the hottest exposed faces of tubes (see Fig 1.10), or as a massive array of deep honeycomb pits (see Fig 1.11). Geometrical/aerodynamic features such as support lugs, fillet weld beads and protruding tube butt welds can give rise to local deep isolated pits.

The worst corrosion occurs on aerodynamically exposed tubes, for example, the leading tubes of a superheater bank, and the rate tends to fall in less exposed tubes. Contraflow (counter flow) superheater banks which combine the hottest flue gases with the hottest steam are more vulnerable to fuel ash corrosion than those designs which have parallel flow.

The corrosion of the tubes is caused by the deposition of sodium and potassium sulphates on the tube surface facing the gas flow where ash build-up determines the heat transfer pattern, as shown in Fig 1.10.

The role of chlorine in the release of sodium and potassium and the formation of their sulphates have been discussed earlier. These alkali metal sulphates formed earlier in the furnace gradually accumulate, by condensation, at the base of the porous ash de-

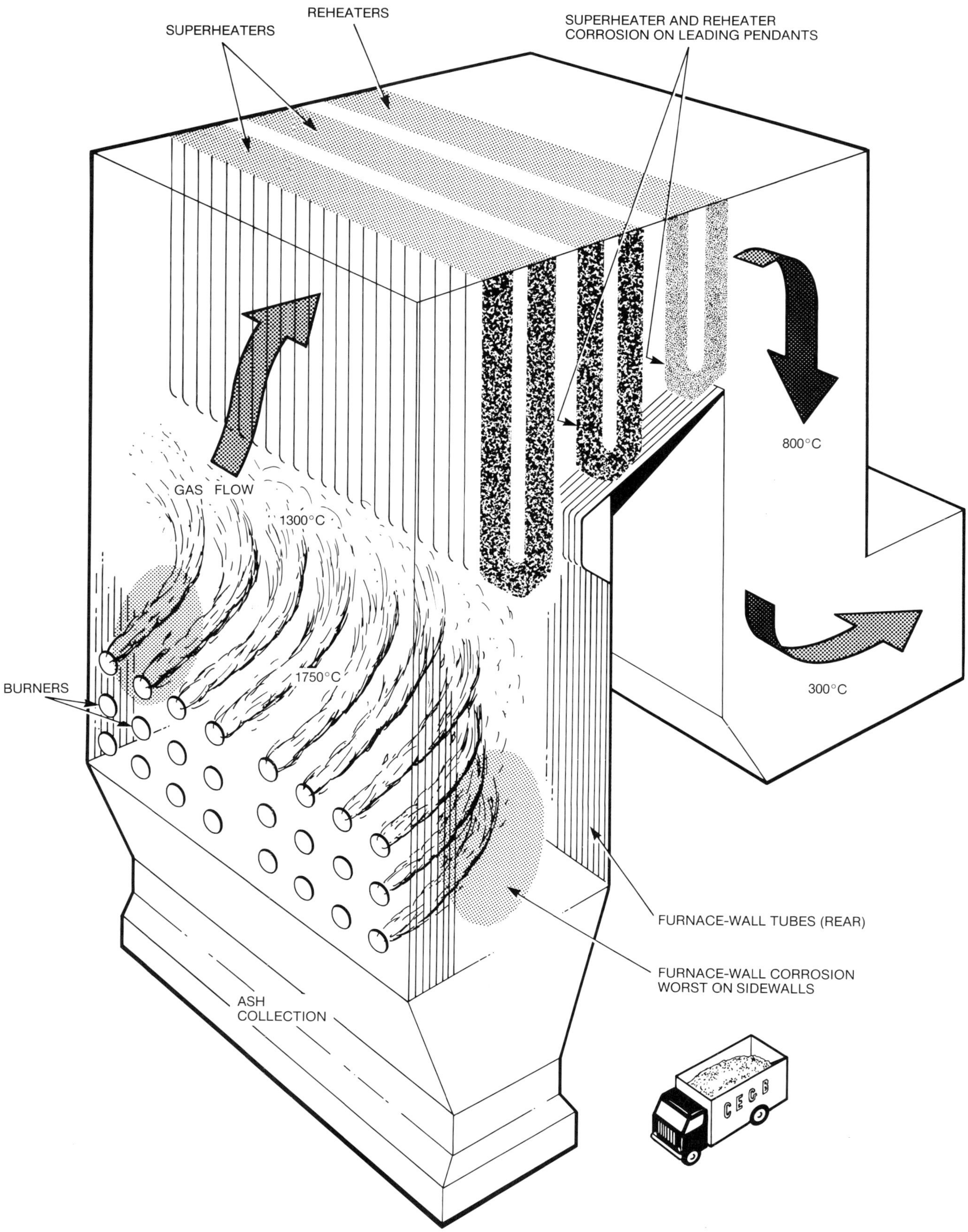

FIG. 1.6 Frontwall-fired boilers showing regions of corrosion

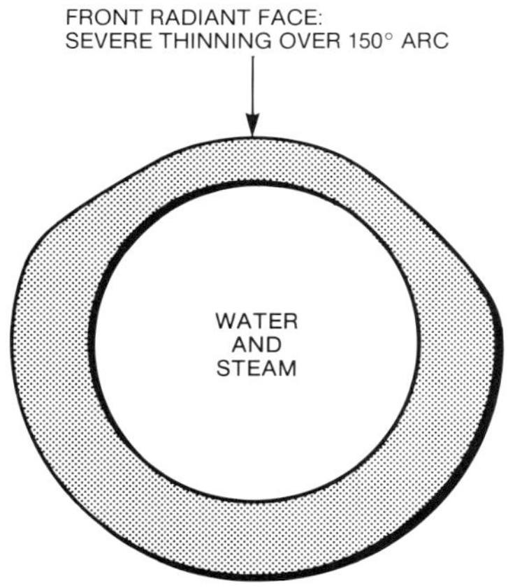

FIG. 1.7 Effect of furnace wall tube corrosion

posit. Such mixed sulphate deposits become molten at tube surface metal temperatures of about 580°C and above, and form a thin layer of highly corrosive liquid next to the protective oxide scale initially formed on the tube surface. The molten sulphate layer contains free sulphur trioxide which stabilises the sulphate and attacks the scale by dissolving the normally protective oxide to form iron, chromium and nickel sulphates.

The stability of the iron and chromium sulphates decreases as the temperature increases, and the temperature gradient through the deposit causes a concentration gradient for the dissolved metal ions. These factors ensure that the dissolution is effectively a continuous process, with iron, chromium and nickel diffusing outward through the molten sulphate layer. At the hotter outer surface, iron/chromium oxides are re-precipitated within the inner ash, but they no longer form a protective layer. This dissolution process results in a linear corrosion rate, as against a parabolic rate for protective oxidation.

The rate of this continuous transport process determines the rate of corrosion, which is therefore highly dependent upon the temperature gradient between the hot gas and the metal surface, i.e., the heat flux established through the tube. Consequently, for a given metal temperature, higher corrosion rates are observed at higher gas temperatures. As metal temperatures increase so do reaction rates, but above a peak temperature of 650–700°C molten sulphate solutions become thermodynamically unstable. This causes reaction (corrosion) rates to fall at higher temperatures, eventually leading to normal high temperature sulphidation/oxidation. This results in the classical *bell-shaped curve* shown in the inset of Fig 1.12.

Operational experience has shown that there is a good correlation between the chloride content of the coal and the rate of corrosion of austenitic superheater and reheater tubes. While it is known that chlorine strongly promotes the release of both sodium and potassium, explaining the correlation with corrosion rate, this correlation is better than the experimentally derived relationship between alkali metal release and chlorine content. It is, therefore, possible that chlorine may play an additional, as yet undetermined, direct role in the corrosion mechanism.

Analysis of operational plant data and carefully controlled probe trials data has resulted in a corrosion rate equation, for austenitic tube materials, which relates the coal chlorine content, tube material, flue gas temperature, tube surface metal temperature, and tube position with respect to gas flow:

$$r = A \times B \left(\frac{T_g}{G}\right)^m \left(\frac{T_m - C}{M}\right)^n (Cl - D)$$

where r = corrosion rate in nano metres per hour, nm/h

T_g = gas temperature, °C

T_m = surface metal temperature, °C

Cl = coal chlorine content, %

A = tube position factor (dimensionless)

B = material factor (dimensionless)

G, M, D, m, n = constants

This equation produces families of curves as schematically illustrated in Fig 1.12.

CEGB data is at present being assessed with a view to developing a correlation for ferritic steels.

Corrosion of superheaters and reheaters in oil-fired boilers is, as discussed previously, again caused by the impurities in the fuel. There are two main differences from coal-fired conditions. Firstly, as the ash content is very much less, then the corrosiveness of the salts is not reduced and secondly sulphate/vanadate deposits melt at lower temperatures. The vanadium/sodium ratio in the final deposits is a good guide to its corrosiveness, i.e., increasing vanadium/sodium ratio causes increased corrosion rates, see Fig 1.13.

As the main steam outlet temperatures for oil-fired boilers have been limited to 540°C, giving metal surface temperatures of above 590°C or below for the high temperature superheaters and reheaters of these boilers, ferritic steel tubes which have sufficient creep strength at these temperatures have been used. Plant trials, laboratory rig tests, and Electricite de France experience have confirmed that ferritic steels have better resistance to corrosion. This is due mainly to their better oxide adhesion characteristics, even in the presence of vanadates, in oil-fired applications.

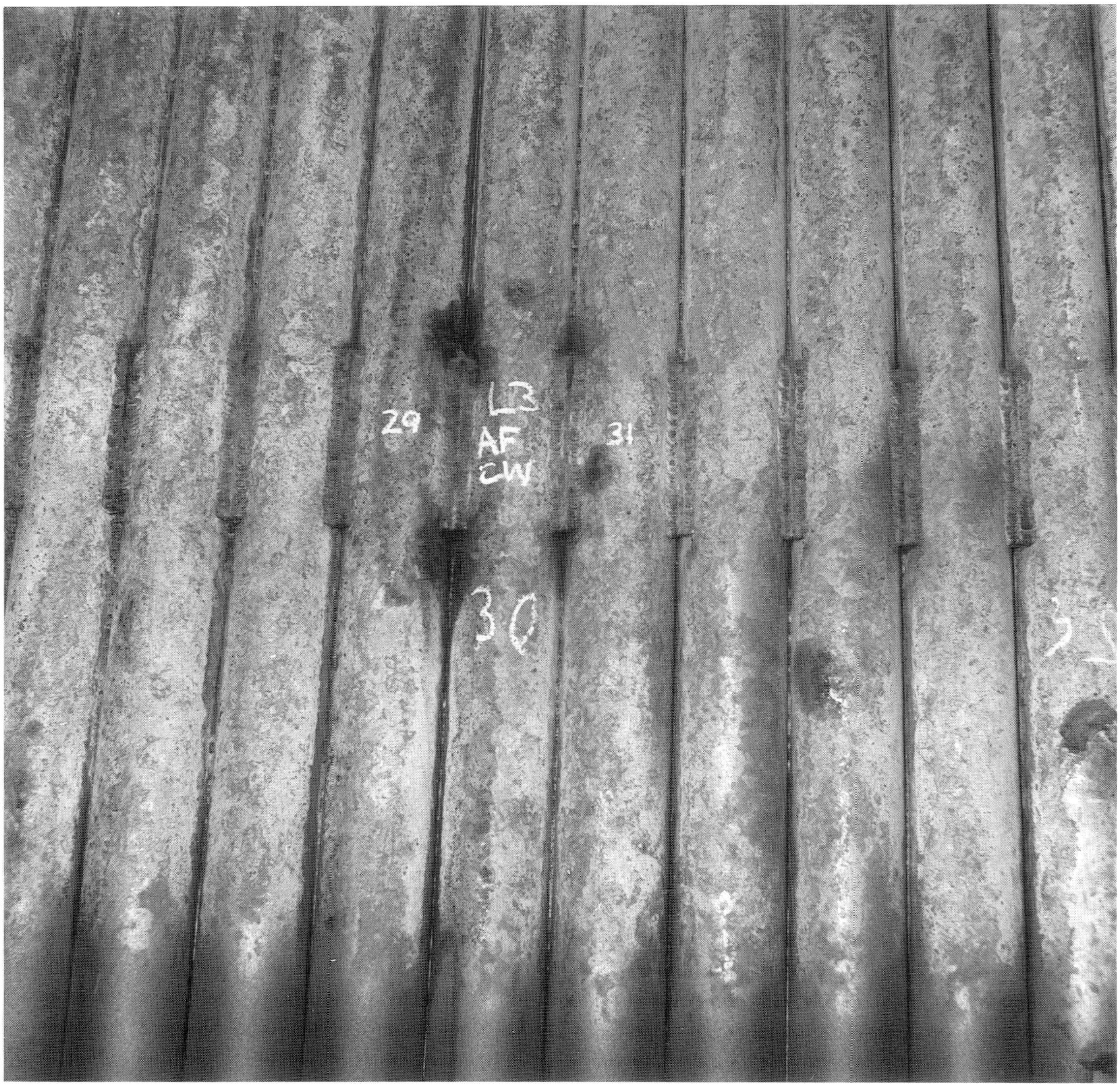

FIG. 1.8 Corrosion of furnace wall tubes at Drakelow *C* Power Station
(see also colour photograph between pp 66 and 67)

A comparison of the corrosion rates of ferritic steels with austenitic steels is shown in Fig 1.14. More specific data has been applied to tube design for all CEGB oil-fired plant from the early 1970s.

2.2.3 Potential solutions to corrosion problems

Furnace wall corrosion problems can be alleviated or eliminated by the application of one or more of the following:

- Elimination of corrosive conditions by combustion system improvements or by burning coal with low, generally below 0.2%, chlorine.
- Use of thicker tubes or faceted tubes, in the corrosion zone.
- Use of more corrosion-resistant tubing, such as co-extruded tube or the use of plasma spray-coated tubing (although this is in the early stages of development at present).

In the case of superheater and reheater corrosion the following remedies may be available or applied:

FIG. 1.9 Failed furnace wall tubes due to fuel ash corrosion at Drax Power Station (see also colour photograph between pp 66 and 67)

- Use of lower chlorine coal either naturally obtained or artificially washed and/or blended.
- By the control of temperature maldistributions, prudent use of sootblowers and control of combustion to reduce differential slagging.
- Use of corrosion-resistant metal shields over the affected areas of tubing.
- Replacement of tubing with more corrosion-resistant tubing, such as co-extruded tubing, or the use of metal-sprayed tubing (when this has been fully developed).

In existing units one of the more cost effective solutions is the use of co-extruded tubing. The tube is manufactured by extruding a composite billet which consists of two close fitting cylinders of the chosen materials. By ensuring that the gap between the cylinders is properly controlled and the surfaces are free of contamination a good metallurgical bond is formed between the two materials during the extrusion process. The inner material is chosen for strength and compatibility with the steam/water environment, and the outer material for its resistance to fuel ash corrosion. Large quantities of co-extruded tubing have been used to combat fuel ash corrosion in CEGB boilers for both furnace wall and superheater and reheater applications with considerable success (increases in tube life of over tenfold have been obtained). The most common combinations that have been used are mild steel inner layer/type 310 outer layer for furnace walls, and Esshete 1250 inner layer/type 310 outer layer for superheaters and reheaters.

For new plant, great attention is paid to ensure that the combustion systems and equipment are designed to avoid reducing conditions in the vicinity of furnace walls, and every effort is made to reduce temperature maldistributions. Actual maximum design temperatures are kept as low as possible, within the limits of the chosen steam cycle, and appropriate corrosion allowances are included in the design of the tube scantlings. Boilers designed and constructed since the mid-1970s have performed considerably better as a result of the above measures, and as yet have not suffered from excessive fuel ash corrosion.

2.3 Gas side erosion

The combustion chamber and high temperature superheater and reheater sections of modern coal-fired boilers do not usually suffer from tube erosion from the impact of flue gas ash particles. This is partly due to the plastic nature of the bonded-slag type deposits which absorb the kinetic energy of the ash particles, and partly due to the nature of the particles themselves. In the convective superheater and reheater areas, where bonded deposits cease to be a problem, the hardness of the ash particles will increase as its temperature drops, thus increasing the erosion propensity.

The majority of tube failures attributable to gas side erosion occur in primary superheater, primary reheater or economiser banks where the gas temperature has fallen below 800°C. In such situations, the erosion rate can be correlated against the ash content of the coal, the ash abrasiveness, and the velocity and temperature of the flue gas. Current practice limits the maximum flue gas velocity to 15 m/s for coal-fired plant, this being the velocity of the gas as it passes between the tubes. Problems occur if high velocities are generated by partial blockage of the gas passes.

The rate of erosion of boiler tubes has been given by Raask [1], viz:

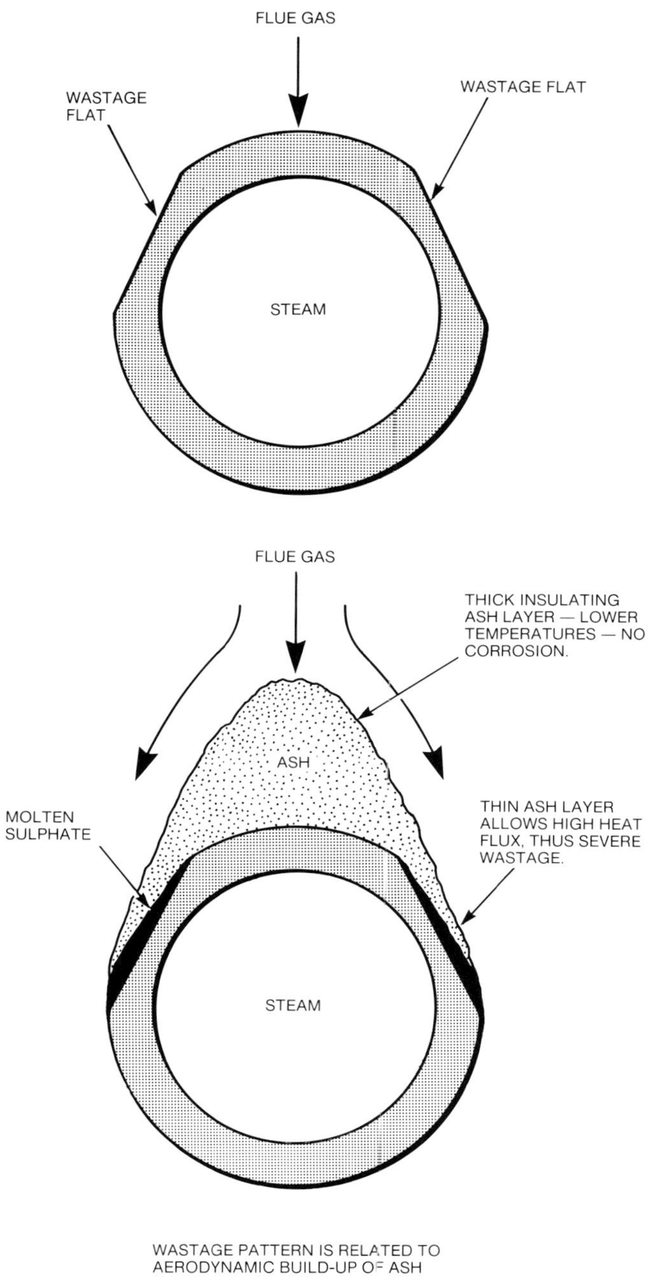

FIG. 1.10 Typical cross-section of corroded superheater tube

$$E = \frac{KI_aR}{T_g}(A/C)\,V^n$$

where E = metal erosion rate, nm/h

K = constant dependent upon tube material

I_a = ash abrasiveness factor

A/C = fuel ash to carbon ratio

V = flue gas velocity

n = index, currently taken to be 3.3

T_g = flue gas absolute temperature

It can be seen from the expression that the most significant factor in the erosion of tubes is that of flue gas velocity, having a power law relationship with erosion rate. However, the abrasiveness of the ash particles is also of fundamental importance so it is worth considering the nature of the ash and the abrasive constituents.

The aluminosilicate clay species make up the bulk of the ash in most coals. The non-abrasive aluminosilicate particles vitrify, agglomerate and spherodise in the combustion process forming a glassy material of Vickers hardness number about 600, thus becoming moderately abrasive.

Quartz mineral contained in coal ash is responsible for erosion wear in fuel handling and milling plant prior to the combustion of the fuel and for a large proportion of the erosion of boiler tubes following combustion.

Quartz is one of the hardest mineral species found in coal ash and in its purest form requires very high ($>2000°K$) temperatures for vitrification and spherodisation temperature. However, in PF-fired boilers, the quartz particles would only be present in temperatures suitable for spherodisation for a very short period of time and consequently complete spherodisation is unable to take place. For this reason the quartz particles retain their irregular shapes and contribute significantly to the abrasive nature of the ash.

Highly erosion-resistant tube materials are not available to the boiler designer. Boiler tubes are fabricated from materials with reasonable ductility. Resistance to erosion would require a hard, brittle tube material which would be unsuitable for pressure part use. For this reason, the pressure parts are required to be designed to allow for a rate of material removal by erosion within their design life.

To keep the rate of material loss by erosion to within design levels, the designer must ensure that superheater, reheater and economiser tube banks remain relatively clean. This is accomplished by the provision of sootblowing equipment and by ensuring that the gas passes do not become blocked by such occurrences as tubes within a given bank falling out of alignment. The latter point is achieved nowadays by the improvement in design of tube-to-tube spacer attachments.

The erosion of tubes by the ash-laden sootblower jet expanding through the tube bank can be a problem in boilers with a heavy dust burden. Palliative measures such as the provision of erosion shields over tubes in close proximity to sootblower jets help to keep the problem manageable between unit outages.

3 Combustion

3.1 Theoretical air

The theoretical air quantity required by any fuel is that which will provide sufficient oxygen to completely

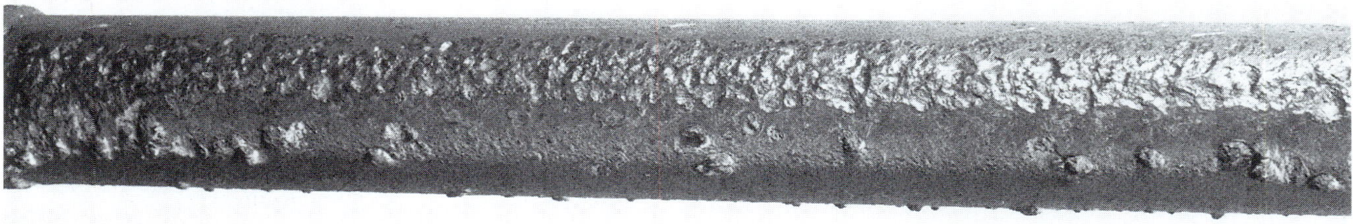

FIG. 1.11 Honeycomb pitting attack

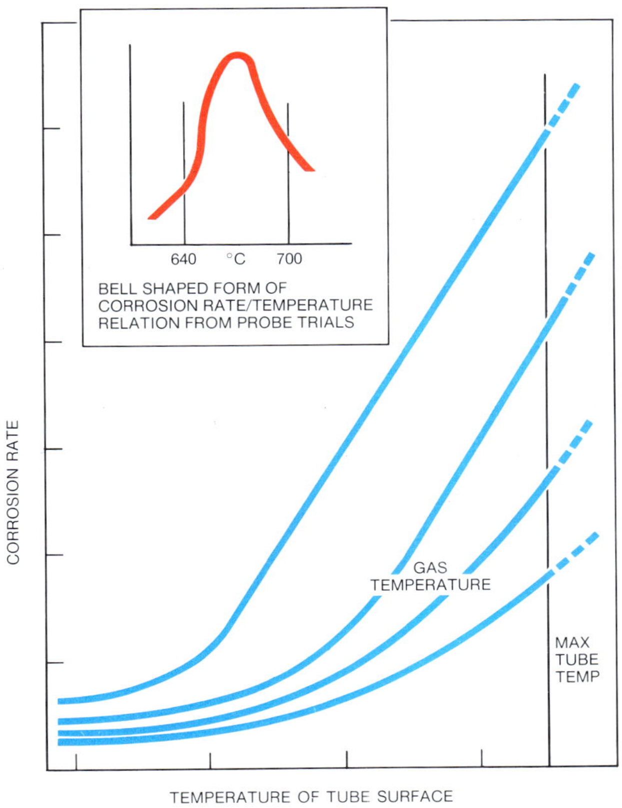

FIG. 1.12 Relationship of corrosion rates with metal and gas temperatures for coal-fired boiler superheater and reheater tubes

oxidise the combustible elements contained in the fuel. This quantity is normally known as the *stoichiometric air requirement*. This theoretical minimum necessary air quantity is obtained by calculation from the known complete (*ultimate*) chemical analysis of the fuel.

Some typical theoretical air quantities for various fuels are given in Table 1.5. It must be emphasised that these are approximate and are given on a dry ash-free basis in the case of solid fuels.

TABLE 1.5

Typical theoretical air quantities for various fuels

FUEL	STOICHIOMETRIC AIR MASS AIR/UNIT MASS FUEL
Bituminous coal	11.18
Anthracite	10.7
Coke	9.8
Lignite	7.5
Peat	5.7
Residual fuel oil	13.85
Distillate fuel oil (gas oil)	14.48
Natural gas (methane base)	17.3

3.2 Excess air

In practice, it is always necessary to supply more air to the combustion system than is theoretically necessary. The reason for this lies in the fuel air-mixing process in any combustion system, as it is not possible to ensure complete and intimate mixing of the fuel with the necessary oxygen at the point of injection. Other factors also have an effect, e.g., the necessary chemical reaction time and the temperature generated by the process. For boiler efficiency reasons it is necessary to contain the excess air to a reasonable minimum. Fan power, stack losses, and unburned carbon-in-ash are the main factors to be considered when optimising the operating condition. The indicator used to express

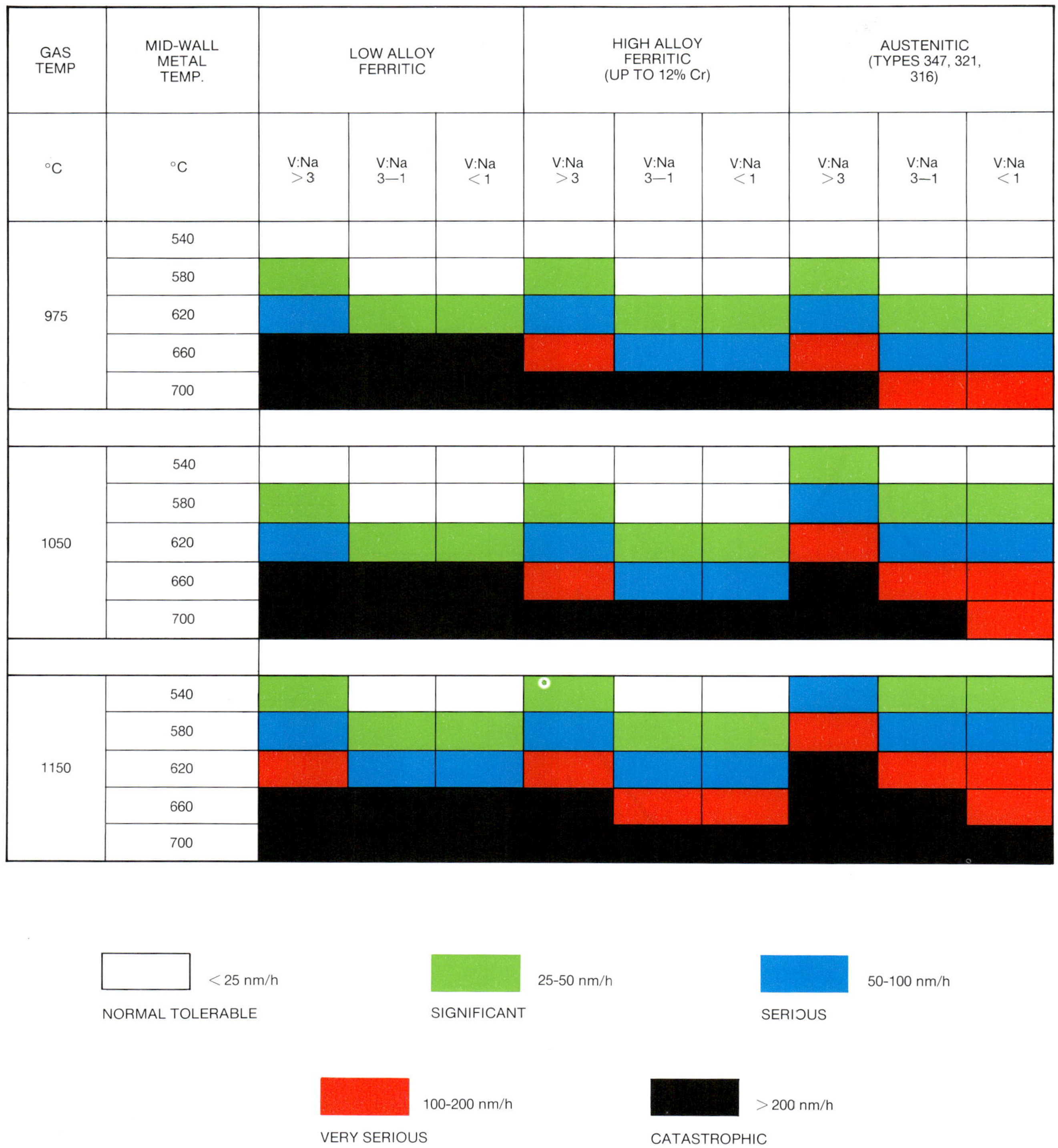

THESE RATES ARE MAXIMUM VALUES, TYPICAL OF LEADING TUBES.

AVERAGE CORROSION RATES FOR NON-LEADING TUBES ARE ABOUT HALF THESE VALUES.

FIG. 1.13 Oil-fired boiler tube corrosion rates

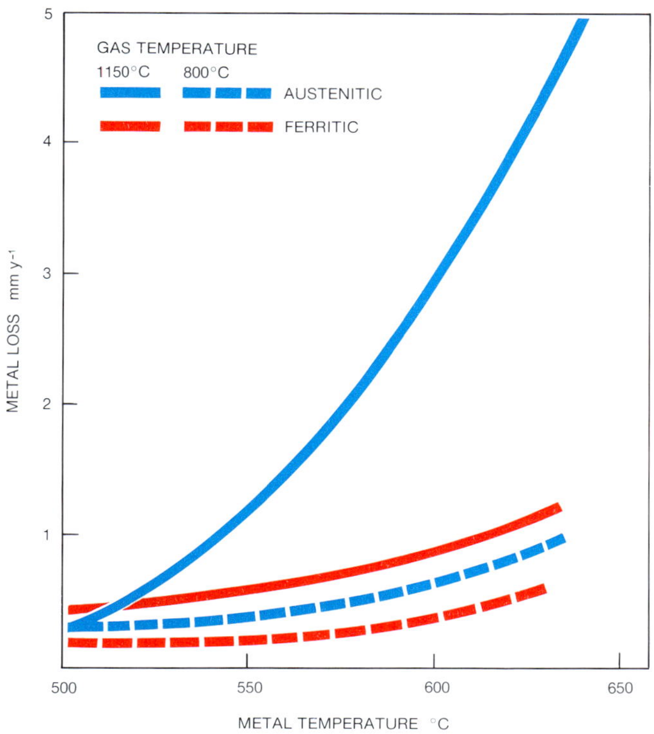

FIG. 1.14 Relationship of corrosion rates with metal temperature for austenitic and ferritic oil-fired boiler superheater and reheater tubes

the excess air quantity is normally the percentage of oxygen (by volume) in the exhaust gas as this is now a normal plant measurement. Carbon dioxide is no longer relevant as an indicator as it cannot provide a reliable indication of minor deviation from a closely controlled *low oxygen* condition.

An alternative expression of excess air is the percentage weight of air supplied to the system over and above the stoichiometric requirements of the particular fuel, e.g., *20% excess air* on a residual fuel oil-fired system means that for each unit mass of oil 16.62 units of air are supplied.

$$\frac{16.62 - 13.85}{13.85} \times 100 = 20\%$$

This condition would equate to approximately 3.7% excess oxygen for this oil.

3.3 Total air

The total air for combustion is therefore made up by the theoretical plus the excess air and can be expressed as mass air/unit mass of fuel burnt, or more usually as the air/fuel ratio. The air/fuel ratio in the residual fuel oil example quoted is 16.62/1.

Therefore from a knowledge of the fuel weight and the excess air requirement it is possible to arrive at a total air weight, and thus to establish fan sizes and other plant parameters. The excess air requirement for any given plant is normally defined by the design of the boiler and combustion system. However, it is normal practice for an allowance of approximately 20% excess air to be made on a large coal-fired unit, and for the design excess air to be as low as 2–2.5% for a large residual oil-fired unit.

Example

The fuel consumption of a 500 MW boiler is 200 t/h burning a bituminous coal which contains 18.5% ash and 8.00% moisture. Thus the theoretical air requirement is:

$$= 200 \times \frac{100 - (18.5 + 8)}{100} \times 11.18$$

$$= 200 \times 0.735 \times 11.18 = 1643.5 \text{ tonnes}$$

Total air required $= 1643.5 \times 1.2 =$ *1972.15* tonnes (20% excess)

As already stated, approximately 20% of the air mass supplied to a coal-fired unit is used for carrying the coal and is fed through the mill at an inlet temperature of up to 300°C to produce a mill outlet temperature of approximately 90°C. This air is known as primary air, and is normally used in an air/coal mass ratio of 1.8–2.5/1.

No primary air is required on an oil-fired unit but the air stream is split on certain applications to reduce emissions.

3.4 Gas weight

The total gas weight is given by the total of air weight + fuel weight (excluding ash), as all the constituents of the fuel, with the exception of a proportion of the ash, pass through the system as exhaust gas.

Thus for the example quoted above the gas weight is:

$$= 1972.15 + 200\,(1 - 0.185)$$

$$= 2135.15 \text{ tonnes}$$

$$\text{Thus the gas/fuel mass ratio} = \frac{2135.15}{200} = 10.67$$

4 Furnaces

4.1 Furnace duty

The modern furnace is designed to have the following:

- A surface area sufficient to reduce the temperature of the furnace gases to a level acceptable to superheater requirements.
- Adequate water circulation in the furnace tubes. This means a tube diameter which has a friction pres-

sure drop which allows sufficient fluid flow to avoid overheating. Natural circulation boilers have larger diameter tubes (with smaller friction drops) than pump-assisted circulated boilers because of the reduced driving force.

- A dimension in the path of the burner sufficient to avoid flame impingement on the opposite wall, but not so excessive that the heat transfer is reduced.

- A width sufficient to accommodate all burners on acceptable pitchings to avoid flame impingement on side and division walls, and interference of one flame with another.

- Overall dimensions and shape sufficient to ensure a gas path which will *fill* the furnace, provide optimum absorption to all parts and a fuel ash particle residence time sufficient to ensure burn out.

- For a coal-fired boiler the furnace together with the platen must have sufficient area to cool the gases below the ash softening temperature, this is to ensure that deposits outside the furnace envelope are solidified.

4.2 Furnace ratings

The dimensions of a furnace vary to accommodate the factors described previously. Typical ratings for coal and oil firing are given in Table 1.6.

The heat release rating can only be regarded as a guide. Projected area rating is indicative of the furnace gas temperature at the nose, whilst volumetric rating is indicative of the residence time allowed for burn out. Plan area together with burner belt height is indicative of the burner belt heat release rates.

These factors are only useful as a comparison between a new design and existing boilers to ensure that operation is within existing experience. A much more detailed examination of the heat transfer and combustion processes must be carried out before a design is considered acceptable.

On coal-fired units a platen transverse pitch of at least 0.6 m is required to avoid ash bridging with UK coals. This pitch is a minimum to overcome the mechanical strength of ash.

4.3 Comparison between coal and oil furnaces

It can be seen from the rating table that oil-fired furnaces are more highly rated than coal-fired furnaces. A number of factors lead to this difference:

- The average oil droplet burn out time is roughly half that of an average pulverised coal particle.

TABLE 1.6

Coal and oil-fired furnace ratings

	COAL				OIL
Type of firing	Opposed	Front	Corner	Corner	Opposed
MCR load	660	500	500	350	660
No. of furnace chambers	2	1	2	2	2
Heat release up to nose — projected area kW/m^2	551.9	544.7	502	433	753
Heat release up to nose — volumetric kW/m^3	211	196.3	228	207	342
Plan area MW/m^2	5.7	5.0	5.05	5.0	8.0
Residence time based on total volume up to nose — seconds	2.4	2.5	2.29	2.28	—
Vertical distance from top burner to nose — metres	12.65	13.4	9.75	11.12	—
Residence time based on volume between top burner and nose — seconds	1.14	1.37	0.972	1.02	—
Platen pitching — metres	0.762	0.616	0.609	0.663	—

- In coal-firing, unlike oil-firing, there is a small percentage of oversize particles which require higher residence times than average particles.

- Heat transfer to the furnace walls is dependent on the amount of ash adhering to the walls and, because a coal-fired furnace has much more ash on the walls than an oil-fired furnace, there is higher resistance to heat transfer in the coal-fired furnace.

The result is that simply to accommodate the combustion process the coal-fired furnace has to be much larger than an oil-fired furnace. The lower heat transfer rates and the requirement to reduce gas temperature below ash deformation temperature in the coal-fired furnace further increases the size and/or surface requirement and, instead of having to provide surface for evaporation duty alone, coal-fired furnaces for high pressure boilers have also to incorporate superheater surface. A size comparison between a typical coal and oil furnace is shown in Fig 1.15.

For coal-firing, special provision has to be made for ash collection at the bottom of the furnace. The furnace hopper slopes at around 60° to the horizontal to ensure free flow of ash, whilst for oil-firing this is not necessary and the slope can be nearer to the horizontal.

4.4 Methods of controlling furnace performance

Apart from controlling the furnace performance by operational means such as firing patterns, selective sootblowing and excess air; there are two other principal methods — tilting burners and gas recirculation.

Tilting burners, which are only employed in corner firing, are positioned in corners of the combustion chamber with the burners firing slightly offset from the chamber wall towards an imaginary circle to which the flame path is tangential. This results in a rotating fireball. In moving the fireball up or down by altering the burner tilts in unison, the heat transfer in the

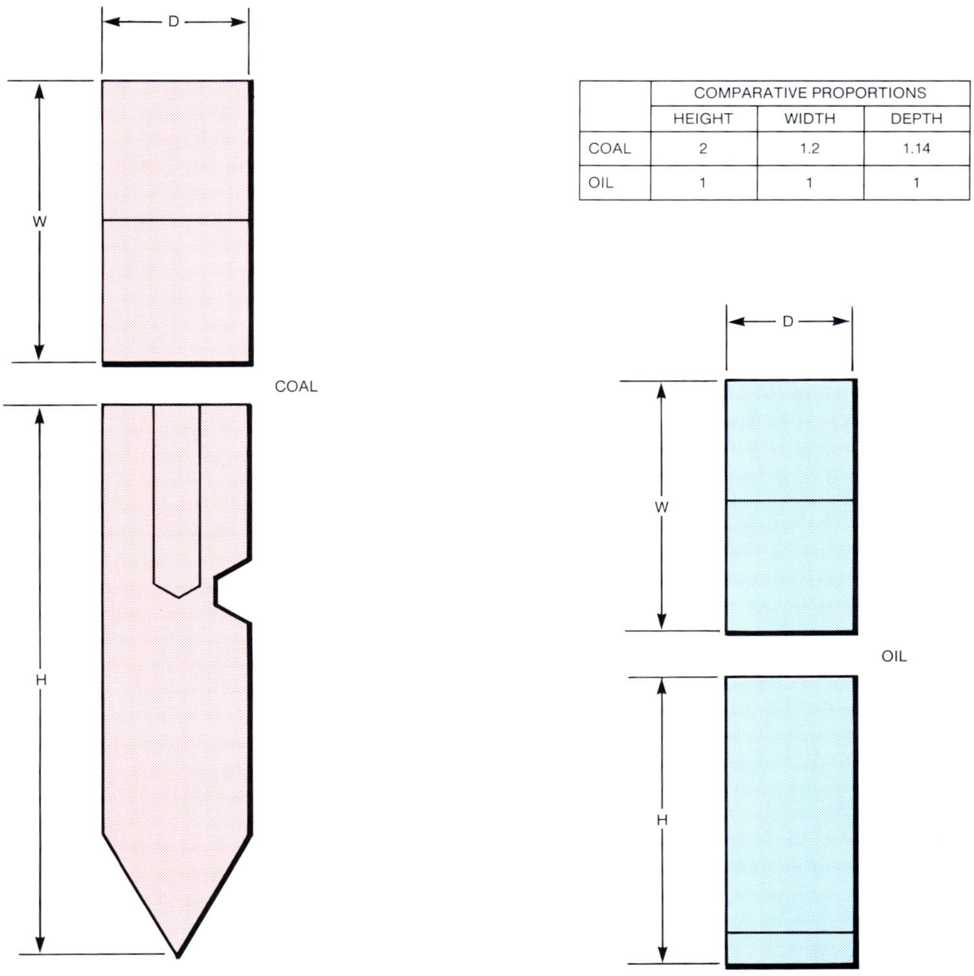

	COMPARATIVE PROPORTIONS		
	HEIGHT	WIDTH	DEPTH
COAL	2	1.2	1.14
OIL	1	1	1

FIG. 1.15 Size comparison of coal and oil furnaces

furnace chamber can be altered. The usual tilt range is $\pm 30°$. With the fireball in the bottom of the furnace, the heat transfer to combustion chamber will increase because the surface area seen by the combustion product is at a maximum. Conversely with the tilt in an upward direction and the fireball at the top of the furnace, the heat transfer is at a minimum.

In the case of gas recirculation, gas is taken from the back end of the boiler and reintroduced into the furnace hopper region. For oil-firing, the gas is recirculated from the economiser outlet, the pressure head being supplied by gas recirculation fans. For coal-firing, the gas may be taken from the same position but to avoid heavy wear rates in the recirculation fans and the cyclone dust separators upstream of the fans, it is preferable to take recirculated gas from after the precipitator. The pressure head can either be provided by gas recirculating fans or by the induced draught fan by provision of a modulating damper at the ID fan outlet.

The recirculated gas reduces the temperatures generated in the combustion process and, as heat transfer in a furnace is mainly radiation, and is therefore a function of flame temperature to the fourth power, heat transfer to the furnace is reduced. The net effect of this process is to reduce furnace exit temperature but, because of the greater gas mass flow, the heat content of the gas leaving the furnace is increased. This increased heat content can be utilised in the superheaters and reheaters because the increased convection heat transfer coefficients, resulting from the greater mass flow, offset reduced gas to steam temperature differences. The overall effect is that increased gas recirculation reduces furnace duty and increases superheater and reheater duty.

4.5 Construction

Evaporator tubes of boilers of ratings up to 120 MWe were allowed to expand within a fixed outer casing. The casing itself was insulated from the evaporator tubes and furnace gases by refractory material fixed to the walls by studs or pins. Problems arose, however, when the refractory materials became detached from the casing and caused jamming of the tube expansion attachments.

Boilers of ratings of 120 MWe and above are arranged with buckstays, casing and insulation supported from the evaporator tubes themselves. This design is known as *skin casing*. Attachments supporting the buckstays must cater for relative differential expansions whilst transmitting pressure forces without causing jamming, tube distortion or seal failures. Initially, skin casing was used with close pitched tangent tubing, but in some of the 500 MWe boilers and all the 660 MWe boilers water walls have been constructed with tubes joined by fins to form a fully welded structure, the membrane wall. The use of membrane walls ensures that the furnace walls are sealed, thus minimising the ingress of tramp air. The outer casing is now a heat shield instead of a pressure shield.

Membrane walls may be manufactured from tubes having integral fins or by welding the fins to plain tubes. Large panels may be fabricated in the manufacturer's workshops and transported to the power station site for final erection. This reduces site erection time and allows for better quality control, since a large proportion of furnace tube welding and inspection can be carried out in works conditions.

Most of the CEGB power stations comprising units in the range 200–600 MWe were constructed in the 1960s and 1970s. Many of these high efficiency, high merit order stations are now nearing their nominal pressure part design lives. Since the early 1980s it has been a CEGB policy to consider life extension refurbishment of these stations where original plant and its operational performance has been judged to justify this approach. Extension of life leads consequently to an increase in thermal and mechanical cycling. This requires a new emphasis to be placed on problems associated with fatigue, for example, the cracking of electrical-resistance-welded membraned tubes in areas subject to thermal and mechanical fatigue stresses due to two-shifting. Experience in this area has led to electrical-resistance-welding of membrane tubing being no longer acceptable.

The boiler must be protected as much as practicable from the effects of implosion or explosion, caused by instability in combustion or draught plant operation. The applicable standard for the design is NFPA 85 'Prevention of furnace explosions in pulverised coal-fired multiple burner boiler-furnaces' [2]. Boiler containment is achieved by a series of buckstays, or girths, which encircle the boiler casing. These beams are connected by vertical stays which give stability to the assembly. Horizontal differential expansions may be accommodated by swivel link connections between the ends of the buckstays at the furnace corners. Originally buckstays were designed to cater for the explosion condition only. However, a series of incidents where loss of firing caused high negative pressures in the furnace and consequent damage has required design for this condition as well as explosions. Buckstay buckling under these conditions is prevented by the installation of vertical restraints to the buckstay outer compression flanges.

In drum type boilers with vertical furnace tubes, the weight of the buckstay system and the ash loads can be taken by the tubes without any increase in tube thickness because the tubes are designed for hoop stress and axial stresses, due to steam pressure, are only half the hoop stress. In once-through boilers with helically wound or horizontal furnace tube systems, the weight of the furnace tubing, the furnace structure and ash loads are additive to the hoop stress. Tube thicknesses can be increased to take some of this stress but in the majority of designs, vertical straps are provided to support the wall at intermediate levels. These straps are either hung from the boilerhouse steelwork through constant load supports or, in later designs, they are attached to vertical tubes in the upper part of the furnace. To reduce differential expansion between furnace wall and straps, the straps are installed adjacent to the tubes and enclosed in insulation.

4.6 Maintenance

Routine inspection of furnaces is necessary to detect signs of trouble in its early stages and in order to correct possible faults in firing, combustion, circulation, sootblowers, etc., before serious damage can result. The procedure adopted during plant overhauls will vary according to design and operating history of individual boilers, but extreme vigilance is necessary through the life of the boiler, particularly in plant where change in operating patterns, such as the adoption of two-shift operation, is made.

When the boiler has been taken out of service, entry to the combustion chamber is made safe by ensuring that the walls and heating surface above the furnace are free of slag. The furnace is scaffolded-up from the hopper slope and rigid work platforms are provided at intermediate levels up furnace walls. The furnace walls are then inspected for areas of high corrosion, erosion due to sootblowing and damage due to falling slag. The integrity of the fin-to-tube welds in membrane wall furnaces is checked and in tangent tube walls the alignment of tubes is checked and, where necessary, tubes are pulled back to the furnace structural steelwork. Tube thickness can be measured by ultrasonic techniques and any tubes which show such thinning that they are unlikely to last until the next major inspection, currently a three year period, are replaced. It is normal practice also to remove a limited number of tubes for internal inspection to ensure that a protective magnetite layer is being maintained. If excessive oxide build up inside the tubes is observed, consideration has to be given to acid cleaning before returning the boiler to service.

Replacement of furnace tubes requires welding and inspection to the original *as built* standards. To ensure this, it is essential to provide good working conditions and have sufficient time in the maintenance schedule for inspection and any subsequent repairs.

5 PF burner types and location in furnace

There are three distinct methods of firing pulverised fuel in a furnace. They are:

(a) Horizontally from one or from two opposed furnace walls — called *wall firing* or *opposed firing*.

(b) Downwards from an arch — called *downshot* firing.

(c) From the corners of the furnace, the coal jets firing tangentially to an imaginary circle in the centre of the furnace — called *corner* or *tangential* firing.

5.1 Wall firing

All boiler plant employing wall (or horizontal) firing has burners situated in the front, back or front and back (opposed firing) furnace walls. A swirl-stabilised form of burner (register) is used which generates the necessary flame stability by means of a combination of secondary air swirl and a flow reversal generated by a central body, the primary air tube, and in some cases an impeller mounted on the oil burner support tube. A divergent refractory quarl is also used to reinforce the recirculation patterns and aid flame stability. It was thought in the past that the quarl aided flame stability as a hot radiant source, but it has now been established that the effect is totally aerodynamic.

An early example of the conventional coal burner is the Foster Wheeler intervane burner (Fig 1.16). In this type of burner pulverised coal transported by primary air is blown tangentially into a central tube and thence into the furnace. The secondary air is admitted to the combustion system through radial inflow swirl vanes, the recirculation generated by the swirl and the quarl being sufficient to stabilise the flame. In this burner about 80% of the combustion air is passed through the secondary air vanes, the additional 20% being admitted as primary air with the coal. The Babcock circular burner is of similar design and operates in a similar fashion, but the coal entry is by means of a 90° elbow with splash plate as opposed to the snail or tangential entry of the Foster Wheeler burner (see Fig 1.17).

Both the Foster Wheeler and the Babcock burners originally incorporated variable-angle swirl vanes, which have been modified by the CEGB to a fixed design. There were several reasons for this; amongst them the inaccuracy and failure of the swirl vane drive, the poor efficiency as a swirl generator and the vulnerability to operational damage.

Windbox pressures of 5–9 mbar are normal with secondary air temperatures of 250–300°C, to give a secondary air velocity at the burner exit of about 36–38 m/s.

Modern designs of coal burner incorporate axial flow swirl vanes which offer a number of practical advantages when compared with the radial in-flow type. This type of swirler is more efficient because it produces the same swirl for less pressure drop and therefore requires less fan power. It can be sited at the entry to the burner away from furnace heat, which increases component lifetimes. It can be more easily adjusted for swirl level, and allows a reduced air register diameter (but not necessarily outlet). This type of swirler also lends itself to the separated and controlled airflow necessary for the reduction of gaseous emissions.

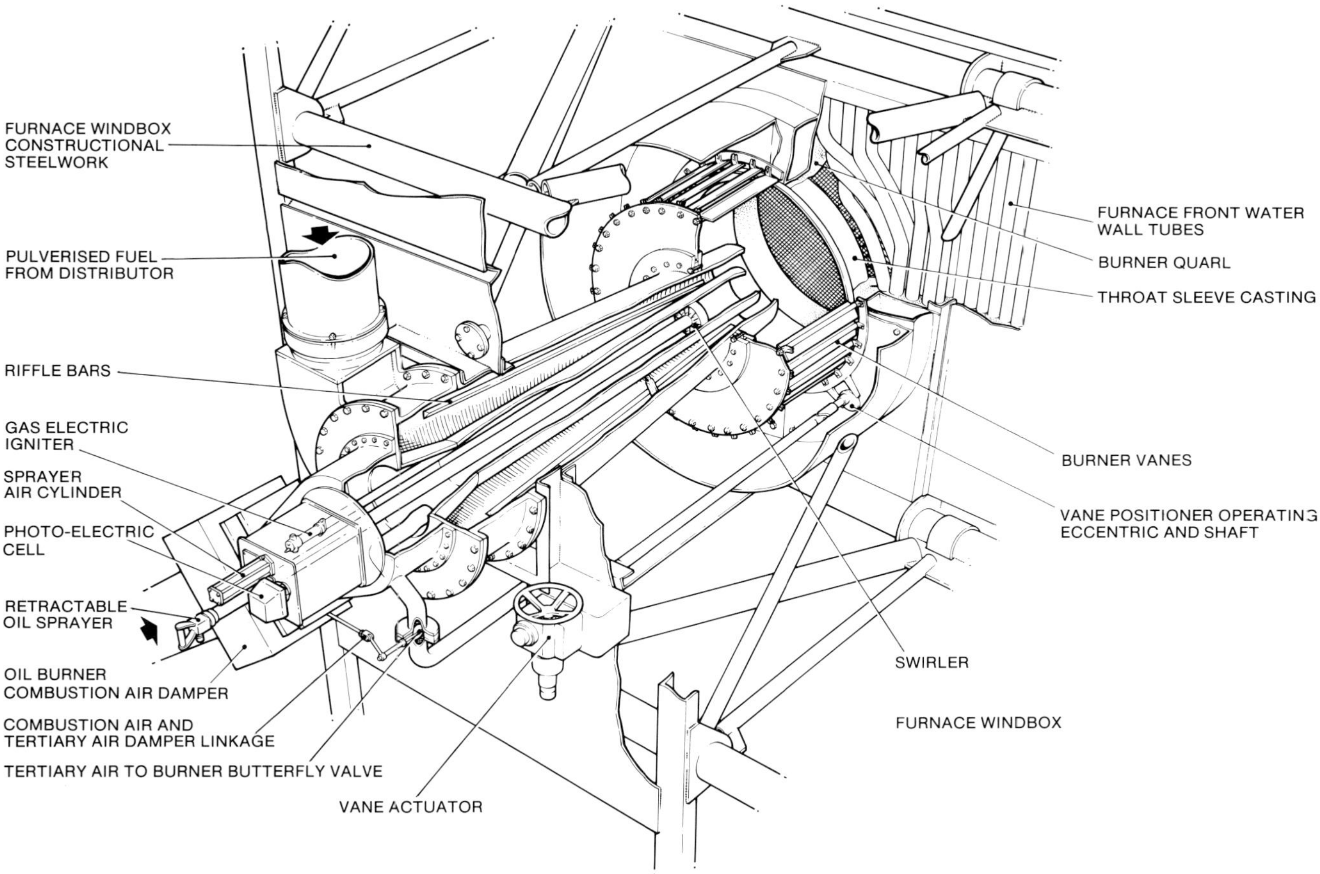

FIG. 1.16 The Foster Wheeler intervane PF burner

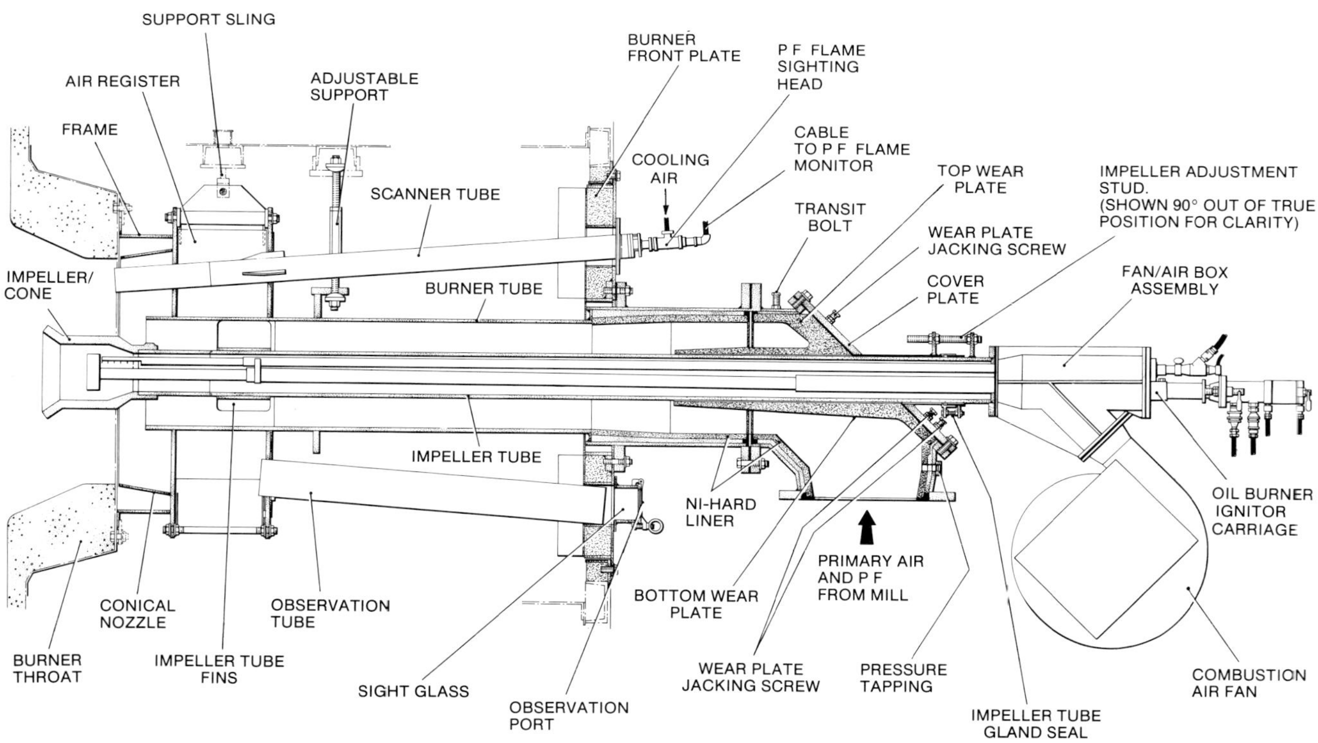

FIG. 1.17 The Babcock PF burner

Burner numbers for a given boiler are largely determined by the size and numbers of mills to be used. In terms of boiler design the burner-to-sidewall, and burner-to-backwall dimensions must be sufficient to avoid flame impingement and the possibility of rapid high temperature corrosion (see Section 2.2 of this chapter). Some typical burner spacings are given in Table 1.7 for various designs of large boilers.

In general there are cost advantages available if burner numbers are reduced, e.g., fewer burners and boiler penetrations, quarls, etc., and fewer burners are more easily fitted into the space available on the boiler walls.

In the past, boiler width or depth has been limited by sootblower travel, this resulting in a shallower, taller boiler than is strictly necessary. However, with modern designs of blower 900 MW boiler designs are completely practicable. In general, turbulent burners in a wall-fired setting will give efficient combustion resulting in carbon in ash of between 1.5% and 4.0%. Such carbon levels are acceptable and do not prevent the sale of the ash for constructional purposes.

5.2 Downshot firing

The vertical or downshot method of firing is adopted where coals of low volatile content (anthracite) are the main fuel. A volatile content of 16% or less is typical.

Low volatile coal requires a long particle residence time for complete combustion and this results in the U-flame concept. Thorough mixing of the coal and air is obtained by dividing the air and coal into alternate streams or by admitting the coal in a thin stream surrounded by air (see Fig 1.18). In this system about 40% of the combustion air is provided at the burner and is projected downwards from the arch towards the furnace bottom, the remaining 60% of the air being fed into the flame through multiple ports arranged in the boiler sidewall. The quantity of air fed through these sidewall ports is varied to adjust the flame characteristics (length, etc.) and reduce slag formation to a minimum.

This system is now only used in the CEGB at Aberthaw Power Station and no more such installations are contemplated.

5.3 Tangential firing

The corner-fired furnace has been developed in this country by International Combustion Ltd. It depends for its operation on the maintenance of a turbulent zone in the centre of the furnace by directing a non-turbulent flame horizontally from each corner of the furnace towards an imaginary circle to which the flame path is a tangent.

The burner itself is of simple construction and mixing of coal and air is obtained by the admission through the burner of alternate layers of coal and air (see Fig 1.19). In order to achieve the necessary fuel penetration and flame stability a windbox pressure of 8–12 mbar is used.

The burner can be tilted through a range of ±30° about the horizontal to maintain steam temperature control. Because of problems experienced with the tilting mechanism, some operators have fixed the tilt and rely on other methods of steam temperature control. However, tilt is useful in a furnace which is arranged to reduce nitrogen oxide (NOx) emissions (see Section 5.6 of this chapter).

Steam temperature control is effected by the change in furnace exit gas temperature which results from the tilt.

5.4 PF ignition systems

Oil burners are always used in CEGB boilers for coal burner ignition, stabilising and furnace start-up purposes.

In the case of wall-fired units each coal burner is fitted with an oil burner mounted in a central support tube (see Fig 1.20). In corner-fired units it is usual to mount an oil burner between each pair of coal burners, e.g., in the twin furnace 500 MW designs there are six mills per boiler and 24 oil burners giving three oil burners per corner associated with six coal burners per corner.

TABLE 1.7

Typical burner spacings

BOILER SIZE MW	NO. OF BURNERS	INDIVIDUAL BURNER INPUT MWth	BURNER SPACING HORIZONTAL (m)	BURNER SPACING VERTICAL (m)	BURNER TO SIDEWALL CLEARANCE (m)
500	48	37	1.3 — 2.4	2.3	2.4
500	24	58	2.5	2.6	2.1
660	60	37	2.59	2.29	3.9

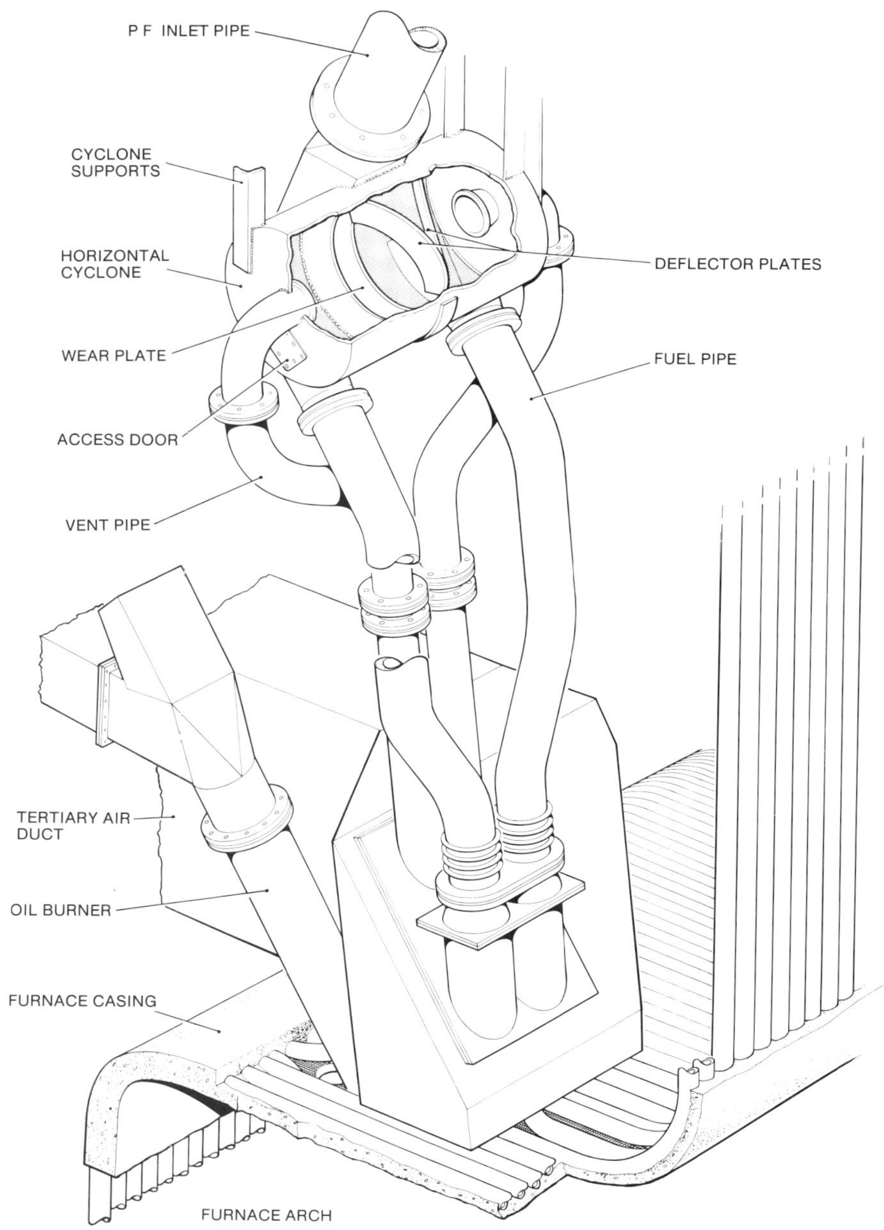

FIG. 1.18 The Foster Wheeler cyclone PF burner

These oil burners are normally designed to meet the requirements of GDCD Standard 166 in modern plant [3]. Each burner is a self-contained unit with individual gas electric ignition systems, burner control gear and individual core air fans, operated by a sequence control unit.

The total oil firing capacity of the installation varies but is usually sized to provide 15–20% boiler MCR. Oil inputs of up to 30–35% have been used at certain times to provide a coal replacement option.

The use of oil burners for boiler start-up is necessary to avoid bringing mills on and off load (called *flashing*) at low boiler loads. This flashing contravenes the requirements of GOM 105 [4].

5.5 Burner windboxes

As burner capacities have increased, windbox dimensions have increased correspondingly to meet secondary air requirements. A difficulty arises on very wide furnaces fitted with side entry windboxes of ensuring that all burners receive the correct proportion of air at all times. Aerodynamic and model studies are now made of the windboxes and their supply ducts, the final designs incorporating flow deflectors, and having flow areas adjusted to give the required conditions. In another form, the windbox system comprises a plenum chamber from which individual ducts lead to groups of burners. An example of this form was at Ferrybridge *C* power station. However, this is being modified to

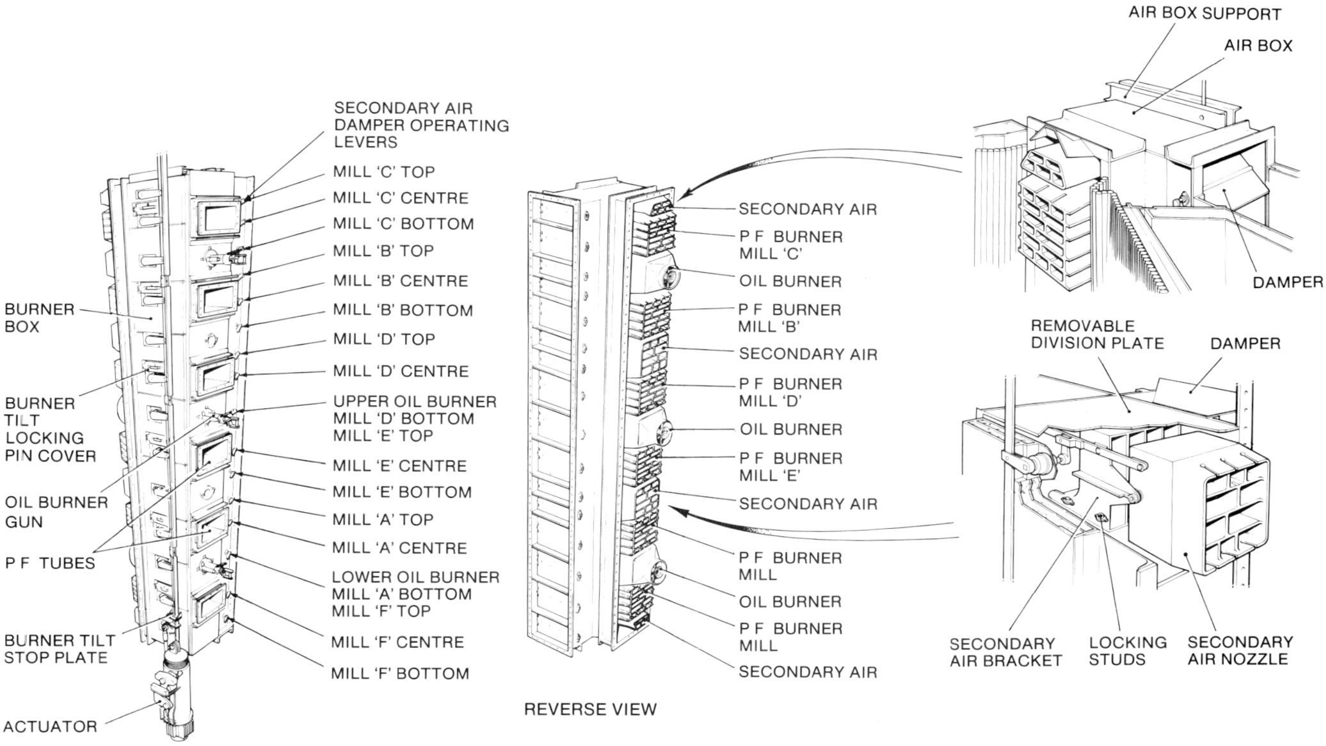

FIG. 1.19 The International Combustion corner-fired burners

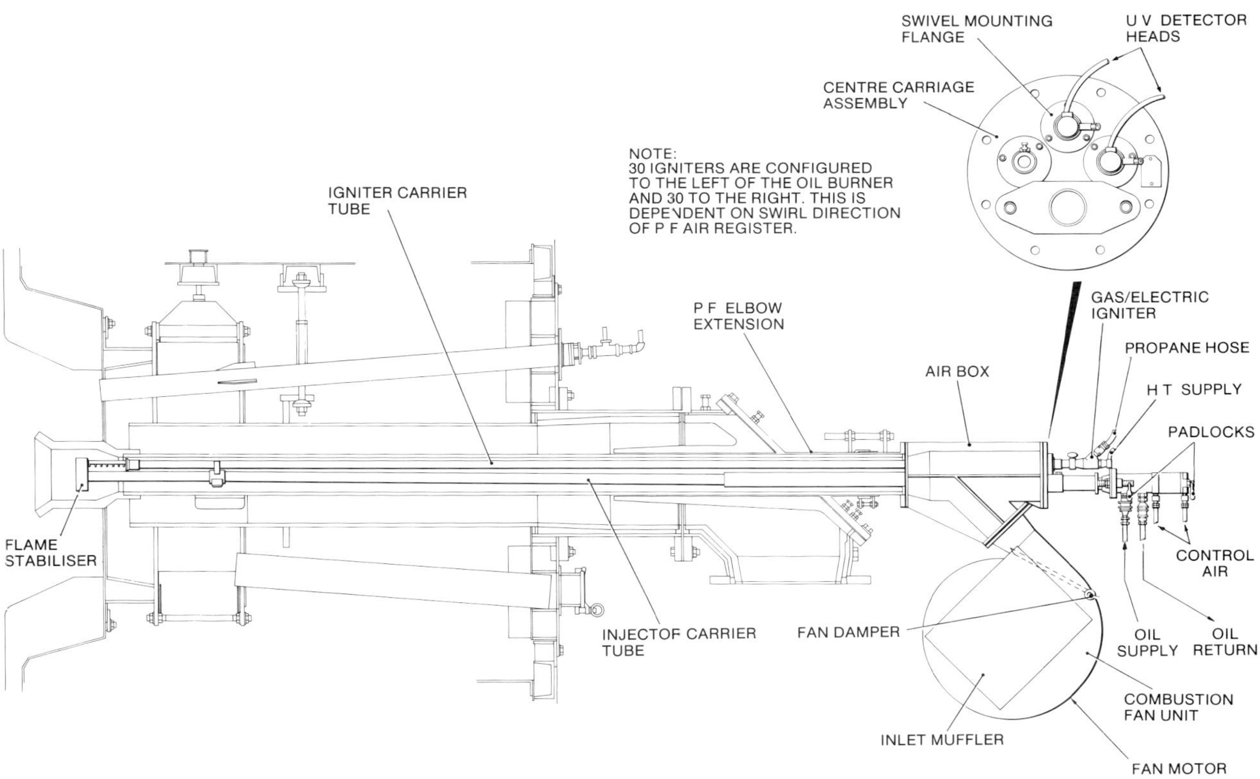

FIG. 1.20 Oil burner

individual ducts to each burner with the intent of improving the burner to burner air distribution.

In practice common windboxes normally achieve ±5%, or a little worse, airflow to individual burners (about a mean), with compartmented windboxes achieving somewhat better than this. There is greater emphasis nowadays on accurate airflow distribution particularly with the modern interest in low NOx burners, which require accurate metering of air and coal on a burner to burner basis in order to obtain the required NOx reduction.

5.6 Reduction of nitrogen oxides

Emissions standards for CEGB power stations are dealt with by Her Majesty's Industrial Air Pollution Inspectorate (HMIAPI). Discussions between the CEGB and HMIAPI indicate that limits on oxides of nitrogen emissions will be imposed probably in accordance with the EEC proposals, when ratified.

It is apparent that any new coal-fired plant ordered by the CEGB will be required to operate with NOx emissions approximately 65% lower than current levels for wall-fired plant, and 40% for corner-fired plant.

This level of NOx emissions may be achievable using a combination of NOx reduction techniques, along with changes in boiler design and operation.

5.6.1 The formation of oxides of nitrogen in combustion plant

The formation of oxides of nitrogen in combustion processes may be explained in terms of the source of nitrogen required for the reaction, that is, atmospheric nitrogen or organically bound nitrogen present in the fuel.

Thermal NOx is formed by the oxidation of atmospheric nitrogen, the mechanism of which was first described by Zeldovich [5]. Tests have shown that thermal NOx levels downstream of the flame front may be predicted by the following equation:

$$[NO] = K_1 e^{(-K_2/T)} [N_2][O_2]^{\frac{1}{2}} t$$

where $[\,]$ = mol fraction

T = temperature

t = time

K_1K_2 = constants

e = Naperian log base

From this equation it can be seen that thermal NO may be decreased by reducing time, temperature and concentration of nitrogen and oxygen.

Prompt NOx, again formed by the oxidation of atmospheric nitrogen but generated in the flame front, where, due to the very short residence time, the Zeldovich reactions are inadequate.

In all cases a reduction in temperature decreases prompt NO levels; however, increasing the oxygen concentration increases NO for fuel-rich flame fronts, and decreases NO for fuel-lean flame fronts.

Fuel NOx is formed by the oxidation of organically-bound nitrogen contained in the fuel. There appears to be no correlation between the quantity of fuel-bound nitrogen and fuel NOx, the fuel nitrogen conversion rate varying widely, depending more on coal rank than fuel nitrogen content. However, experimental results suggest there may be some correlation between fuel NOx and fuel nitrogen/oxygen ratio.

From the foregoing, the most important criteria relevant to the control of both thermal and fuel NOx for coal firing may be summarised as follows:

- Coals with the lowest fuel nitrogen and fuel oxygen/nitrogen ratios generally will produce the lowest NOx.
- Fuel NOx may be minimised by controlling the quantity of air available to the fuel during the early stages of combustion.
- The thermal NOx contribution to total NOx may be reduced by operating at low excess air and limiting peak gas temperatures throughout the furnace.
- Total NOx may be reduced by splitting the primary air/PF into two streams, one fuel-rich and one fuel-lean, the resulting average NOx level being lower than that generated from a single flame at the normal boiler operating burner zone air fuel ratio.

Most current NOx reduction systems utilise one or more of the foregoing techniques.

5.6.2 Proven NOx reduction technology

Burner zone heat release rate (BZHRR)

The burner zone heat release rate (the net heat input to the burner zone divided by the effective projected heat transfer surface) affects flame temperature, thereby determining the amount of thermal NOx generated.

On units equipped with highly turbulent burners, a 30% decrease in BZHRR may possibly result in a similar reduction in NOx levels. However, the affect may not be so marked when low NOx burners are installed, due to the lower peak flame temperatures associated with these units.

Although only having a small affect, a reduction in BZHRR may be an added benefit when furnace dimensions are altered to facilitate the incorporation of NOx reduction systems such as low NOx burners and reburning.

Overfire air (OFA)

The utilisation of overfire air is a well known technique for NOx reduction. The units' overall excess air level is unchanged, but a significant proportion of the air is injected above the burner zone, the burners themselves operating substoichiometrically. This has the effect of reducing the availability of oxygen to the flames during the early stages of combustion and lowering peak furnace gas temperatures.

However, there is a practical limit to the amount of overfire air which may be used, since operating the lower section of the furnace substoichiometrically can lead to increased slagging and tube wastage problems, particularly with British coals, as well as increased carbon in ash levels. These problems may be offset to some extent by the incorporation of air ports adjacent to the furnace walls, to protect the water tubes from the corrosive gases by producing a protective boundary layer of air.

Nevertheless, practical NOx reductions of around 20% are obtainable from the introduction of overfire air.

Low NOx burners — first generation wall-fired

The first generation low NOx burners generally incorporate air staging, and in some cases stratification of the pulverised coal, to achieve significant reductions in NOx emissions.

Air staging is accomplished by the utilisation of either dual registers, often with variable swirl generating components, or by the introduction of tertiary air through ports adjacent to the burners in the furnace wall. This has the effect of reducing the oxygen available to the flame during the early stages of combustion, hence controlling the generation of oxides of nitrogen.

The concentration of pulverised coal into discrete streams results in an additional delay before oxygen from the combustion air is available to a proportion of the fuel.

The variable geometry (swirl vanes, coal nozzle position, etc.) allows the burner to be optimised with respect to stability, flame length and shape, and NOx level generated for each particular installation. As for overfire air, boundary air ports may be required to protect furnace walls.

Examples of first generation low NOx burners are the Foster Wheeler Energy Corporation 'controlled flow/split flame' (CFSF) and Riley Stoker 'controlled combustion venturi' (CCV) burners.

Reductions in the emissions of oxides of nitrogen of approximately 60% compared to highly turbulent burners are claimed for low NOx burner systems, however these claims have yet to be substantiated on UK plant. It is estimated that a combination of these items may result in reductions of NO emissions in the range 40–70%.

Low NOx burners — first generation corner-fired

As described previously, NOx is produced within the furnace combustion system by the oxidation of both atmospheric nitrogen and the nitrogen chemically combined with the fuel. In order to limit the NOx output it is necessary to reduce the oxygen available to the nitrogen in the primary combustion process (by fuel-rich operation) and subsequently to control the overall gas temperature to limit the further oxidation of nitrogen while completing combustion (fuel-lean operation). In the low NOx concentric firing system (LNCFS) developed by Combustion Engineering (USA), the fuel-rich stage is achieved by directing the fuel and secondary air streams to different parts of the flame stabilising vortex, giving a fuel-rich region in the centre of the furnace and an air-rich region close to the furnace walls. In addition the general level of excess oxygen in the primary combustion zone is lower than normally employed, the balance being introduced as overfire air high in the furnace.

Thus the main NOx reducing parameters are met by the following measures in a corner-fired unit:

- Reduction of secondary air quantity injected through the fuel nozzles
- Diversion of the secondary air injected through the auxiliary secondary air nozzles towards the furnace walls
- Injection of the remaining secondary air through ports above the top fuel nozzles on the original firing circle (overfire air).

A CEGB 500 MW corner-fired boiler has been converted to this system and NOx reductions of up to 40% obtained without increase in carbon-in-ash losses and an improvement in the slagging characteristics of the boiler.

Reburning

Reburning is based on the principle that under certain conditions, namely high temperature and low oxygen concentration, hydrocarbon-based radicals strip oxygen from NO molecules, the remaining nitrogen ions combining to produce molecular nitrogen.

The reburn hydrocarbons are combusted with additional air which is added above the reburn zone, at a lower temperature, which restricts the generation of further oxides of nitrogen.

The mechanisms may be represented by the following overall equation:

$$2NO + 2C_nH_m + \left(2n + \frac{m}{2} - 1\right)O_2 \leftrightharpoons N_2 + 2nCO_2 + mH_2O$$

In practice the boiler main burners are operated as normal and reburn fuel is injected into the furnace above the burner zone (and OFA ports if fitted). At the point of injection the furnace temperature should be in excess of 1200°C.

The reburn fuel (pulverised coal) is transported by an inert carrier gas (usually recirculated flue gas), and a minimum residence time of 0.1 seconds is required. Generally, as the reburn fuel fixed carbon to volatile matter ratio increases, the efficiency of the deNOx process is reduced.

Oxidant is injected through additional air ports above the deNOx zone, and again, a minimum residence time of 0.1 seconds is necessary to complete carbon burn out.

Oxides of nitrogen are generated in the *combustion completion zone*, the actual concentration being dependent on the fuel nitrogen content of the reburn fuel and temperature. (Increasing fuel nitrogen increases the amount of NO produced.) Nevertheless, net NOx reductions of 50%, regardless of inlet NOx concentration, are claimed.

Disadvantages of the reburn process include:

- Probability of increased tube wastage problems.
- The extra furnace height required to accommodate the necessary residence times.
- The additional cost and complexity associated with the reburn system.
- The possibility of increased carbon-in-ash levels.

Reburn technology is marketed by several Japanese companies, for example, the Mitsubishi Advanced Combustion Technology (MACT) and Hitachi In Furnace NOx Reduction (IFNR) processes.

Reburning technology has yet to be demonstrated on UK plant, and there is little reported data regarding the use of PF as the reburn fuel.

Miscellaneous

There are various other NOx reduction techniques which have been disregarded due to the minimal decreases in emissions obtainable, their effect on boiler operation, or the associated costs due to additional plant or efficiency penalties, for example:

- Low excess air combustion.
- Flue gas recirculation.
- Reduced air preheat.
- Water injection.

5.6.3 Future NOx reduction technology

Low NOx burners — second generation

Second generation low NOx burners generally depend on the principles of both fuel and air staging. Air staging is similar to the first generation low NOx burners, whereas fuel staging is achieved by splitting the primary air/PF flow into two streams.

The Mitsubishi Pollution Minimum (PM) burner, as applied to tangentially-fired boilers, operates with adjacent fuel-rich and fuel-lean flames in conjunction with separate gas recirculation (SGR). When installed with overfire air the manufacturers claim NOx emission levels as low as 150 PPM.

Foster Wheeler Energy Corporation is developing a fuel staging burner, from which it believes NOx reductions of up to 85% compared to highly turbulent Foster Wheeler intervane burners may be achievable. This burner has a very substoichiometric gas pilot stabilised flame along the burner axis, which is surrounded by the main flame. The internal flame produces partially-combusted products which then pass through the main flame, apparently resulting in a reburn effect similar to IFNR/MACT.

The fuel staging burner is said to have a low load capability which may result in the reduction of the amount of auxiliary fuel required during low load operation, with associated cost savings.

6 PF milling plant

6.1 Choice of plant

Both of the UK boiler makers have their own designs of mills and because of this they usually offer these for their own boilers. To avoid split contractual responsibility the CEGB has up to now accepted this situation and it has resulted in a variety of mill types for similar design requirements. The only exception has been in the case of low volatile coals for which tube ball mills have been preferred. This is because these coals need to have their fineness of grinding maintained for burner combustion reasons, and this is better done with tube ball rather than vertical spindle mills.

Tube ball mills are also more suitable than vertical spindle mills for grinding coal that has a very high and abrasive ash content. Grinding this coal in a vertical spindle mill will result in frequent major mill outages for replacement of the grinding elements, whereas it is easy to carry out on-load replacement of the grinding ball charges on tube ball mills. These do have the disadvantage of being more susceptible than vertical spindle mills to the adverse effect of high moisture coals.

All types of mills are capable of grinding to the same fineness if they are correctly sized and provided that the mill and furnace requirements are correctly specified there is no reason why the mill maker and the boiler manufacturer need to be the same firm.

Assuming identical grinding performance and neglecting commercial considerations, the choice of mills depends on power consumption, maintenance cost and plant layout problems. These features are dealt with in greater detail in later sections.

6.2 PF milling plant and associated equipment

Plant used by the CEGB is almost exclusively either tube ball mills or vertical spindle mills; this choice results from the properties of coals burnt.

A general survey of the various types of mills and the theory of grinding is given by Lowrison [6].

6.2.1 Tube ball mills

Some tube ball mills are operated under suction whilst the others are pressurised. Suction mills use an exhauster fan to lift the milled coal out of the drum and this results in heavy erosive wear of the exhauster impeller and liners. A pressure type mill has air supplied by the primary air fan, which is handling comparatively clean air and hence not subject to a high rate of erosive wear. However, a pressurised mill requires efficient drum trunnion seals to prevent outwards leakage of pulverised fuel.

To obtain satisfactory cascading of the ball charge, the mill drum is rotated at about 70% of the speed at which the ball charge would be held against the inner surface by centrifugal force.

Tube ball mills have the advantage of on-load replacement of the grinding balls but the disadvantage of high specific power consumption at part load. Whilst the tube ball mill itself is more reliable than a vertical spindle mill, its reliability can only be that of the mill group, i.e., coal feeder plus mill plus fan, etc. The CEGB operates tube ball mills with a capacity of up to approximately 75 tonnes/hour. A general survey of the theory of tube ball mills is given by Rose and Sullivan [7].

A typical tube ball mill is shown in Fig 1.21.

6.2.2 Vertical spindle mills

This type of mill is used by the CEGB in greater numbers than any other. Examples of it are the Babcock *E* and the NEI/International Combustion LM type. The Babcock mills operate under pressure as do the later versions of the LM mills, whilst the older LM mills operate under suction. The CEGB also operates other makes of vertical spindle mill such as MPS and Berz, but only in small numbers.

On pressurised mill systems the primary air fan, which provides the pulverised fuel transport medium, is situated before the mill and therefore handles clean air and is not subject to erosive wear as is an exhauster fan. This is the chief advantage of the pressurised mill system, however, the mill does need sealing air which is usually provided by a separate fan at a pressure higher than that of the mill interior.

A disadvantage of the pressure type mill is that it must be absolutely airtight in order to avoid pulverised fuel leakage to the atmosphere. Conversely the standard of sealing on a suction mill need not be so high, but it must not be allowed to deteriorate too far as the inwards leakage, being cold air, will make it difficult to dry the wetter coals. This leakage air is also unmeasured as regards its quantity, and if excessive under certain conditions produces a high air/coal ratio which may be explosive should there be an ignition source.

Vertical spindle mills used by the CEGB range in size from 10 tonnes/hour to 57 tonnes/hour.

As these mills have become larger, the size of the grinding balls and rollers relative to the size of the grinding rings and tables has increased and the result has been an increase in the life of these items. There has also been a trend from loading the grinding elements by springs to the use of either gas or hydraulic cylinders.

An outline of the theory of vertical spindle mills is given by V. C. Marshall [8].

A Babcock type 10E mill is shown in Fig 1.22; this has a nominal output of 40 tonnes/hour.

An NEI Lopulco LM 45/3P mill is shown in Fig 1.23; this has a nominal output of 45 tonnes/hour.

6.2.3 Coal feeders

The choice of feeder is determined by the flow properties of the coals. Mechanised mining produces fine coal which sometimes has ash with a clay content that is sticky when wet. The resulting coal stoppages and loss of plant availability led to an examination of the entire problem of handling coal between the bunker and the mills. The result was the adoption of bunkers having a large rectangular outlet in combination with drag link feeders. Prior to this, table feeders were fitted.

The principle of the drag link feeder (Fig 1.24), is that the coal is scraped from the bunker outlet and dragged along the feeder discharge table by the upper strand of the chain. It falls from the end of the table, between the strands of the chain, on to the bottom casing, where it is pushed in the opposite direction by the return strands of the chain into the mill inlet chute. Provided the chain scraper bars are spaced at not less than approximately 300 mm, difficult coals can generally be handled. Control of the coal depth is by means of a gate, and the rate of feed is controlled by a variable-speed motor drive.

6.2.4 Power consumption

Mention has already been made of the power consumption of various mill types and Table 1.8 gives a comparison of typical power consumptions between different mill types.

6.2.5 Air quantity

All mills are designed so that under optimum running conditions there is a definite relationship between the quantity of air and coal passing through (the air/fuel ratio), this ratio being under automatic control throughout a limited load range.

Factors governing the choice of air/fuel ratio for a given mill layout are principally:

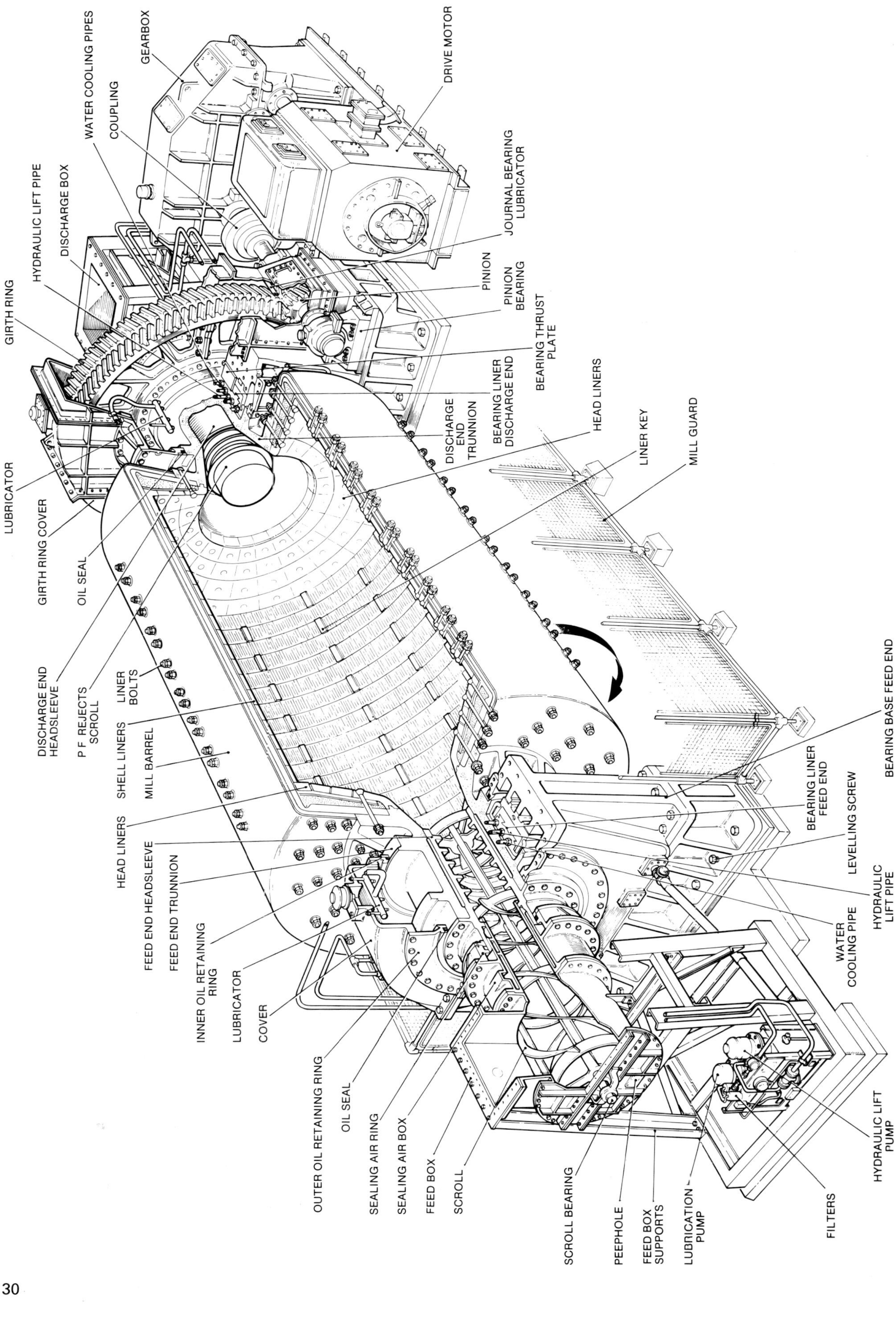

Fig. 1.21 Tube ball mill

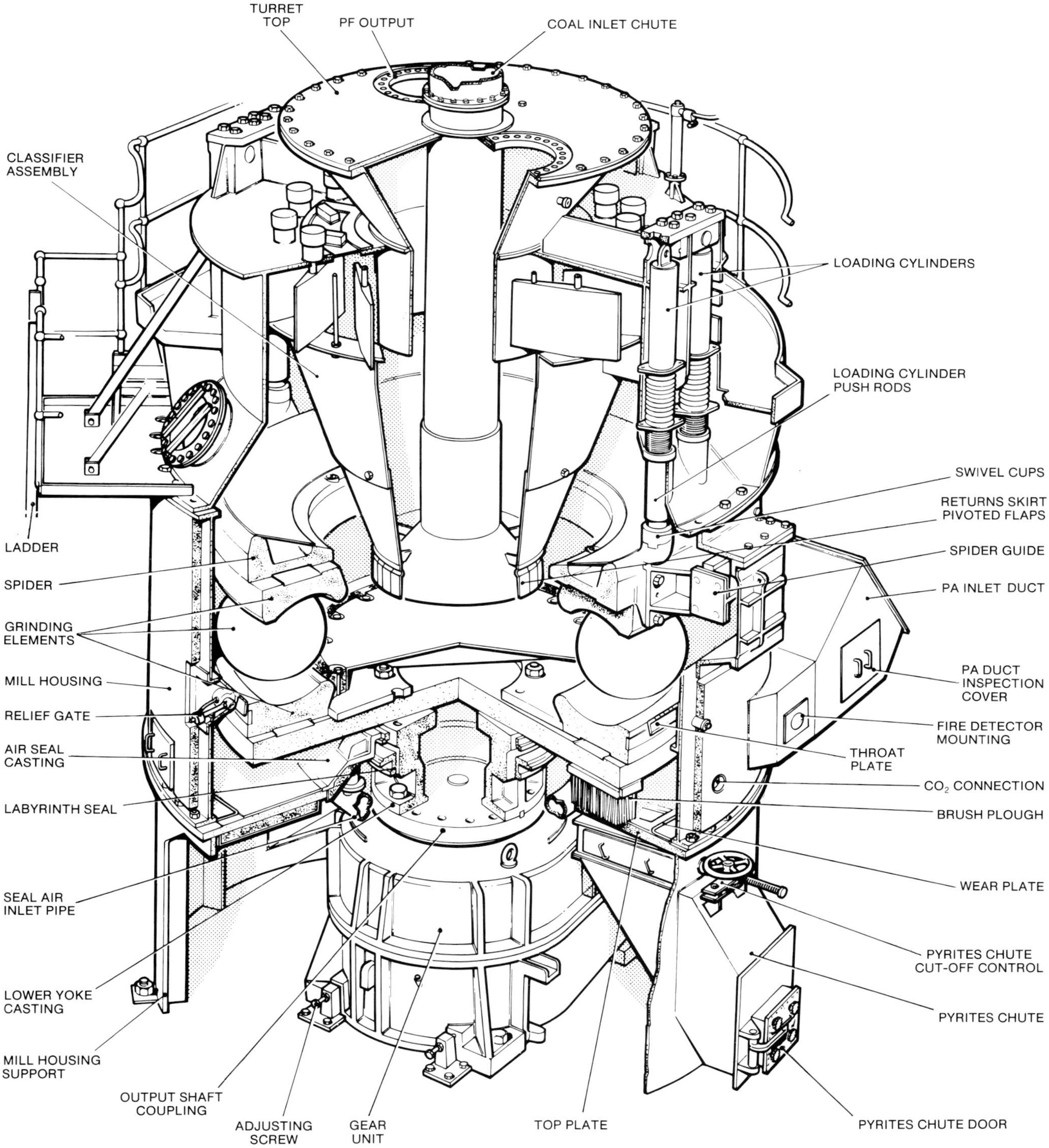

FIG. 1.22 Babcock type 10E vertical spindle mill

- The aerodynamic flow pattern required within the mill.
- Coal drying requirements.
- Fuel pipe velocities to ensure that coal particles remain airborne.
- The design of the coal burner.

On medium-speed mills the ratio is approximately 1.7 to 2.2:1 and on the tube ball mill about 1.2 to 1.5:1, i.e., less air is used. Since the incoming air performs the function of coal drying, higher mill air inlet temperatures are required by tube ball mills compared with medium-speed mills when handling coals having equal moisture contents. In order to maintain the required heat input for drying, it is not unusual

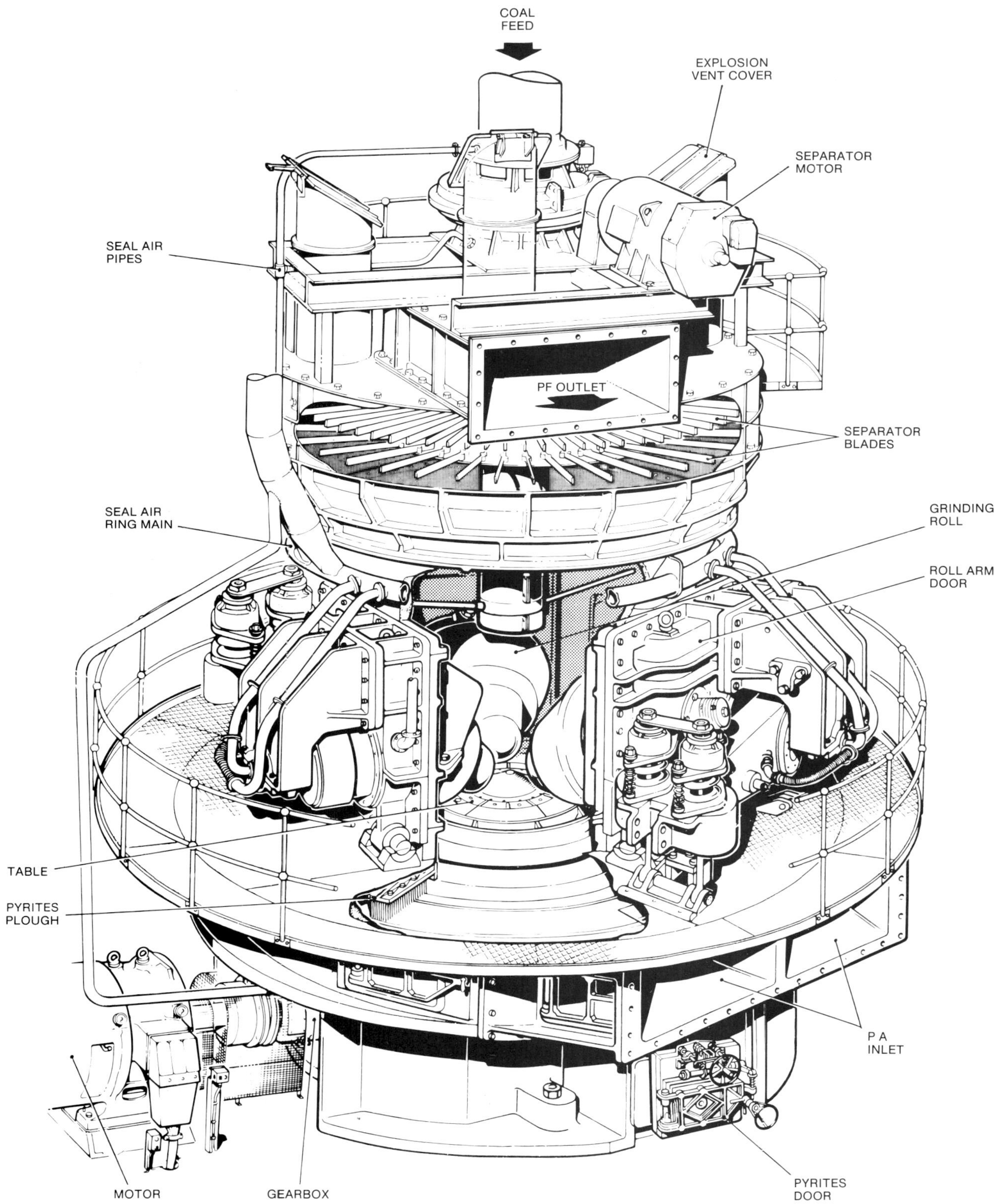

FIG. 1.23 Lopulco LM45/3P vertical spindle mill

to see differences of 80°C between vertical spindle and tube ball mills handling high moisture coals.

6.3 Mill/burner arrangements

The ideal mill/fuel pipe/burner arrangement is one in which an even distribution of coal is obtained across the furnace width at any load and with any combination of mills. It is a condition which is difficult to achieve, mainly because of the unavoidable differences of layout between each fuel pipe. In practice, in order to provide the best coal distribution

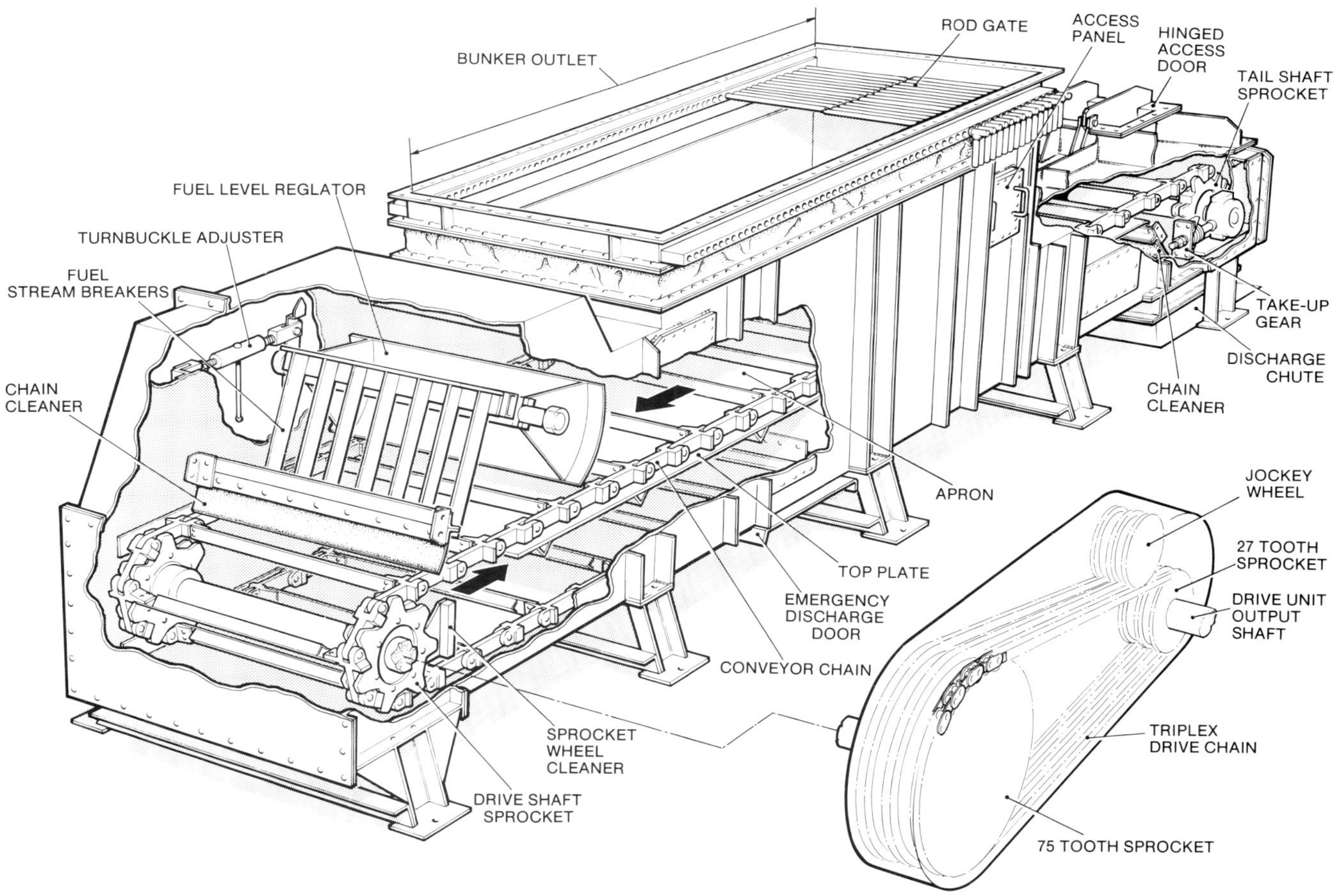

FIG. 1.24 Drag link coal feeder

TABLE 1.8
Typical power consumption of PF mills of approximately 40 tonne/hour capacity

MILL TYPE	MILL	FEEDER	PA FAN	TOTAL
Tube ball (pressure)	12.5	0.2	9.0	21.7
Vertical spindle (pressure)	8.0	0.2	10.8	19.0

in the furnace with different mill combinations, it is necessary to compromise on geometric layout and make up the deficiency as far as practicable by using riffle boxes or other types of distributors.

The performance of any riffle box, however well designed, is affected by the quality of pulverised fuel dispersion at its inlet. Current practice is to try to position riffle boxes so that they are immediately preceded by a vertical section of pipe in which the pulverised fuel can disperse evenly.

Various examples of mill/fuel pipe/burner arrangements are now considered, and for the sake of simplicity these only deal with a small number of mills and burners such as are found on some of the CEGB's older and smaller boilers.

The problems and layout principles involved remain the same for larger boilers.

Figure 1.25 shows a typical arrangement for front-wall firing in which eight burners are supplied by four pressure type mills, any three of which will enable the boiler to steam at full load. It will be noted that particular care has been taken in the layout of piping to ensure that a minimum of unbalance of coal input occurs across the furnace width when any one of the

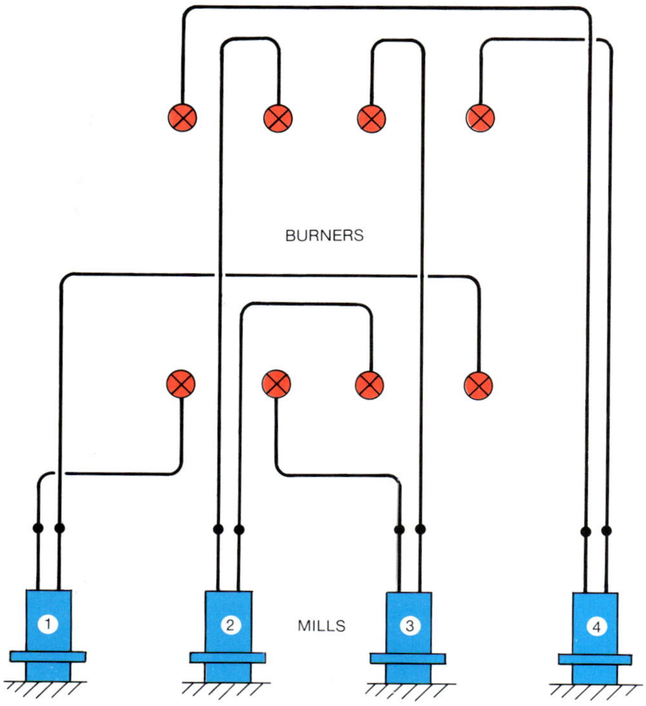

FIG. 1.25 Diagrammatic arrangement of fuel pipes from four mills to eight frontwall burners

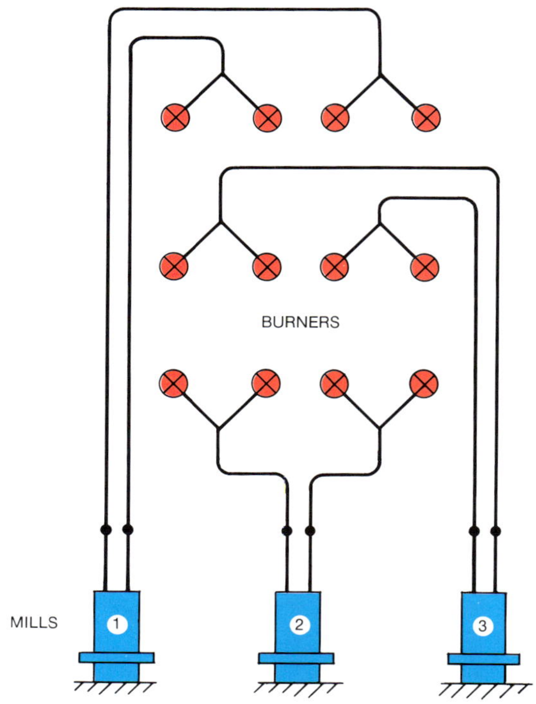

FIG. 1.27 Diagrammatic arrangement of fuel pipes to give a coarse steam temperature control

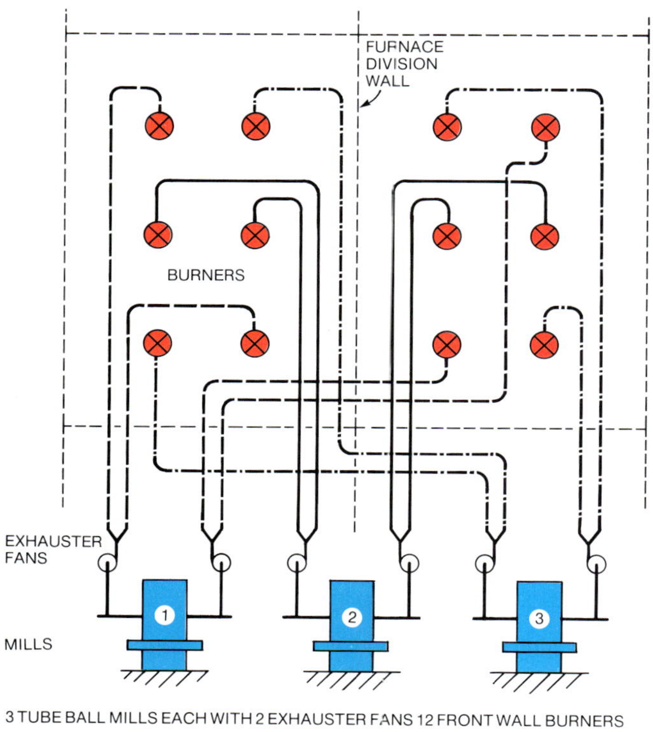

FIG. 1.26 Diagrammatic arrangement of fuel pipes to a divided furnace

four mills is out of service. Where a vertical spindle mill having two or more outlets is employed it is not unusual to find an uneven concentration of fuel in each of these outlets so that, for instance, on a two outlet arrangement a higher concentration may be found in, say, the right hand outlet. The arrangement shown in Fig 1.25 would, therefore, produce a higher concentration of fuel in the right hand half of the furnace. An effort could be made in the layout to prevent this by arranging the right hand outlet from two mills to serve the left hand half of the furnace and vice versa, but the success of this arrangement can only be assured if all mills are in operation each with equal outputs, which defeats the object of balanced firing with any mill out of service.

A further complication in frontwall firing is introduced where a furnace division wall is included. Figure 1.26 shows an arrangement in which the layout of fuel piping is such that whichever mill is out of service equal heat input to each half of the furnace is maintained with the remaining mills.

It is not unusual to find in frontwall firing that each mill is arranged to serve one complete horizontal row of burners. As well as providing a uniform distribution across the furnace this arrangement also provides a measure of coarse steam temperature control (Fig 1.27). Provided there is sufficient mill capacity to give full duty with one out of service, a variation in total furnace heat absorption (and, therefore, total heat in the gases leaving the furnace) can be obtained by raising or lowering the zone of combustion by excluding the top or bottom row of burners. This principle is analogous to steam temperature control by tilting burners in the corner fired-furnace although it is not nearly so positive.

Figure 1.28 shows an arrangement in which each mill feeds eight burners in a divided corner-fired fur-

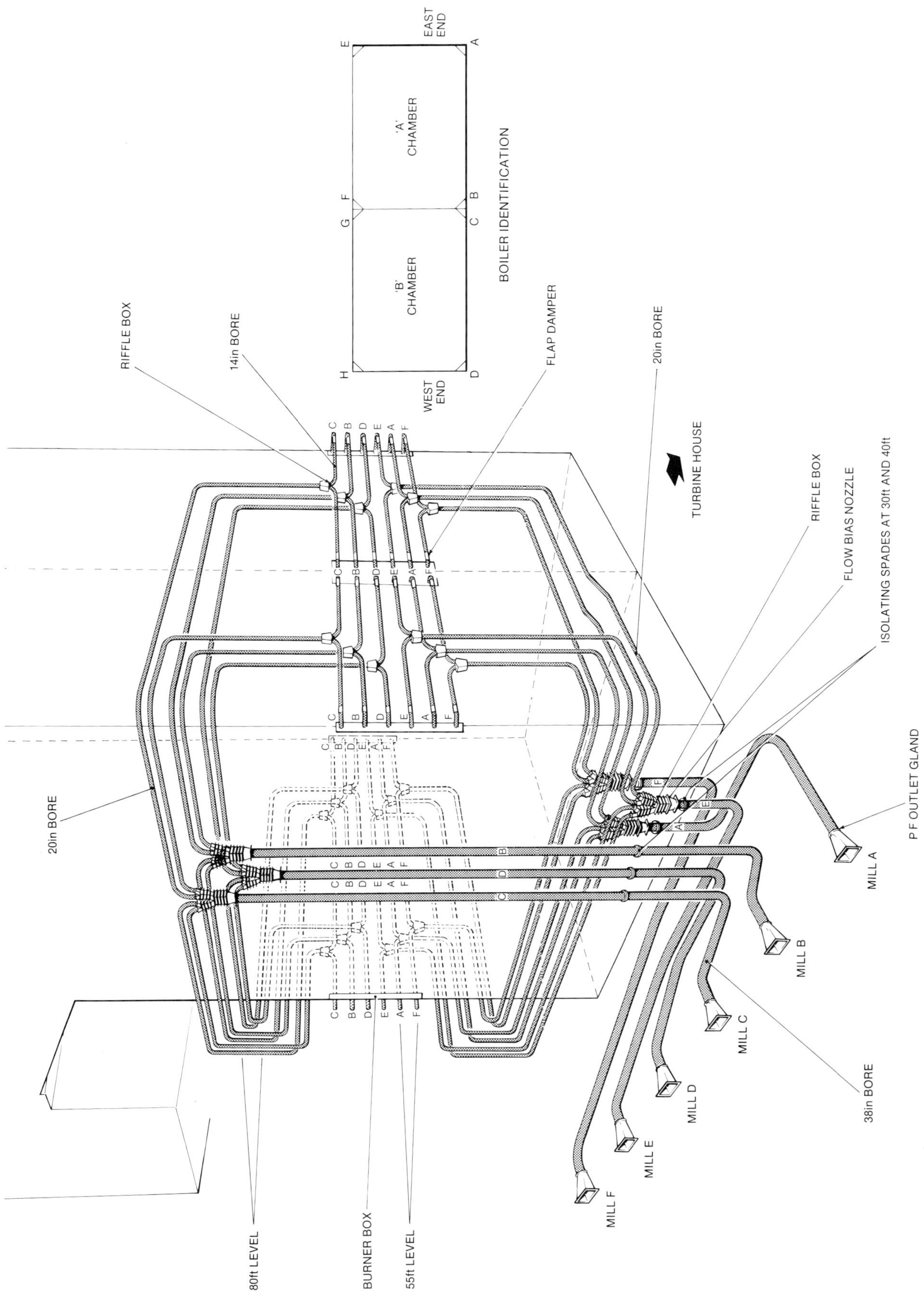

FIG. 1.28 Arrangement of fuel pipes to a divided furnace — eight burners per mill

nace. It will be seen that there are three successive subdivisions between the mill and the burners and a total of seven riffle boxes per mill. Obviously the greater the number of divisions the greater the risk of uneven fuel distribution. Adjustments to the riffle boxes may improve the distribution over a limited load range but a more satisfactory solution would probably need modifications to the layout of the piping. Whilst the preceding remarks serve to show how maldistribution can occur in a corner-fired arrangement, it can occur equally in other forms of firing where it is necessary to subdivide the coal flow between burners.

On large boilers where up to 10 mills and 60 burners may be employed the problems become complicated, but nevertheless the same essentials are required and arrangements must be such that even fuel distribution shall be possible whatever the combination of mills in operation.

6.4 Factors affecting milling plant performance

The performance of the milling plant is affected by a number of factors mainly associated with the properties of the coal being ground.

6.4.1 Hardgrove Index

The Hardgrove Index is a measure of the ease with which coal can be ground. It is obtained by grinding a sample of coal under controlled conditions in a miniature vertical spindle mill. Whilst it is a guide to the power needed to grind the coal it is not necessarily indicative of the coal's abrasive properties. The test machine is motor driven and grinds a 50 gramme sample by turning through 60 revolutions. The test method is given in BS1016 Part 20 [9].

In general, British bituminous coals have a Hardgrove Index within the range 45–60 but those with a low volatile content have a higher Index. For example, some of the coals from Kent and the South Wales semi-bituminous coalfields have Hardgrove Indices in the region of 80–100, these being the softest coals in the country. The converse applies to the older coals outside the bituminous range. Pure anthracites for instance have a low volatile content and are very hard with an Index in the order of 35–40.

From the foregoing a mill designed to handle a coal having a Hardgrove Index of 50 will have a greater output when grinding coal that has an Index of 65, and conversely a reduced output if harder coals are handled.

Since the majority of bituminous coals have a Hardgrove Index in the range of 45–60 it is common practice for mill manufacturers, when considering this type of fuel, to design their plant to give full mill duty with an Index of 50 and to provide a curve to show the increase or decrease in output with softer and harder coals. A typical curve is shown in Fig 1.29.

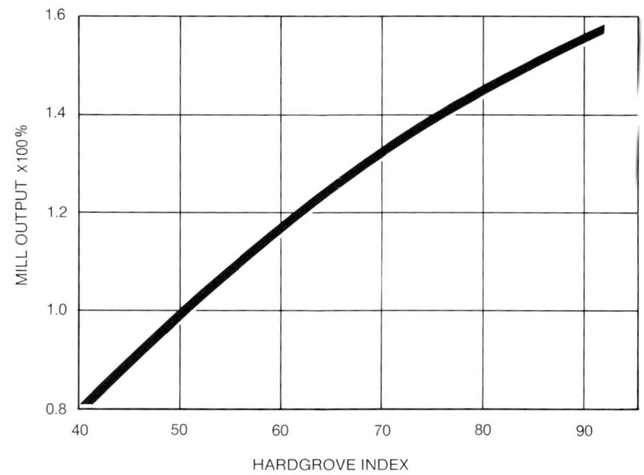

FIG. 1.29 Relationship between mill output and Hardgrove grindability index

In properly sized equipment it is unnecessary for limitations in mill output to occur through a change in Hardgrove Index, since the plant should be designed to handle the harder coal and will automatically give an increased output with the softer coals, other things being equal.

The resistance of coal to breakage depends on whether it is broken by impact (fast compression) or by crushing (slow compression). This may be additionally affected by the form of the coal, particularly its susceptibility to crack propagation. Furthermore, difficulties arise in achieving a representative sample, particularly of untreated coals, and of ensuring that test results in the machine correspond to those which will be achieved on a full-sized mill. A knowledge of the coals to be handled is therefore essential if the true performance of the mill is to be established.

6.4.2 Moisture content

The total moisture content of raw coal is made up of inherent (i.e., that locked within the structure of the coal) and free or surface moisture.

Drying of coal is carried out by the passage of hot air through the mill and an adequately sized mill should be able to extract the whole of the free and up to half the inherent moisture of the wettest specified coal. The output of a vertical spindle mill is less sensitive to increases in moisture than a tube ball mill and this difference is shown in Fig 1.30.

Hot air supply arrangements for the mills are divided generally into two main types. A boiler will have either:

- Main airheaters which supply hot air to the mills and the furnace windbox and which can have the air exit temperature boosted for high moisture coals by means of an economiser gas bypass or:

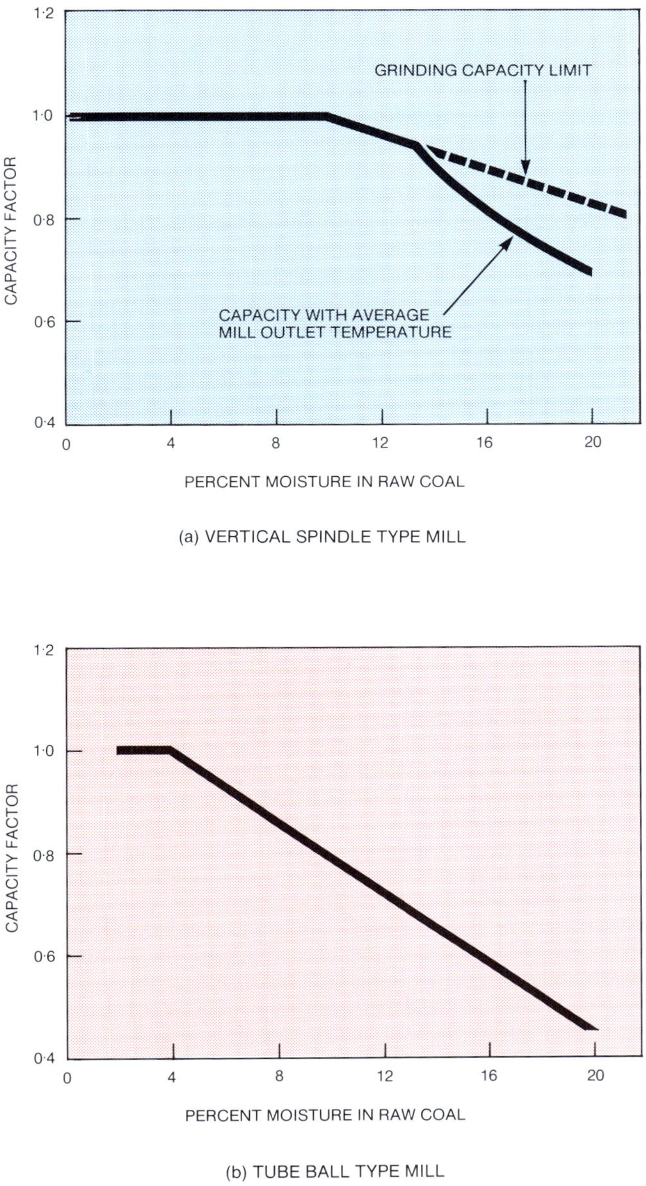

FIG. 1.30 Variations in mill output with total moisture content in raw coal

- Separate airheaters which serve only the mills. These heaters are usually sized to deal with the wettest coal without the use of an economiser bypass.

An alternative system employed for suction tube ball mills on a few boilers is to draw off hot gas from the bottom of the furnaces to supplement the hot air from the main airheaters when handling the wettest coal.

6.4.3 Fineness of milled product

When handling the majority of UK bituminous coals, a mill and its classifier will be designed to produce a pulverised fuel fineness such that 70% of it will pass through a 200 mesh sieve. For the majority of bituminous coals this fineness is sufficient to maintain satisfactory combustion and any increase in fineness will waste mill power. Where coals that have a lower volatile content are being ground, the fineness required is increased typically to 85% passing a 200 mesh sieve.

Mill output and fineness are interdependent and an increase in output can be obtained only at the expense of a deterioration in fineness, assuming that the power input to the mill remains the same. This practice is not recommended unless the coarser pulverised fuel provides satisfactory combustion.

Mill output should never be quoted without relating it to fineness of grinding, Hardgrove Index, and moisture content.

A curve showing the variation in mill output with fineness is shown in Fig 1.31.

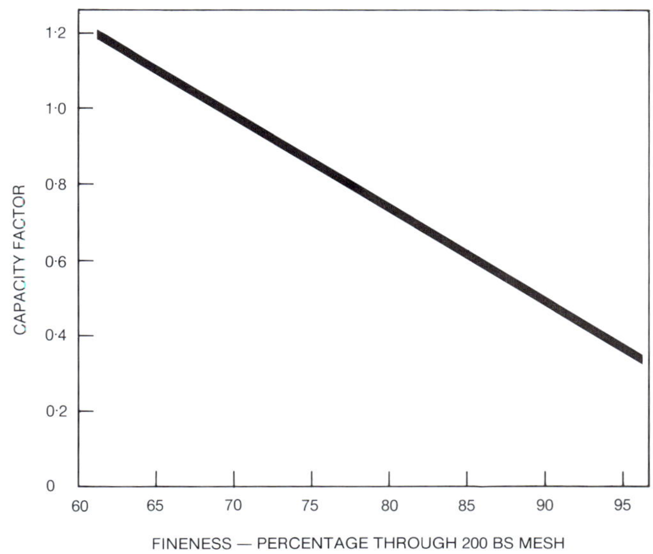

FIG. 1.31 Variation in mill output with fineness of product

6.4.4 Size of raw coal

The size of coal fed to a mill obviously affects the amount of work that it has to do. Mill manufacturers design their mills for a chosen top size of coal and in practice this is rarely exceeded on CEGB plant. Mill performance correction factors are available but it is seldom found necessary to refer to them once the mills have been commissioned.

6.4.5 Mill wear

There are several methods of measuring coal abrasiveness, one of which was devised by Yancy, Gear and Price [10]. It consists of measuring the weight loss of four mild steel paddle blades, which rotate for a given number of revolutions in a vessel that contains a specified weight of coal. The method of testing is given in BS1016 Part 19 [11].

The balls in a tube ball mill decrease in size as they wear down and the balance in the quantity of large, medium and small balls must be maintained by adding a fresh charge of balls at regular intervals. The larger balls are needed to crush the bigger pieces of coal whilst the small balls are necessary to produce the finer particles of pulverised fuel.

The table segments and roller tyres of ring-and-roller vertical spindle mills wear preferentially in zones of high local coalbed pressure. This results in abrasion producing excessive local clearances between the tyres and the table and leads to a reduction in mill output.

Local wear clearances of this type do not appear in the balls-and-rings or balls-and-race vertical spindle mills, however, wear does reduce the ball diameter and whilst on the larger mills this does not appreciably reduce the output there is an increase in the grinding power needed per tonne of coal.

The general question of mill wear is discussed in Section 6.6.2 of this chapter.

6.5 Standby capacity

The majority of pulverised fuel milling plant in power stations is of the vertical spindle medium-speed type, and wherever plant of this type is in use periodic adjustment of roller setting and replacement of worn parts, etc., is necessary at intervals. Unless spare capacity is available, load reduction is then inevitable.

With this in mind the CEGB in preparing its specification for boiler plant, insists that maximum boiler output should be possible with one PF mill out of service where such mills are to be installed. By this arrangement mill adjustment can be carried out at almost any time and major items of mill overhaul, such as replacement of worn parts, can be carried out in normal working hours.

CEGB boilers with tube ball mills up to now have not had a spare mill specified, as it was considered unnecessary because of the ease with which the ball charge can be topped up whilst the mill is running. Experience has shown that the availability of the mill is only as good as the reliability of its auxiliaries such as feeders and damper actuators, and any future tube ball mill installations will have a spare mill.

Mill standby capacity can be further augmented by oversizing the oil lighting-up burners so that in total they can produce up to say 30% of boiler MCR duty. This will, if required, release more than one mill for maintenance.

6.6 Operation and maintenance of PF plant

6.6.1 Operation

The effect of hardness and moisture content of the raw coal upon mill performance and the necessity to obtain an optimum fineness of milled product in order to achieve economy in milling power have already been considered in Section 6.4 of this chapter.

Another factor affecting milling power is the output required of the mill. As the output of the mill decreases, for instance, the power required/tonne of coal milled will increase, since in the extreme, power is required to turn the mill when giving no output (see Fig 1.32).

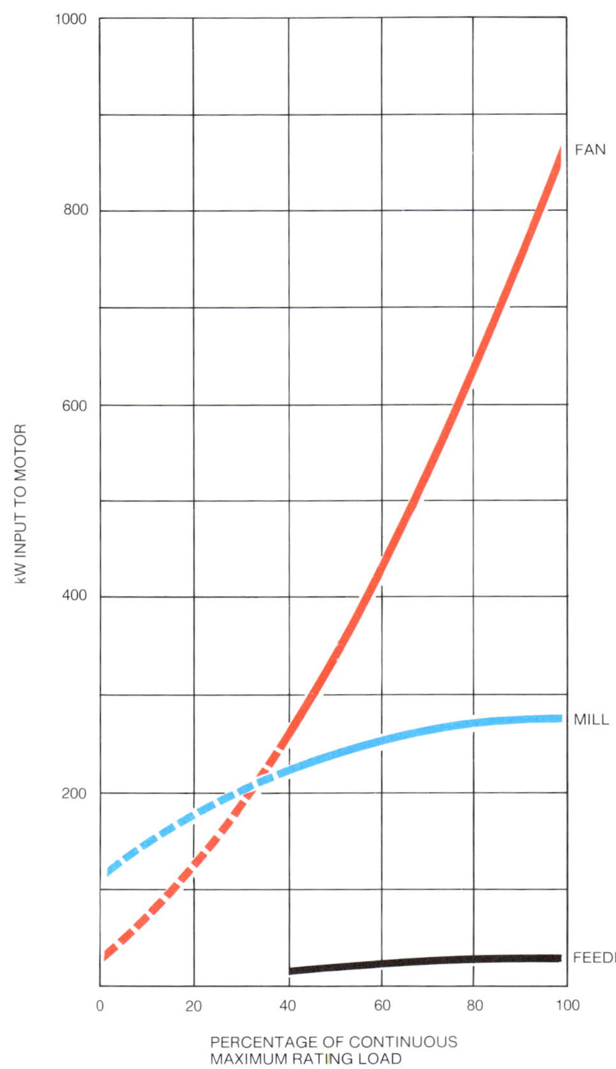

FIG. 1.32 Typical power input requirements with increasing mill output

Maximum economy of milling plant power is thus obtained when all mills are operating at their maximum output. In plant where full boiler output is obtainable with one mill out of service it is, therefore, the usual practice to operate in this manner rather than with the full complement of mills partially loaded. Furthermore, where partial mill loading is necessary to meet a reduced boiler output, it is good practice to load each mill to the same degree, since secondary air flow conditions would need to be changed on those burners served by a mill not conforming to this pattern of operation.

The satisfactory day-to-day running of milling plant is obtained largely as the result of experience. One of the targets that should be aimed at is to obtain equal distribution of PF to all the burners fed by a mill, thus avoiding local reducing conditions in the furnace. Should this arise there is a risk of damage to the tube walls local to the burners through wastage.

In order to avoid excessive settlement of PF in, and possible blockage of, fuel pipes, it is necessary to maintain a minimum average transport velocity of 20 m/s. The normal operating velocity will be between 24 m/s and 27 m/s. A speed higher than this will cause accelerated erosion of the PF piping.

On properly designed plant, the PF transport air leaving the mill is normally at constant temperature for a given load and should only vary over a narrow range whatever the moisture content of the coal; this assumes that the properties of the coal are within the design specification.

In cases where the moisture content is low, less heat is required in the mill to dry the coal and therefore the outlet temperature rises. Tempering air is added to control the outlet within the range 60–80°C. Should the outlet be less than 60°C, there is a risk of moisture in the pipes due to the air having reached its dewpoint; PF may then stick to the inside of the pipes.

The higher the mill outlet temperature the higher the boiler efficiency, and therefore mills should be operated at the top end of the temperature band wherever possible. The figure selected must take account of its effect on possibly accelerating spontaneous combustion of coal deposits in PF piping, and also not be so high that it causes deterioration of the rubber joint rings in the pipe flexible couplings.

The necessity to carry out adequate purging of a furnace when ignition is lost cannot be stressed too strongly. On no account should an oil burner be inserted for ignition of the coal until the induced draught fans have been run a sufficient length of time to ensure that no pulverised coal remains in the furnace. This rule should apply under any condition of operation, whether the boiler is on-load or just being lit-up.

6.6.2 Maintenance

As already indicated in Section 6.4.5 of this chapter, the major items of heavy wear maintenance on vertical spindle mills are balls, rings, roller tyres and table segments, plus in the case of suction mills the exhauster blades.

Wear on tube ball suction mills is chiefly confined to the exhauster fan. Obviously the mill drum liners wear but they are not a maintenance problem in the way that the grinding elements in a vertical spindle mill are; this is because they have an appreciably longer life. The rate of wear of these items depends on the abrasiveness of the coal ash content and on the proportion of ash in the coal. Quartz and pyrites in the ash will increase its abrasiveness.

The life of mill grinding elements depends on the mill type as well as the coal quality. Typical figures for large NEI Lopulco mills grinding average UK coals are about 6000 hours for the roller tyres and 12 000 hours for the table segments, whilst the balls and rings of the Babcock Type 10E mills are usually changed after 16 000 to 20 000 hours. When examining these figures, it should be remembered that grinding elements are sometimes changed before they are worn out because it is convenient at that time to have the mill off-load for the maintenance of other items.

There are a few UK coals which are highly abrasive and which double the wear rate. In the case of the Babcock Type 10E mill this is dealt with by allowing the original set of 10 balls to wear down from 768 mm diameter to 654 mm diameter and then fitting a second set of 12 balls of reduced diameter. In this way the maximum life is obtained from the grinding rings; these have a longer life than the balls.

The majority of CEGB stations with type 10E mills, which are grinding average coals, change the balls and rings at the same time because of operational convenience.

Tube ball mills require periodic make-up of the ball charge, and when grinding average UK coals a make-up of approximately 0.14 kg of balls per tonne of coal is required.

The main area of wear in the exhauster fan is, as would be expected, the impeller blading. To deal with this the blades are often protected by a weld deposited wear-resistant facing, generally of high chrome alloy. Wear of exhauster blades is a constant problem and best avoided by having a pressurised milling system that uses a primary air fan.

The CEGB incurs significant expenditure replacing the wearing parts of milling systems and consequently there is a strong incentive to extend the life of these items. When prospective alternative materials become available they are evaluated, but it is not easy to draw conclusions because of the length of time needed to establish the wear life and the difficulty of comparing like with like; this is because of differences in plant operating regimes and fuel properties.

6.6.3 Costs

There can be a temptation to write a boiler fuel specification that embraces a wide range of coal properties, particularly gross calorific value, just to be on the safe side. In practice it is unlikely that such a wide range of coals will be burnt and the result is that more milling plant than necessary will have been bought. Careful attention to the fuel specification can secure major economy in the capital cost of the plant.

On a cost per tonne basis, large mills are cheaper than small mills, but this does not necessarily imply that a boiler should have a small number of very large mills. The eventual decision must take account of such factors as step changes in heat input as mills are brought into or taken out of service, the effect of the turndown of individual mills on the overall turndown of the boiler, availability on the market of reliable large mills, the time needed to carry out maintenance and the cost of spare parts.

7 Oil burner types and location in furnace

There are three basic types of oil burner atomiser in use today and these may be combined with either the axial/radial inflow type swirl-generating register (Fig 1.33) or the simple axial flow register combined with a bladed swirler type stabiliser (Fig 1.34).

The three types of atomiser are:

- The mechanical (or pressure) atomiser usually known as the *pressure jet*. This may be installed as a simple pressure jet or as a spill return pressure jet, the difference is described in Section 7.1 of this chapter.
- The steam atomised burner, which may be internally mixed *steam atomised* or externally mixed *steam assisted*.
- The air atomised burner. This type exists in similar subdivisions to those of the steam atomiser.

Figure 1.35 shows the fuel oil and air flows in the axial/radial inflow type swirl generating register.

The purpose of the air register is to supply a controlled quantity of air in a clearly defined flow pattern to the oil spray and to ensure correct mixing of oil and air. This must provide a stable flow over the full range of operation and also give maximum combustion efficiency.

Flame stability is created by the swirl flow generated either by the bladed swirler or by the radial/axial swirl vanes contained within the register air passages. The swirl promotes a low pressure area downstream of the burner exit which induces a gas recirculation pattern, thus providing a stable ignition source and flame. The proportions of swirl and axial airflow are critical and must be optimised in order to achieve the best balance of stability and combustion performance.

The oil burner register normally incorporates a sleeve damper to shut off the airflow to the burner when not in use. This damper may allow a small proportion of air leakage in order to keep the burner components within their design temperatures.

It has been recognised for many years now that a low oxygen operating regime is imperative for heavy oil-fired boilers in order to reduce acid formation in the combustion system and the subsequent acid smut emission.

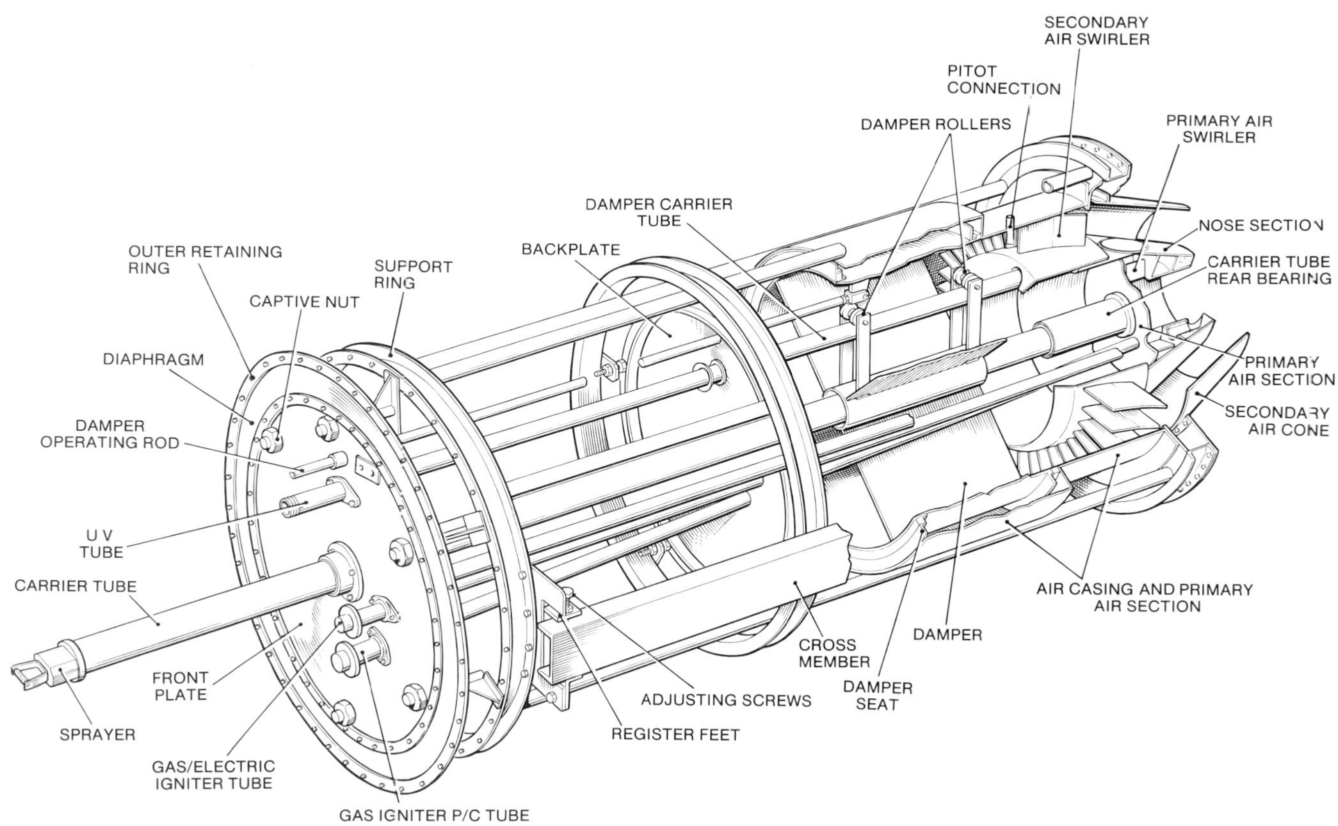

FIG. 1.33 Oil burner with axial/radial inflow type swirl-generating register

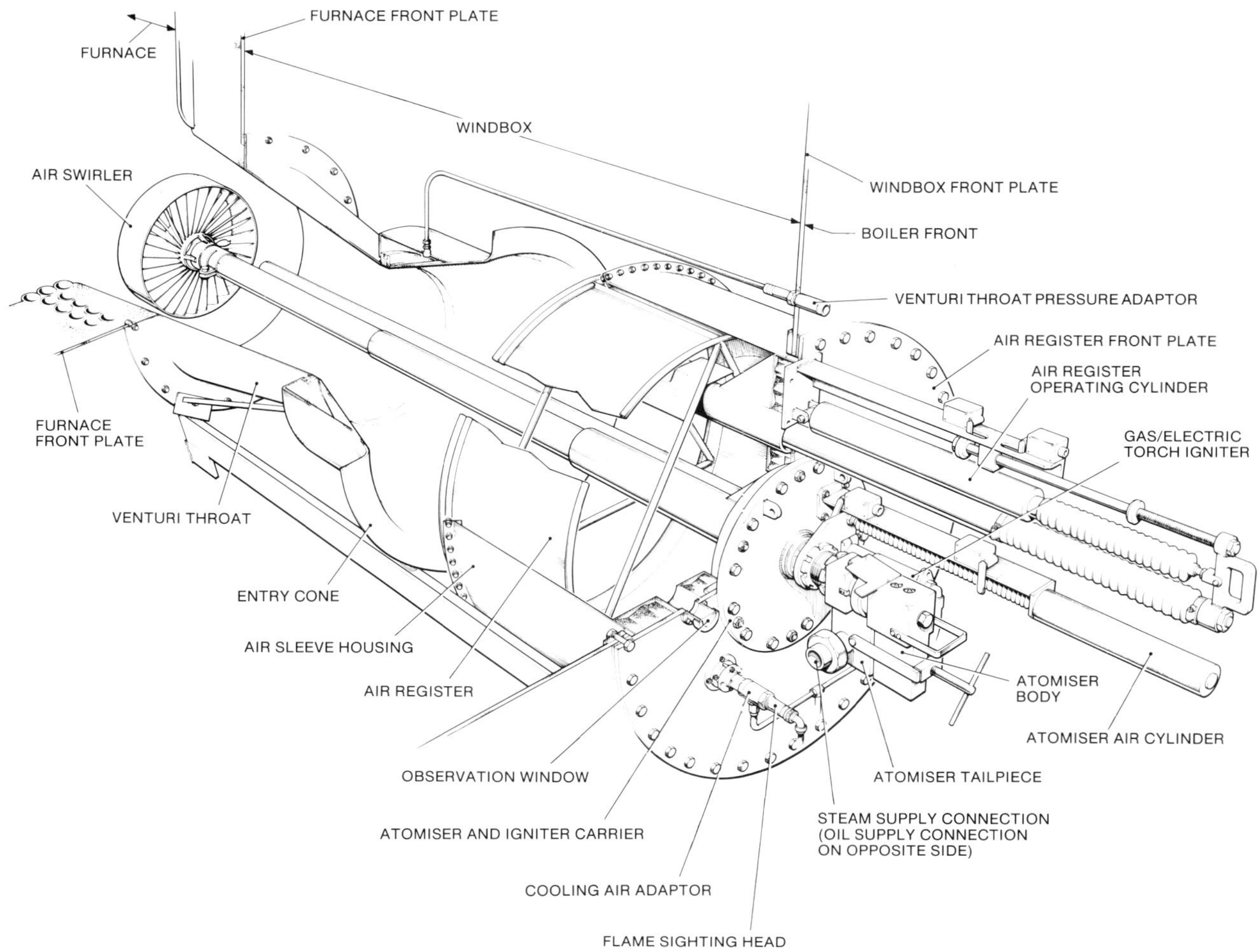

FIG. 1.34 Oil burner with axial flow register and bladed swirler type stabiliser

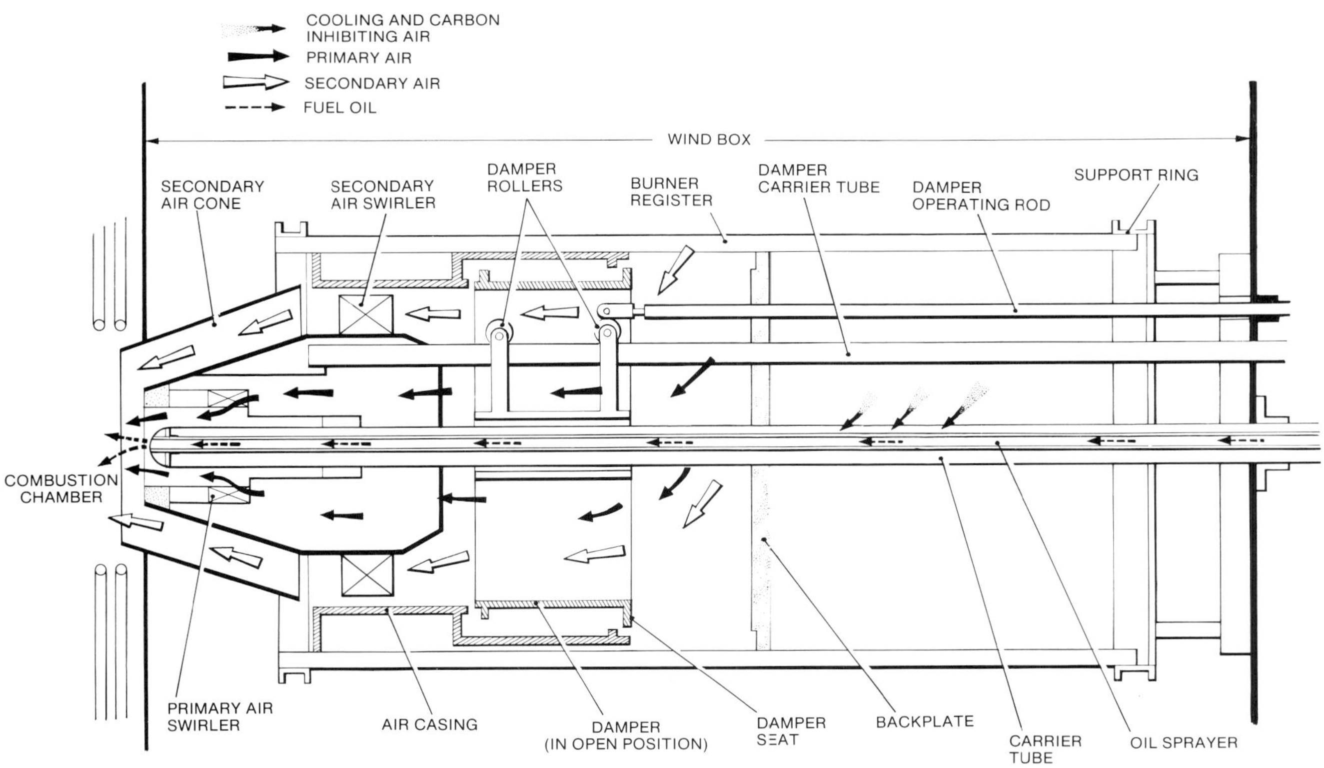

FIG. 1.35 Burner fuel and air flows

7.1 Mechanical atomising burner

This is also known as the pressure atomising burner. In this system the oil is supplied at high pressure from the oil pumping and heating equipment. The pressure at the burner is usually designed to be within the range 31 bar to 83 bar, the design pressure depending on the burner throughput and the atomising quality required. For oil burners of up to 76 kg/s a pressure range of 31 bar to 45 bar is sufficient.

In the simple pressure jet, all the oil passes through a distribution plate and subsequently through swirl slots or holes before entering a swirl chamber tangentially. This swirl chamber then tapers down through an exit cone to a circular sharp-edged orifice. In this fashion the oil leaves the orifice as a hollow conical thin sheet which breaks into droplets of varying sizes. All the pressure energy is thus converted into overcoming internal losses and generating the swirl velocity in the oil.

The flow through a simple pressure jet is proportional to the root of the applied pressure (within fairly broad limits), each nozzle having a factor (flow factor) which is notionally constant and is derived from the flow that results from a given pressure. Using consistent units:

$$\text{Flow factor}(c) = \frac{\text{flow}}{\sqrt{\text{pressure}}}$$

From this relationship it is apparent that the turndown on a pressure jet is very restricted as atomisation, quality degrades rapidly with pressure. If, for example, 24 bar is assumed as the minimum pressure to obtain satisfactory atomisation, the turndown on a small (19 kg/s) atomiser with a full load pressure of 38 bar will be:

$$\frac{24}{38} = 0.79$$

Thus the turndown flow is limited to 79% of the full load value, and the only way to extend this is to increase the design pressure with repercussions on capital and running costs of the installation. For these reasons, turndown on a pressure jet burner is normally limited to 1.4:1.

The spill pressure jet has been used widely in the last three decades to partially overcome the problem of restricted maximum. In this type of system the input pressure is usually kept constant at or around the maximum value, and a central bleed flow from the swirl chamber (the spill return) increased from zero (full load) to a maximum flow (maximum turn down), see Fig 1.36. This control is effected by valves (spill valves) which control the pressure in the spill system. Such valves may either be integral with the burner bar or external in a control valve block or the pipework system. The burner illustrated also has the facility to 'tip recirculate' when off-load (see Section 8.1 of this chapter).

The oil viscosity required at the burner should be between 12 cSt and 18 cSt for satisfactory atomisation. Experience on large oil-fired units shows that decreasing the viscosity (by oil temperature increase) offers worthwhile gains in combustion performance and therefore lower solid emission levels. For a heavy residual fuel oil, temperatures at the burner of up to 140°C are normal; BS2869 refers and the correct temperature to obtain the required viscosity on a range of fuel oils can be obtained from this document [12]. The operational aspects are discussed further in Volume G.

7.2 Steam atomising burner

In this burner the atomising process is created by the oil being broken up into droplets by a high velocity steam jet directed across the oil outlet jet. The basic form of this atomiser is often known as the Y-jet. Such atomisers contain up to 15–20 discrete oil/steam jets depending on the atomiser capacity required. Fuel flow through these atomisers is roughly directly proportional to the applied fuel pressure. The steam and oil are transmitted to the atomiser by a burner bar consisting of concentric tubes (see Fig 1.37). One end of these tubes must be sealed by a gland system to allow for differential expansion — since the steam and oil are at different temperatures. The steam is normally carried by the centre tube.

In another version of the steam atomiser the steam is fed down the outer tube, oil down the inner. This is achieved by means of complex drillings in the atomiser (the skew jet) originally developed to improve the atomisation. The steam conditions used are usually a pressure of 7–11 bar with a small amount of superheat. The oil pressure is varied for flow control purposes and is usually a maximum of 17 bar at the burner.

As the atomisation does not rely principally on the energy in the oil, but on that in the steam, the turndown available is much greater than in the pressure jet type of oil burner. A ratio of 5:1 is readily available, and, if the steam pressure is kept constant, atomisation improves with turndown.

Steam atomisers use 4% or more (weight for weight) steam to oil and normally require an auxiliary boiler installation for supply. This results in a considerable capital and revenue penalty on such systems.

The externally mixed *steam assisted* burner is intended to overcome some of the turndown restrictions of a pressure jet. Thus, complete annular or discrete steam jets are directed to penetrate the oil cone issuing from a normal pressure jet, and used at turndown, the steam will provide better oil atomisation.

7.3 Air atomising burner

This burner system uses air in a directly similar manner to the steam described in the previous paragraphs. Air

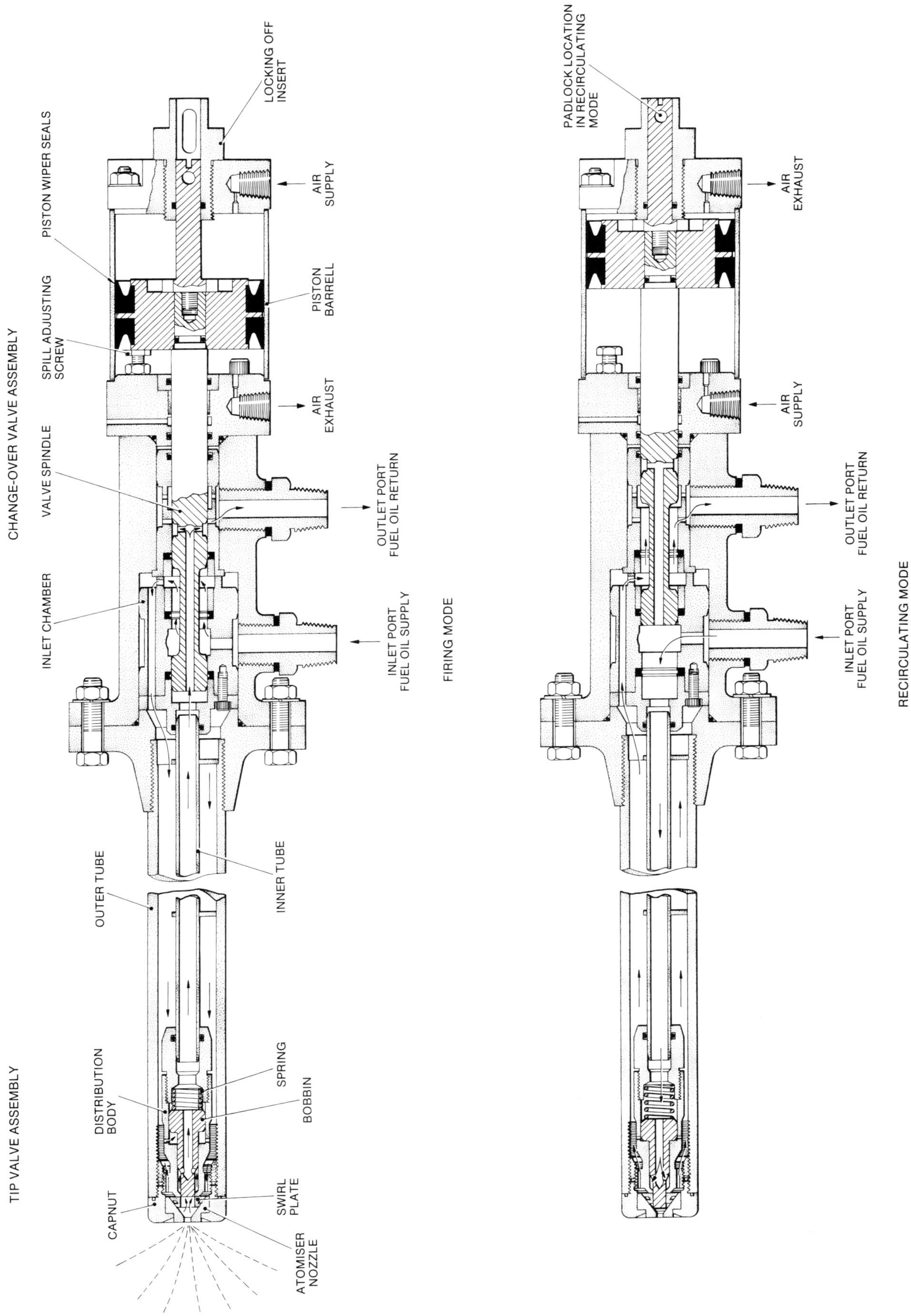

FIG. 1.36 Spill pressure jet fuel oil injector

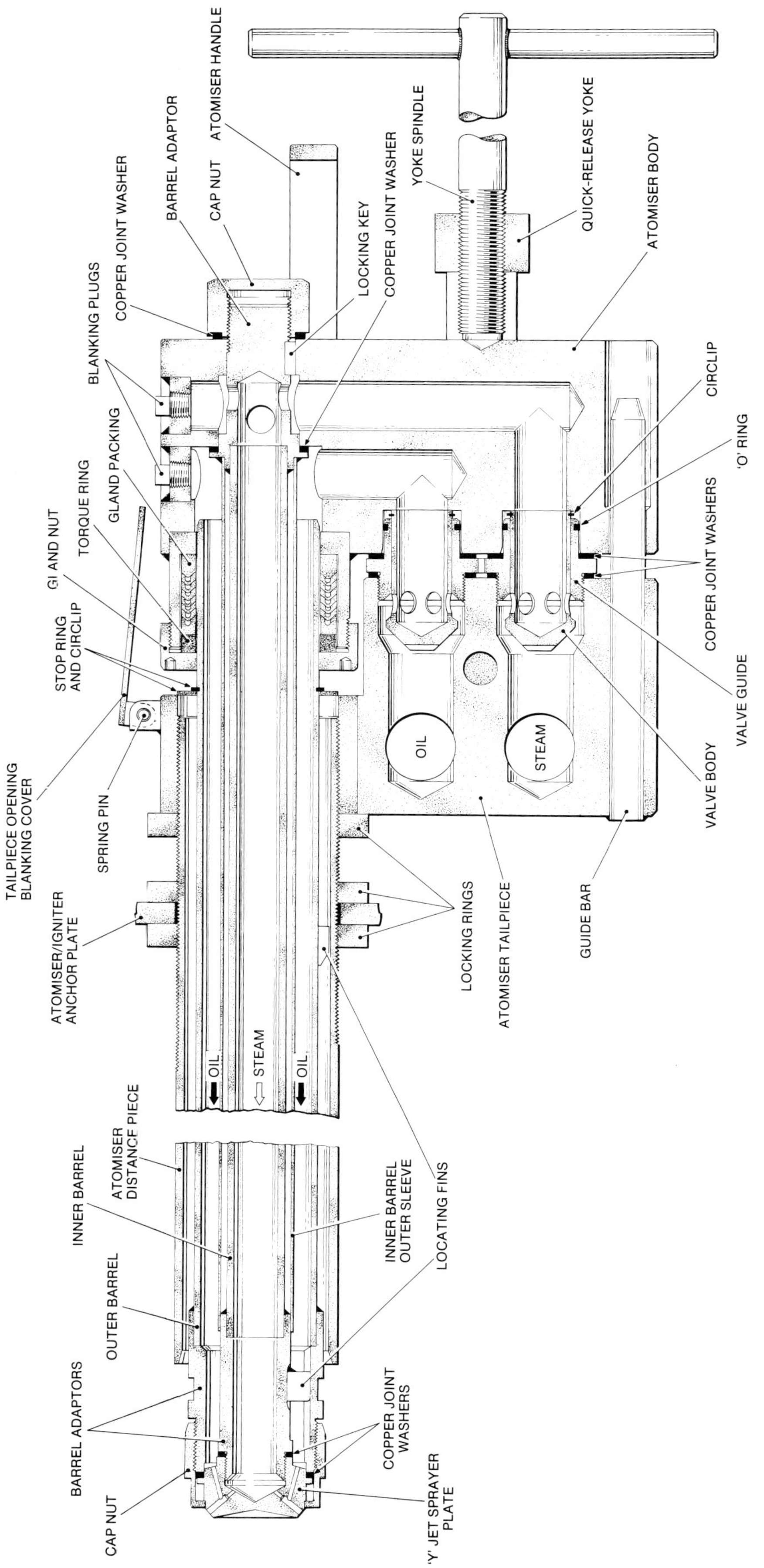

FIG. 1.37 Atomiser assembly

at up to 22 bar is used and considerable compressor capacity is required. Air does not possess any particular advantage when compared with steam and the choice is really one of cost — the capital cost of the compressors and their running costs versus the capital cost of the auxiliary boilers and their operating costs, mainly fuel.

Of course air does have a cooling effect, particularly when operating with residual fuel oil, and this could be construed as an operational disadvantage.

Air-assisted atomisers are similar in principle and operation to steam assisted atomisers.

7.4 Location in furnace

Burner arrangement in an oil-fired furnace follows a similar pattern to that in a PF furnace, with wall, opposed and corner fired units in service.

7.4.1 Wall fired

Wall-fired units use either of the burner types described, with any of the atomiser types mentioned. A burner pitching of 1.8 m horizontally and 2.4–2.5 m vertically has been used in the largest CEGB (660 MW) oil-fired units and found satisfactory. A sidewall clearance of up to 3.6 m has been used.

7.4.2 Corner fired

In a corner-fired unit the type of oil burner is normally of the air tube and bladed stabiliser type; again combined with any of the atomiser types mentioned, more normally with spill pressure jets or simple pressure jets.

7.5 Oil burner igniters and controls

7.5.1 Oil burner igniters

Many types of oil burner igniters have been used in the past ranging from single high tension (HT) or high energy (HE) spark to a small auxiliary oil jet itself ignited by a high tension spark. With the increasing use of residual fuel oils a simple spark was found to be totally inadequate, being positionally sensitive, prone to fouling and failure, and not giving sufficient energy to ignite the oil. The gas/electric igniter was developed to overcome the problems of the HT spark and this type of igniter using propane (or butane) has been in almost exclusive use for the past 25 years. The gas is normally ignited by an HT spark but HE has been used. Various types of gas igniter are in use, for example:

(a) Simple gas, in which gas is fed down the igniter and lit and stabilised on the end of the igniter. This type tends to be sensitive to windbox conditions.

(b) Premixed, in which a combustible mixture of gas and air is fed down the igniter and ignited and stabilised at the firing end of the igniter. This type of igniter develops an intense bunsen-like flame which has been shown to be too small for some installations. The air may be supplied from the windbox or from a compressed air system.

(c) A combination of (a) and (b). In this form, to overcome the failings of the other types, a small premixed flame is used to stabilise a large neat gas jet providing a high energy flame which is effectively independent of environmental conditions. This type of igniter is suitable for the largest type of oil burner.

7.5.2 Oil burner controls

Oil burner installations are now exclusively semi-automatic within the CEGB. Semi-automatic means that the operator starts the firing sequence manually and the sequence then proceeds automatically to establish the oil flame. Oil burner controls have progressed through discrete relay logic to the current full microprocessor/computer control.

A normal ignition sequence would be as follows, and would be fully monitored by the control gear on the operator pressing the START button:

1. Check that associated boiler interlocks are all cleared.
2. Check no flame indication from flame monitor ('dark check').
3. Insert burner (if appropriate).
4. Burner purge — 60 seconds (if appropriate).
5. Prove atomising steam on (if appropriate).
6. Energise electrical ignition circuit.
7. Open gas valve.
8. Prove gas flame lit.
9. Open oil valve and continue to prove gas flame lit.
10. Prove oil flame and continue to prove gas flame lit.
11. Shut off igniter.
12. Prove oil flame still lit.

From this sequence it can be seen that a flame monitor system for both gas and the oil flame is essential. These monitors must be separate. Monitors are available in many forms; ultraviolet (UV), infra-red flicker, visible light correlation, cross correlation and many others. The spark gap ionisation proving system is particularly suitable for premixed gas igniters.

If the oil flame is not proven within a period of time governed by preset timers in the control system, the burner is shut down and a fault signal given. It is normal now to incorporate a sequence diagnostic step monitor in the control gear to identify the fault area and thus enable rapid diagnosis and repair.

One of the main boiler interlocks (see step *1* of the ignition sequence) is that of furnace purge. The boiler must be purged with air before any light-up attempt and also after the last burner is shut down on ceasing operation. The normal purge rate is a minimum of 50% MCR airflow for a minimum of four minutes giving at least five complete air changes. This is essential, in order to reduce the risk of boiler explosion caused by a build-up of combustible vapour.

Step *3* in the sequence (insert burner) is necessary if a retractable burner system is installed. The designer has three options when choosing a burner type; a retractable burner, in which the burner is retracted from the high heat zone when out of service; a tip shut-off system combined with a spill return pressure jet in which the burner bar is fixed in position and cooled by recirculating oil; and a fixed burner, either pressure jet or twin-fluid in which the atomiser is cooled by the purge fluid steam or air.

It is normal to purge a pressure jet or twin-fluid burner of fuel oil into the furnace on burner shutdown using steam for a residual fuel oil system (air may be used for a distillate oil system). The igniter must be fired during this sequence.

7.6 Oil burner windboxes

The general comments made in Section 5.5 of this chapter apply, however, oil burner windboxes are now always modelled in order to ensure air distribution between burners of better than ±3%. The best air distribution obtainable is essential if low oxygen operation is to be maintained without any deterioration in performance. As stated elsewhere 'low oxygen' implies an excess oxygen at the furnace outlet of 0.5% or less. Most oil plant is now operated at the carbon monoxide (CO) threshold which is the point at which CO starts to rise above 100 PPM in flue gas. This is usually less than 0.5% oxygen.

It is possible that individual windboxes with individual airflow control would provide a slightly better performance but the accuracy, reliability and maintenance demands of such a multiplicity of flow devices has so far precluded their use on CEGB plant.

7.7 Oil burner performance and maintenance

When firing residual fuel oil it is essential that the oil is fed to the burner at the correct input conditions. It is also essential that the geometry of the air register is maintained because, despite the overall flow control, the final airflow and mixing of the fuel and air to produce the required performance are governed by this geometry.

There are three main factors by which oil burner performance is defined:

(a) Performance at light up, i.e., combustion efficiency in a cold/cool environment and the facility to prevent the escape of unburnt oil or vapour to the economiser/airheater, etc., during a prolonged boiler warming phase (GOM 68 refers) [13].

(b) The ability to operate on low excess oxygen in order to reduce the formation of sulphur trioxide and thus prevent acid smuts [14].

(c) The ability to operate at low excess oxygen without exceeding the HMIAPI limits on solids in flue gas (currently 115 mg/m^3).

In order to maintain this performance careful maintenance is necessary, particularly on the factors affecting the oil atomiser. The air register geometry relies on primary design and installation and repair, if damaged, it requires a boiler outage, but atomisers deteriorate throughout normal use and the following procedure should be followed for any normal maintenance process.

7.8 Testing and maintenance of oil atomisers

The components of an oil sprayer (atomiser), namely the orifice plate, swirl chamber and distribution plate, are precision parts manufactured to close tolerances. This is necessary to ensure consistent atomiser performance.

In operation, these components may suffer from blockage or wear, both of which are reduced to a minimum by the use of hardened orifices, adequate cooling and good fuel filtration. Thus, in a well designed installation, the sprayer is capable of prolonged use without maintenance. Other burner arrangements may not be so satisfactory and frequent attention may be required to prevent malfunctioning.

The following maintenance system is suggested based on the premise that dismantling is wasteful in time and increases the possibility of damage to the sprayer components.

An important aspect of any maintenance system is to establish the degree of attention required. Therefore, at regular intervals, the frequency of which will be determined by subsequent experience, a sample number of burners should be removed for cleaning and calibration. The selection should be undertaken in rotation so that all burners receive equal attention.

7.8.1 Pressure jet and spill atomisers

Initially the burner barrel should be flushed clean of fuel oil. This task should be completed on a closed circuit system utilising gas-oil or kerosine as a flushing medium. If the burner has been removed from the boiler for some time, a facility for barrel heating will be required on a tip shut-off burner system. The flushing operation should be of about 5 min duration. The clean barrel is then installed on a sprayer test rig, on which gas-oil is supplied to the burner at a selected pressure (marked on the pressure gauge).

The gas-oil flowrate indicated on the flowmeter should be noted and the spray visually examined for cone angle and uniformity (absence of streaks). Providing the flowmeter reading is within $\pm 2\%$ of the design fuel throughout, when corrected for temperature, and there is no apparent leakage from the cap nut or any other intermediate connection, the burner may be set aside for return to the boiler.

If leakage is apparent, an attempt should be made to reseat the sprayer components by tightening the cap nut. If this cures the leakage, the barrel should be returned to the sprayer test rig and retested.

If leakage persists, or the required flowrate is not achieved or the spray appears non-uniform, then the sprayer should be dismantled and inspected for deposits in the swirl passages or between the plates. Any deposits should be removed using copper tools, or a heated fluidised bed, and the sprayer reassembled using new seals where applicable and retested.

Should sprayer performance still be unsatisfactory, the complete sprayer assembly should be discarded and replaced with new components and retested.

If a series of burners installed on the spray test rig give consistently high or low measurements of flow rate or a new sprayer assembly is out of limits on flow rate, the calibration of the rig pressure gauge and flowmeter should be checked.

By using gas-oil or kerosine for flushing and testing, the burners with satisfactory performance may be returned to service without the need for an air or steam purge.

7.8.2 Steam or air atomisers

The oil passages should be flushed clean using the closed circuit system, as for pressure jet and spill atomisers. Since the designs of twin fluid atomisers at present in use in the CEGB have the cone angle physically set by the method of construction, the test rig may be simplified to deal only with the problems of wear and incorrect assembly.

The cleaned barrel is installed on the sprayer test rig, the oil and steam/air inlets being connected to metered oil and air systems, respectively. Adjacent to each inlet connection should be a pressure gauge.

The desired flowrates should be established from a series of sprayer tests on a number of new atomiser assemblies. Once the nominal 'as new' flowrates have been deduced, the oil and air flowrates through used burners, when corrected for temperature and pressure, should be within $\pm 2\%$ of the average 'as new' value.

The oil passages should be tested using similar techniques for pressure jet and spill atomisers, that is, the flowrate through the sprayer is measured at a preset oil pressure and compared with the nominal 'as new' value. Should the flowrate be out of limits, refurbishment is limited to cleaning or replacement of sprayer components.

The steam or air passages are tested using air as the test medium. The air supply should be adjusted to a selected pressure (marked on the pressure gauge). The airflow through the atomiser may be read off the air flowmeter installed on the test rig.

While the outer annulus is under air test, the cap nut should be examined for leakage with a soap solution.

As for pressure jets, the sprayer should only be dismantled if suspect. Cleaning should only be undertaken with soft tools or a heated fluidised bed, and care should be exercised when checking orifices with *go* and *no go* gauges.

7.8.3 Cleaning record

It is useful to record the results from sprayer tests with particular reference to the position of the burner in the boiler, the type of fault, if any, and the time of removal from the boiler, to allow operating time to be established.

From this data the required frequency of sprayer examination may be determined and the location of possible aggressive environments within the boiler may be identified.

7.8.4 Safety aspects

Sprayer test rigs should be constructed in compliance with GDCD Standard 164 Appendix A [15].

8 Arrangement of oil handling plant

Oil handling plant is installed on coal-fired and oil-fired boilers. On the former it is needed for lighting-up and stabilising the PF fired furnace, whilst on the latter it is used for the main burners. The majority of boilers have plant that is designed for a range of oils from light to heavy and generally the problems are the same whether it is a coal or an oil-fired boiler.

The handling and storage equipment may be divided into two parts:

- Off-loading and storage area (discussed in Volume A Chapter 2).
- Pumping and heating arrangements.

8.1 System

As the size of coal-fired boilers has grown, it has become common practice to increase the capacity of the oil lighting-up burners beyond that required for coal ignition, to a capacity that enables the turbine to be synchronised on oil alone. This also gives a moderate amount of replacement generating capacity in the event that an excessive number of mills are out of service due to unforeseen circumstances. The oil firing capacity of some of the 500 MW and 660 MW coal-fired boilers is now as much as the 100% MCR output of some of the smaller 100% oil-fired boilers.

Whilst for editorial reasons the oil tanks and off-loading equipment are dealt with in Volume A it must be remembered that a change to any part of the complete system (i.e., from tanks to burners) cannot be made in isolation without considering the overall effect.

The basic design requirements of the whole system are:

- To keep the oil in the storage tanks at a temperature at which it will flow by gravity into the transfer pump.
- To raise the oil to a temperature which will give a viscosity that is suitable for atomising into a spray at the burners. This temperature must be controlled not just to maintain efficient atomisation but also to present the oil flow meters with a fluid of constant viscosity. In the case of oil-fired boilers this is necessary in order to control the ratio of oil to combustion air so as to achieve combustion with low excess air.
- To raise the pressure of the oil to that necessary to give efficient burner atomisation over a specified load range.

For small single boilers it has been common practice to install *duplex* combined heating and pumping, i.e., two 100% heaters and two 100% pumps, all mounted on a single common baseplate.

Where only one boiler is under consideration such an arrangement is ideal, but where there is more than one boiler the extent of standby becomes rather excessive should a duplex unit be chosen for each boiler. In order to reduce this excessive standby, interconnection between one boiler and another is adopted and *simplex* units may be provided for each boiler. A simplex unit consists of one 100% heater and for 100% pump on a common baseplate. A simplex arrangement for three boilers is shown in Fig 1.38.

An arrangement such as this is typical for units up to 120 MW but as boiler sizes have increased it has become desirable to increase the number of heaters and pumps per boiler to obtain greater flexibility of operation. A typical arrangement for a 500 MW oil-fired boiler is shown in Fig 1.39, in which one boiler is served by four 25% heaters plus one 25% spare, all in parallel and placed on the suction side of three 50% pressure pumps which are also in parallel. The heaters in such an arrangement would each have a capacity of about 32 tonnes per hour and the pumps 64 tonnes per hour.

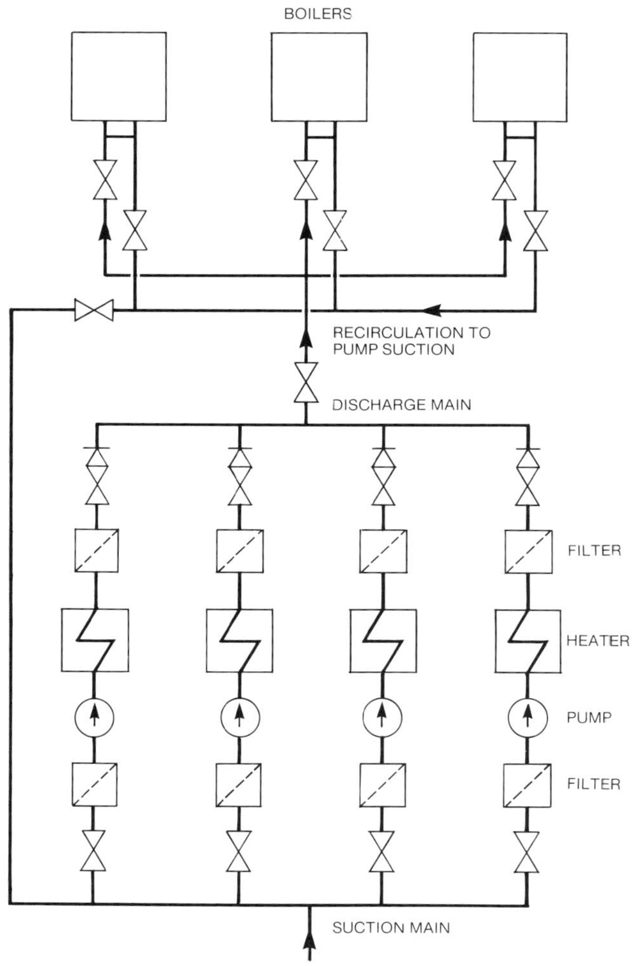

FIG. 1.38 Diagrammatic arrangement of four simplex units serving three boilers

On current designs which use residual oil, continuous recirculation is used to maintain flow through the distribution piping; in addition, heat tracing is provided to make up the loss through the piping insulation.

Often a separate pump is used for recirculation; this draws oil from the heater discharge, circulates it up to the burners and discharges back to the heater suction. By this means the only oil at a temperature below that required for burner atomisation is a small quantity in the burner itself. Some designs of burners even avoid this by recirculation which is carried out right up to the burner tip (see Fig 1.36).

Where there is a limitation in the quantity of steam available from the main boilers for oil heating or where large quantities are required, it is generally economical to install auxiliary low pressure oil-fired shell boilers of the *economic* type.

8.2 Pumps

There are two main types of pressure pumps, centrifugal and positive displacement screw. Both are me-

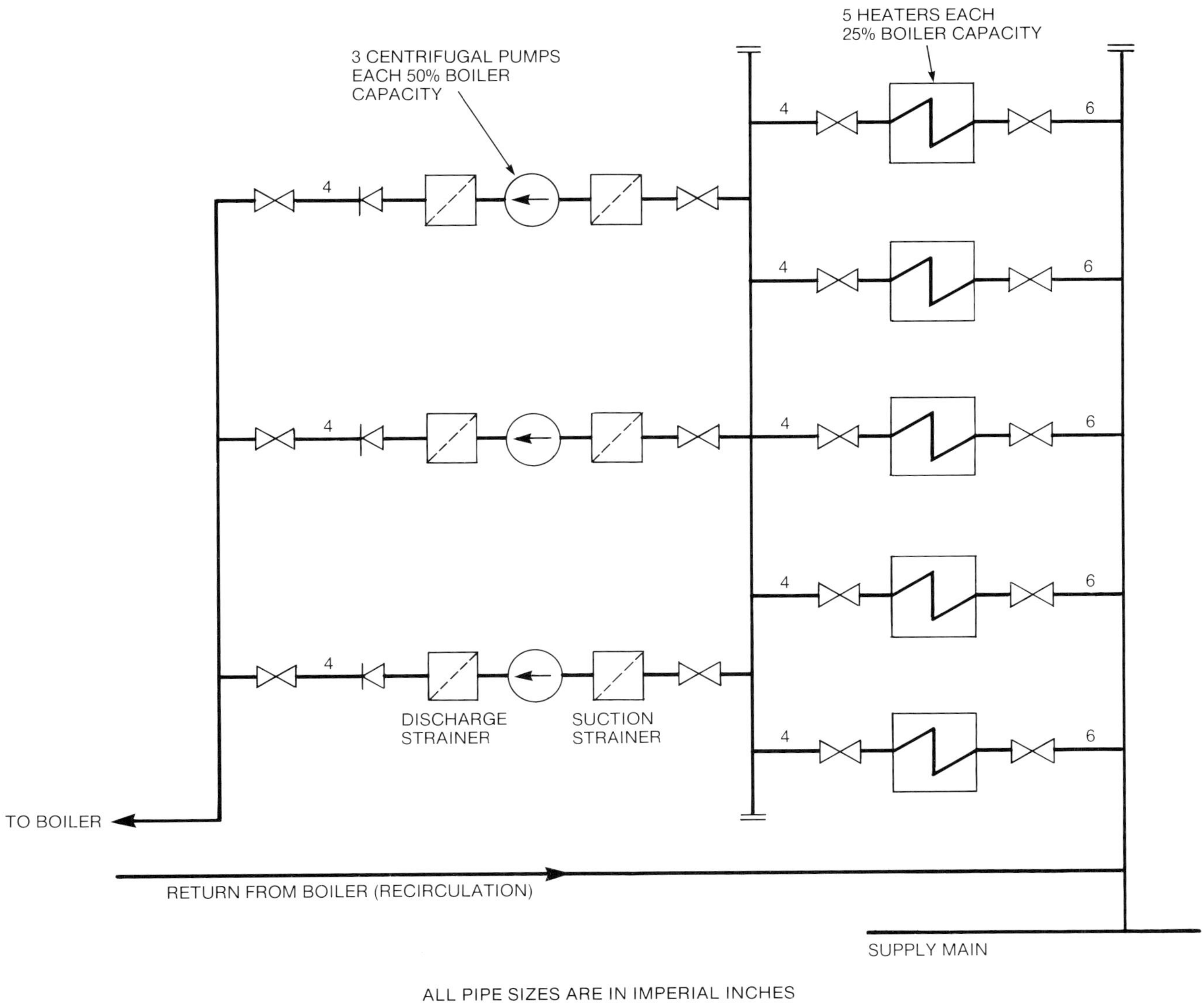

FIG. 1.39 Typical arrangement of oil-firing equipment for a 500 MW boiler

chanically reliable and there are no clear advantages for either type.

Centrifugal pumps are used extensively in the petro-chemical industry and perform satisfactorily; there is no shortage of manufacturers and prices are competitive. The control of these pumps is simple and is achieved by increasing the pressure drop in the system by closing in a throttling valve; the pump output quantity then decreases in accordance with the pump performance curve.

Positive displacement screw type pumps deliver a fixed quantity of oil at a pressure, the maximum value of which, if uncontrolled, would only be limited by the power of the pump motor and the containment pressure of the system. In practice a spill valve on the discharge side of the pump is used to control the quantity and pressure of the oil going to the boiler. The excess oil is spilled back to the oil storage tanks or some other point upstream of the pump inlet.

Centrifugal pumps are more efficient when pumping low viscosity oil and are normally located after the heaters, whilst positive displacement pumps are better when handling thicker oils and are therefore located before the heaters.

Typical centrifugal and positive displacement pumps are shown in Fig 1.40 and 1.41 respectively.

8.3 Heaters

The medium for heating the oil can be electricity, hot water or steam and the choice will be determined by installation and running costs. Most of the heaters installed by the CEGB use steam and are of the shell and tube type.

If the pressure pumps are located before the heater then the steam will be at a lower pressure than the oil and any leakage will be of oil into the steam/

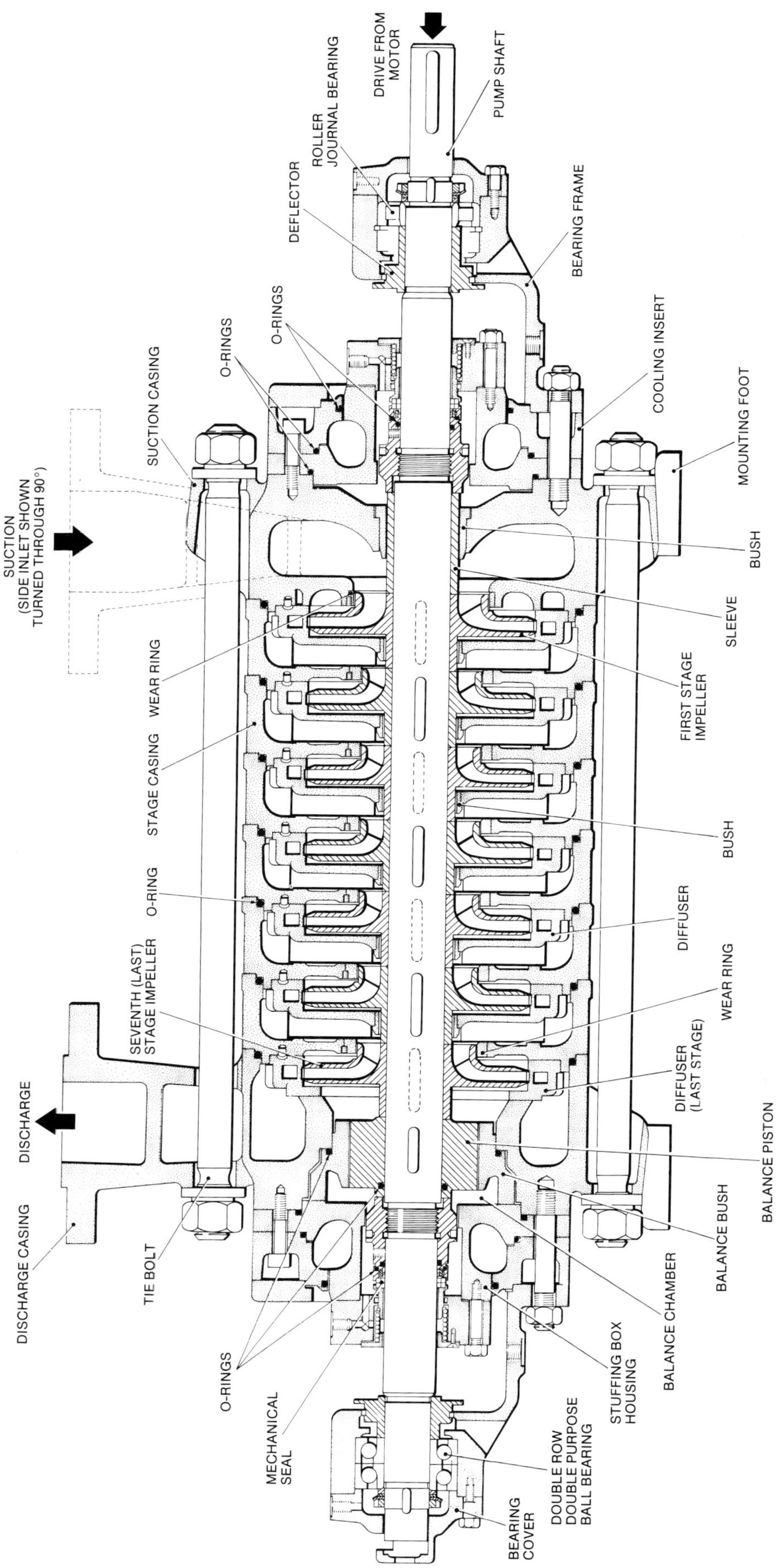

FIG. 1.40 Typical multi-stage centrifugal type fuel oil pump

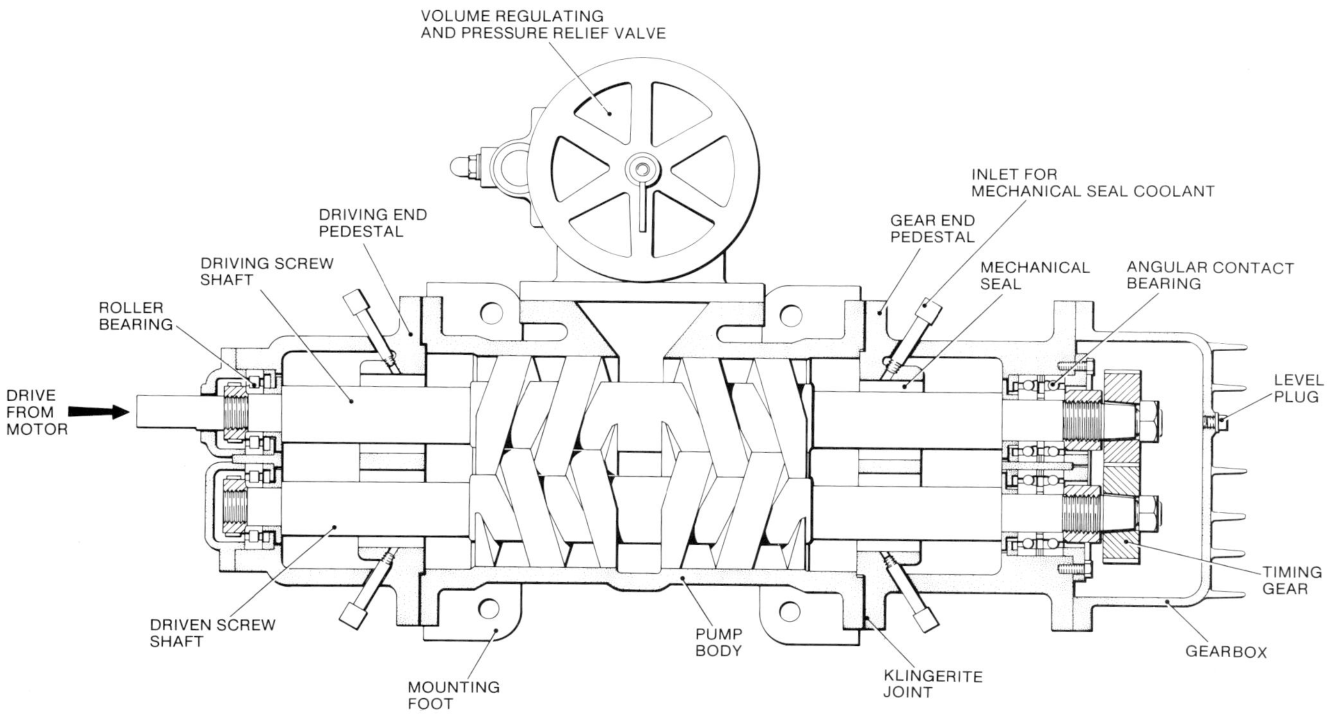

FIG. 1.41 Typical positive displacement screw type fuel oil pump

condensate system, which is undesirable because oil may get into the auxiliary boiler feedwater. If the pressure pumps are after the heater then the situation is reversed. There are no strong deciding arguments and the choice is usually based on cost.

The heater design presents no particular problems provided that attention is paid to:

- Avoidance of steam impingement (steam/oil heaters).
- Adequate oil velocity to reduce oil side deposits, combined with smooth surfaces to which sludge will not stick.
- An efficient condensate removal system (steam/oil heaters).
- Plant spacing and lifting arrangements to give easy removal of heater tube nests.

A typical steam fuel oil heater is shown in Fig 1.42.

8.4 Piping

Choice of the diameter of oil piping is a compromise between reducing pumping power by opting for large pipes, and reducing capital cost with small pipes. Oil velocities are usually within the range 1.5 m/s to 2.0 m/s.

Guidance on the design and construction of piping systems is given in BS3351 [16].

Most of the fuels burnt by the CEGB are of the residual type and therefore require some method of limiting heat losses from the pumps, piping, valves and strainers. This is done by insulation augmented by heat tracing; the objective is to keep the temperature up to a level which will allow the oil to be pumped after it has been static for some time. This temperature is not necessarily that needed for burner atomisation. As previously mentioned the design of current oil handling systems is one that employs constant recirculation of oil through most of the piping. This keeps it clear and tracing is only used as an emergency back-up if needed. However, constant trace heating is used on limited sections of piping where there are unavoidable dead legs.

Under normal boiler firing conditions the heat loss between the heaters and the boiler front will be kept to an acceptable level by the piping insulation and trace heating does not need to be in operation.

The trace heating medium can be electricity, steam or hot water and the choice rests on an economic assessment of capital, running and maintenance costs. Electric tracing is relatively simple to install and has a low capital cost but operating costs for continuous use may be higher than using steam or hot water. Steam tracing systems of any size have a large number of steam traps which collectively need significant maintenance.

Convenient sections of the oil handling system should be provided with isolating valves so that wherever possible maintenance can be carried out on main plant items without bringing the oil system off-load. Sections containing high pressure oil (13.8 bar and above) should be provided with double isolating valves. To reduce

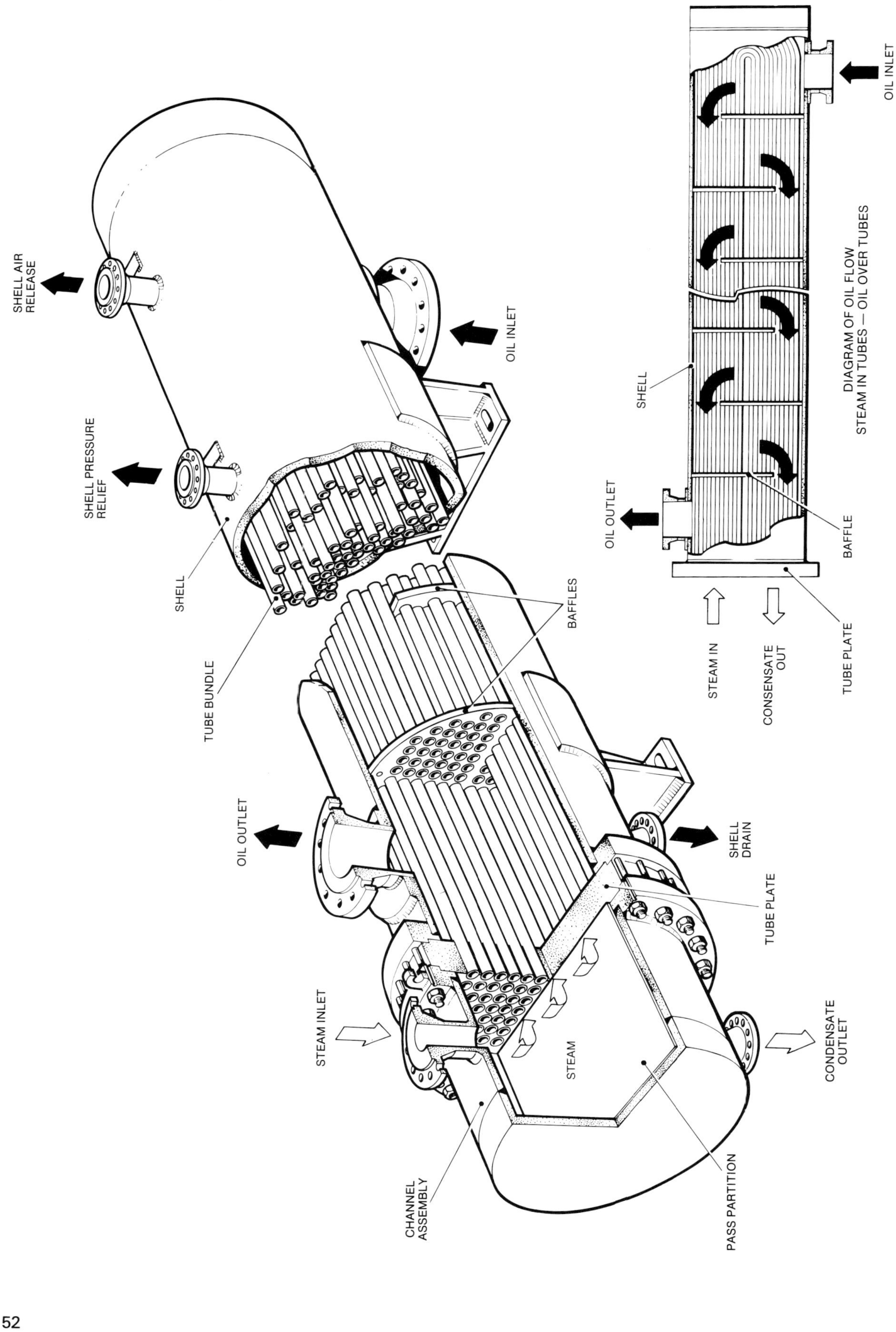

FIG. 1.42 Typical steam fuel oil heater

fire hazard, oil piping joints should be welded wherever possible. All sections of piping that can be isolated should be fitted with relief valves to allow for expansion of trapped oil. For the same reason the bodies of valves and strainers should not be made from cast iron.

9 Future trends

The CEGB can see the need for a series of coal-fired stations in the future both to meet increased demand and to replace units reaching the end of their life. After an exhaustive examination of the risks and advantages of advancing from the current 660 MWe units, it has been recommended to move to 900 MWe units, using slightly enhanced steam conditions but not supercritical pressures. The availability of large mills that have accumulated many hundreds of thousands of hours of operation allows the new boilers to be designed with only six or seven mills and a correspondingly small number of burners which eases the logistics of maintenance in this area.

It is not anticipated that oil-fired plant will be ordered in the foreseeable future, but the flexibility of the present oil-fired units and fluctuations of oil prices give little incentive to convert the oil-fired stations to coal firing. The compact designs developed do not allow modification without loss of output and, if conversion were ever required, it is likely to involve replacement with new coal boilers rather than modification of the existing boilers.

The trend in fuel supplies is towards coal with more corrosive constituents and this could limit steam temperatures to those presently employed. Further development of coextruded tubing could lead to higher steam temperatures.

Heavy fuel oil will be used exclusively for start-up and flame stabilisation of coal-fired boilers, and this will require pumping and heating systems similar to those installed at oil-fired stations. As oil prices increase in comparison to coal prices, there will be pressures to reduce the consumption of oil at coal-fired stations. Considerable progress has already been made and a side benefit of the development of low NOx firing systems may well be reduction of oil consumption for flame stabilisation, as most NOx systems require very intense and therefore stable initial combustion.

10 References

[1] Raask, E.: Mineral Impurities in Coal Combustion: Hemisphere Publishing Corporation, US: 1985

[2] NFPA 85: Prevention of Furnace Explosions in Pulverised Coal-Fired Multiple Burner Furnaces

[3] CEGB, GDCD Standard 166: Oil Burners (Light-up/part load): Issue 2: September 1980

[4] CEGB, GOM 105: The Operation of Pulverised Fuel Plant to Avoid Explosions, Issue 1: January 1983

[5] Zeldovich, T. B.: Acta Physica Chem USSR Vol. 21, pp 577: 1946

[6] Lowrison, G.: Crushing and Grinding: Butterworths, London: 1975

[7] Rose, H. E. and Sullivan, R. M. E.: A Treatise on the Internal Mechanics of Ball, Tube and Rod Mills: Constable & Co. London

[8] Marshall, V. C.: Comminution: Institution of Chemical Engineers: 1975

[9] BS1016, Part 20: Determination of Hardgrove Grindability Index of Hard Coal: 1981

[10] Yancy, H. F., Gear, M. R. and Price, J. D.: An Investigation of the Abrasiveness of Coal and its Associated Impurities: Transactions — American Institute of Mining, Metallurgical and Petroleum Engineers, 190, pp 262–8: 1951.

[11] BS1016, Part 19: Determination of Index of Abrasion of Coal: 1980

[12] BS2869: Specification of Fuel Oils for Oil Engines and Burners for Non-marine use: 1983

[13] CEGB, GOM 68: The Prevention of Air Heater Fires: Issue 3: March 1983

[14] Brooks, W. J. D., Holmes, L. and Leasons, D. B.: Power Station Oil Burner Systems for Low Excess Air Operation: Journal of Institute of Fuel, Vol. 38 pp 218–237: May 1965

[15] CEGB, GDCD Standard 164: Oil Burner Atomiser Tips, Issue 2: October 1979

[16] BS3351: Specification for Piping Systems for Petroleum Refineries and Petrochemical Plants: 1971

CHAPTER 2

Boiler unit — thermal and pressure parts design

Contents

1 Introduction

This chapter deals with the boiler unit as a heat exchanger, with regard to both its performance and construction. In order to obtain a satisfactory form of analysis, it is convenient to regard the boiler as composed of a number of integrated heat exchangers. The fluid on the hot side of the heat exchanger is the mixture resulting from the combustion of the fuel in air; this expression covers the substances from the initiation of the combustion reaction, through the completion of combustion to the physical process of cooling the gas. The fluid on the cold side of the heat exchanger is water substance, entering as liquid and leaving as steam, an imperfect gas.

2 Disposition of heating surfaces

At first sight of a sectional side elevation of a modern boiler (Fig 2.1) it may seem that although the gas flow is quite simple, the water and steam flow path is unduly complicated or even random. In fact, the disposition of the various parts of the cooling surface is carefully thought out to make the most economic use of natural physical phenomena. It is fundamental

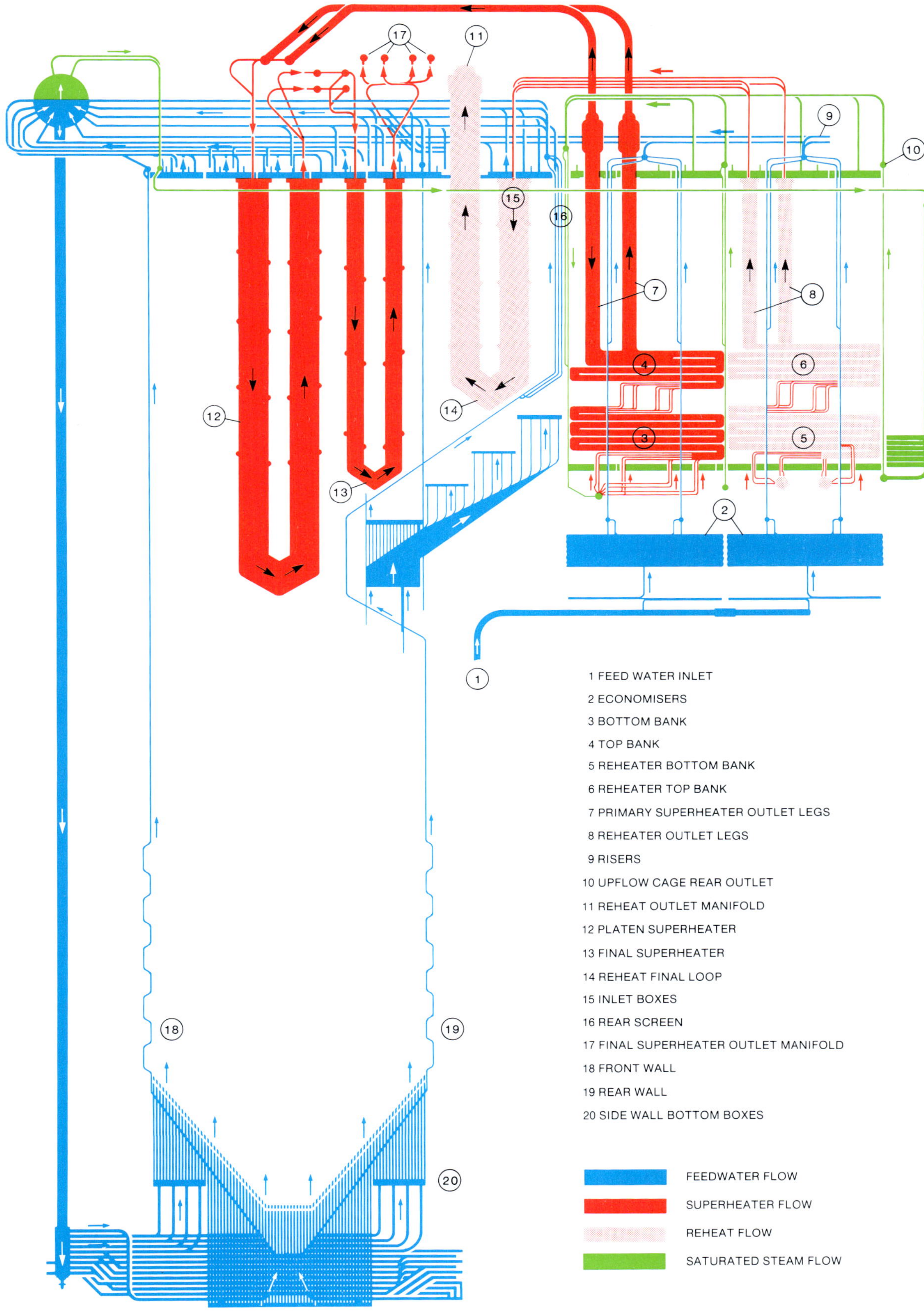

FIG. 2.1 Section through a 660 MW boiler

to the exchange of heat between substances that there must be a temperature difference between the substances and, furthermore, that the rate of exchange varies with the temperature difference. This suggests the use of a counterflow heat exchanger, that is to say hot steam heated by hot gas and cool water heated by cool gas. In part, this holds good, but in part is over-ridden by consideration of the pressure parts which are utilised to separate the fluids. The tubes used for this duty will take up some temperature between that of the fluids they separate. It shall be seen later that where water or slightly superheated steam is the cold side fluid, the tube metal temperature is relatively near to this fluid temperature, but where the coolant is more highly superheated steam, the tube metal temperature is a relatively higher increment above the already high coolant temperature. Now, since a higher temperature duty for a tube requires the use of a higher alloy and hence more expensive steel, the high steam temperature surfaces need to be carefully located so as to limit the tube metal temperature and yet near enough to the furnace to keep the amount of heating surface to a minimum.

There are three main zones within the overall heat exchanger:

- The furnace — where the high temperature gases of combustion can be used for heating water and steam with low to medium superheat.
- The convection zone — where medium temperature gases can be used for heating steam with medium to high superheat.
- The heat recovery zone — with cool gases, heat can only be abstracted effectively by cool fluids, such as feedwater and steam with low superheat.

Within these zones, there is scope for variations on the theme within which the designer must provide for absorption of the correct proportion of heat in all the stages as well as the correct total absorption. The balance must be maintained when firing any fuel that has been specified for the boiler, and under varying load conditions.

It must also be noted that the proportioning of heat distribution varies with the thermodynamic cycle used. As pressure and degree of superheat increases and reheat is introduced, latent heat becomes progressively a smaller proportion of the total heat transferred and the furnace has to provide superheat duty as well as evaporation.

3 Heat transfer and fluid mechanics

3.1 Furnace heat transfer

The rate of heat transfer in the furnace is complex and is dependent on a number of interacting parameters. It is usually calculated using a validated computational algorithm to give steady state gas temperatures and radiative fluxes. The three-dimensional system geometry of the furnace is divided for computational purposes into a number of cells. For each of the cells, the following have to be identified:

- Mass flow across each cell face. These are obtained in generalised form from model tests generated within the program.
- Heat release within the cell. Heat from each flame is released progressively along its length; the rate can be established from measurements on single flames and in generalised form is introduced into the program.
- Wall emissivities and reflectivities. These can be obtained from both plant and laboratory measurements.
- Flame emissivity. This is made up of three components:

  (a) The non-luminous radiation from triatomic gases CO_2 and H_2O.

  (b) The radiation of the soot particles. These particles originate from the thermal decomposition of the hydrocarbon components of the volatile matter.

  (c) Radiation of solid coal, char and ash particles.

- Resistance to wall heat absorption due to deposits. This is simulated by specifying the furnace tube fluid temperature and the wall heat transfer coefficient. It allows the surface temperature appropriate to the local heat fluxes to be calculated: for oil-fired boilers the coefficient is high whilst for coal firing it is low and varies considerably with operating conditions.

Having defined the gas flow into and out of each zone, the rate of heat generation together with the data on the radiation characteristics of the furnace gases and walls, the thermal and radiation energy balance is calculated within the code. The resultant equations are solved simultaneously for the gas temperature within each zone and the net radiation across each zone boundary. The output generated for such a program is illustrated in the prediction of heat flux for a 500 MW oil-fired unit, see Fig 2.2.

3.2 Furnace water/steam circuit

3.2.1 Drum boilers

In a drum boiler, water flows from the drum via downcomers and feeder pipes to the bottom of the furnace

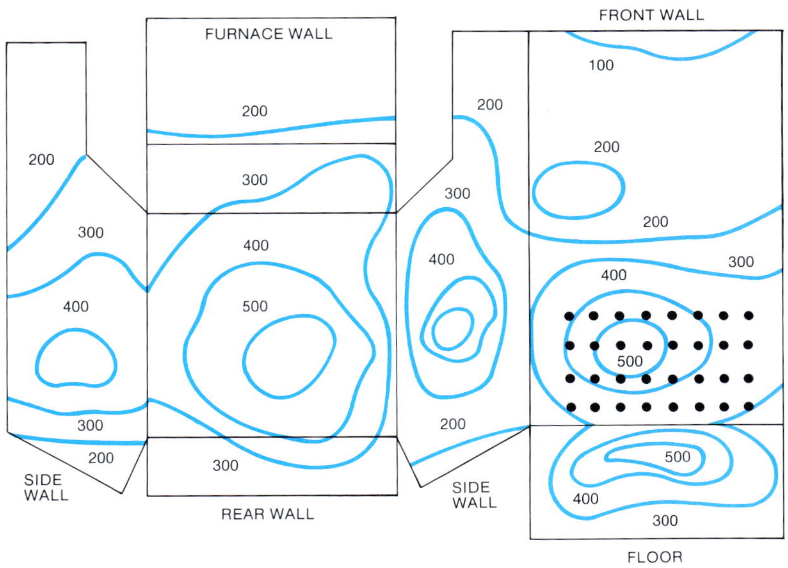

FIG. 2.2 Predicting heat flux distribution in a 500 MW oil-fired furnace

where it feeds the evaporator tubes forming the furnace envelope. Within the furnace, tubes are heated and the water is gradually changed to a water/steam mixture of increasing dryness along its length. This mixture is returned to the drum where dry steam is separated from the mixture. Since the water/steam in the furnace tubes is of a lower density than the water in the downcomers, a static head difference is generated and this results in a downward flow in the downcomers and an upward flow in the furnace tubes. With natural circulation boilers, the flow is generated solely by this density difference, with assisted circulation boilers the flow receives additional assistance from the pumps. The pumps are usually located at the bottom of the downcomers.

Calculation of overall circulation rates and individual tube conditions requires knowledge of distribution of heat input to the furnace walls; this information is then applied to groups of tubes and circulation rates are then obtained by balancing the driving forces, i.e., static head difference against the frictional and acceleration resistance to flow. This process is normally carried out by a computer.

While the calculation process for natural and assisted circulation boilers is similar, and their vertical tube design is similar, there are differences in their characteristics.

Natural circulation furnaces are always designed so that flow rate increases with increasing heat input and they can therefore tolerate variations in heat input around the furnace perimeter (natural circulation characteristic). To achieve economic designs, however, the circuits are arranged to broadly match heat input, larger numbers of tubes being fed from each downcomer in low heat input areas than in high heat input areas. Tubes are generally of 63 mm outside diameter.

Assisted circulation furnaces, to achieve design economy, normally have small diameter tubes (44 mm) and the number of downcomers is reduced. The tube flows have less natural circulation characteristics, i.e., flow rate does not increase so greatly with heat input as with natural circulation boilers and more elaborate matching of flows to heat input is required. In most designs, tubes are individually orificed to achieve the necessary matching. Circulation commences as soon as the circulating pumps are started and this eliminates the temperature differences that are formed during shutdown (although seen, these temperature differences have not proved a significant operational problem in natural circulation boilers).

With assisted circulation boilers, downcomers are commoned into the manifold at the pump inlet side and discharged into a common discharge manifold before distribution to the furnace headers. Generally, one spare pump is provided and the manifolds are sized to allow full load operation with any one pump out of service. Load has to be reduced if two pumps are out of service. In CEGB stations, load limitation due to pump unreliability has been minimal.

Drum boilers are dependent on the difference in density between steam and water and, as pressure increases, these densities converge until at and above the critical pressure they are equal. Drum boilers are therefore restricted in pressure to somewhat below critical pressure, the limit of economic designs being somewhere around 183 bar.

3.2.2 Once-through boilers

Once-through boilers can operate at any pressure below or above the critical as furnace flow is imposed by the feed pump. While water flows in the furnace tubes of natural and assisted circulation boilers remain generally constant from 5–100% CMR, flow in a once-through boiler is proportional to load. At low loads, inadequate furnace protection is provided by once-through flow and in most designs the minimum furnace flow is maintained above 25–30% CMR flow at low load and start-up either by recirculation or by dumping. In most modern designs, a separator vessel is provided at the furnace exit and circulation pumps recirculate water from this vessel through the economiser and furnace tubes, effectively creating an assisted circulation boiler at low load and start-up.

The need to maintain flow stability and cooling at minimum once-through flow results in high furnace pressure drop at CMR and in turn a forced flow characteristic at high loads, i.e., increased heat input to a tube reduces flow and hence increases steam quality or superheat at furnace outlet. To minimise this effect, once-through furnaces generally have spirally-wound tubes to average the heat input per tube. To achieve acceptable part load velocities tube diameters are smaller than in drum boilers.

With the furnace tubes in an inclined or near horizontal attitude they are intrinsically weaker in the vertical direction than vertical tubes of the same geometry. The tubes have therefore to be thickened relative to their pressure-containing requirement to withstand the furnace ash loads, loads which can be carried without reinforcement in vertical tube boilers. To achieve an economic design, many once-through boilers are supported at intermediate levels.

Vertical tube designs have been developed for once-through boilers but this has generally been for supercritical pressures as multi-pass designs can be developed without the separation problems in intermediate headers which can occur at subcritical pressures. Tubes generally are smaller than in drum boilers, because of the need to maintain high velocities at part loads and often additional support has to be given to the walls to support ash loads.

3.2.3 Water/steam heat transfer

In a furnace operating at subcritical pressure, water enters the evaporator tubes subcooled and leaves as a steam/water mixture. The heat fluxes in a furnace generated by the combustion process are extremely high and if the heat flux is allowed to reach a critical value there is a drastic fall in the waterside heat transfer coefficient. This is known as critical heat flux, departure from nucleate boiling, burn out or dryout. The phenomenon responsible for the fall in heat transfer coefficient is the replacement of a layer of liquid and vapour adjacent to the wall with a layer of vapour alone which has a very much lower conductivity. Tube failure can then result from two mechanisms:

(a) The low heat transfer coefficient combined with heat flux raises the metal temperature to levels where creep life is rapidly consumed.

(b) Local concentration of dissolved solids at the steam/water interface at point of dryout leading to rapid corrosion.

Critical heat flux is dependent on a number of parameters:

- Operating pressure.
- Steam quality, i.e., degree of wetness, subcooling, etc.
- Type of tube, rifled or plain bore.
- Tube diameter.
- Flux profile around the tube.
- Steam/water flow in the tube usually expressed as mass flow per unit area.
- Inclination of the tube — vertical tubes have a higher critical heat flux value compared to inclined tubes, due to buoyance effects.

It has not been found possible to find a theoretical equation to encompass all these factors and therefore the design information is obtained from rig work which simulates the actual conditions to be found in any particular furnace under consideration. The CEGB has carried out extensive tests at its Marchwood Engineering Laboratories. Figure 2.3 illustrates the type of flows found. These depend, amongst other things, on the magnitude of flow and steam quality. At a low quality, bubbly flow occurs with the steam travelling as disconnected bubbles in a stream of water. Under this condition, the critical heat flux is related to peak heat flux at the crown of the tube. At higher qualities the flow becomes annular, the steam travels along the tube as a continuous stream, carrying with it some water droplets, while the remaining water flows as a thin film. With annular flow the critical heat flux is dependent on the total heat input to the tube and not

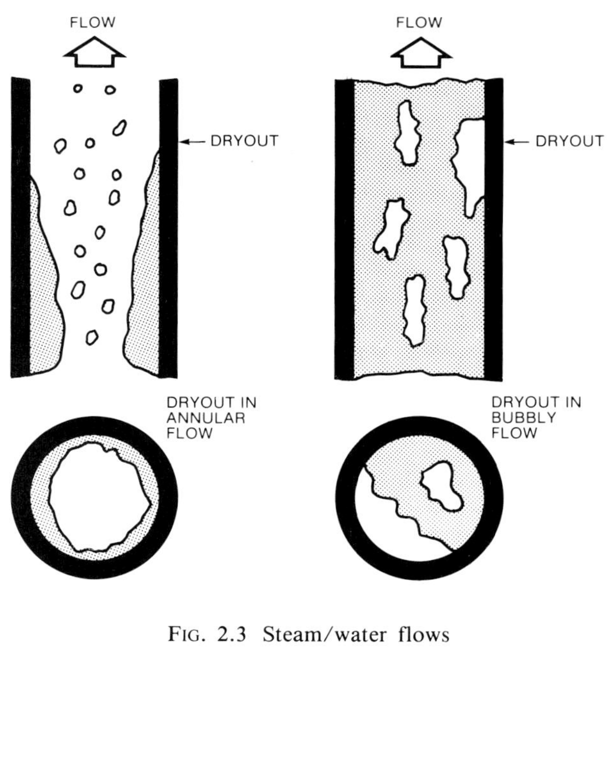

FIG. 2.3 Steam/water flows

the peak heat flux at the crown. In bubbly flow, dryout occurs when nucleate boiling ceases and a steam blanket occurs, whilst for annular flow the depletion of the annular film is greater than the re-deposition of water droplets from the core. There are other flow patterns which fall between the two described.

Once the designer has the required critical heat flux data from the rig, an assessment can be made of the vulnerability to dryout for any particular furnace design. This is carried out by comparing the designer's estimate of heat flux along the length of the tube (see Section 3.1 of this chapter) with the relevant critical heat flux data, see Fig 2.4. For the design to be acceptable the critical heat flux must always be greater by a margin than the heat flux generated in the combustion chamber.

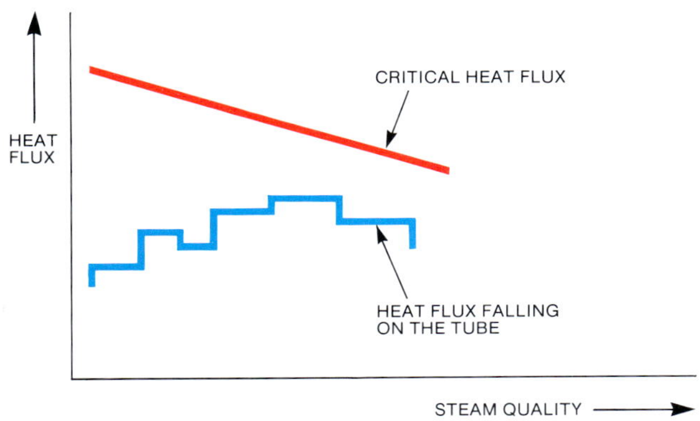

FIG. 2.4 Heat flux v steam quality

3.3 Heat transfer in tube banks

The steam raising surface within a boiler may receive heat by three main processes:

(a) Convection, due to the flow of high temperature gases over or around the tubes.

(b) Non-luminuous radiation from certain of the gases contained in the fuel combustion products.

(c) Direct radiation from the furnace.

The mechanism of direct furnace radiation heat transfer is described in Section 3.1 of this chapter. The following is a discussion of the methods of calculating heat transfer due to convection and non-luminous radiation. The interaction of the outside film coefficient and the heat transfer mechanism inside boiler tubes, to produce an overall heat transfer coefficient, is also described.

3.3.1 Non-luminous radiation

Non-luminous radiation is the term used to describe radiation of heat from hot gases which have left the furnace environment and are no longer part of the flame. As these gases pass over boiler tube banks, the radiant heat transfer may be visualised by considering that at any instant in time the tube bank will contain a volume of hot gas which is stationary and radiating heat to each tube within the bank. This form of heat transfer is also known as intertube radiation.

Of the many products of combustion of a boiler fuel, the only gases which possess radiant properties worthy of consideration for this analysis are carbon dioxide (CO_2), water vapour (H_2O) and sulphur dioxide (SO_2). Other gases such as hydrocarbons and alcohols are known to exhibit radiant properties, but these are not normally encountered in British coal or oil combustion products. Gases with symmetrical molecules such as hydrogen, oxygen and nitrogen have been found not to show significant radiant properties at temperatures prevailing in boiler plant.

The main factors affecting the contribution to heat transfer of the radiation from individual gases are the partial pressures exerted by the gases and the thickness of the gas layer. The gas layer thickness is also known as the beam length, and is a function of the dimensions of the void between the tubes or tube banks which is occupied by the gas. In other words the larger the volume of gas, the greater will be the heat radiated from it.

Other factors influencing the transfer of heat from the gas to the steam tubes are the temperatures of the gas and tube surface, and the emissivity of the tube wall. The relationship between these quantities is described by the Stefan-Boltzmann Law [1].

The calculation of radiation from non-luminous gases is usually based upon the method established by H. C. Hottel [2]. This method, although considered to be exact, is extremely laborious and is impractical for use even by digital computers. An empirical method was derived by Sharan [3]. This method predicts values in close agreement with the exact method and in addition to being much less time-consuming, presents a form of equations to represent the parameters which permit their direct use in mathematical computer modelling of boiler plant.

Further empirical relationships have been developed, from which correction factors for a range of surface emissivities and temperatures were obtained.

When considering non-luminous radiation from cavities to adjacent tube banks, it must be considered that not all the incident radiation may necessarily be absorbed by the tube bank. Where the tube bank is small and of insufficient depth to absorb all incident radiation, some may pass straight through the bank. The portion of incident radiation absorbed from an adjacent cavity is therefore a function of the depth of the tube bank and of the free space between the tubes. Various empirical exponential relationships have been developed to describe the decay of incident radiation through tube banks in different parts of the boiler.

Resistance to radiation heat transfer, in the form of fuel ash deposit build up on the tubes, causing an increase in effective radiation receiving temperature, is accounted for by an effectiveness factor applied to the calculated radiation heat transfer coefficient.

3.3.2 Convection heat transfer

Boiler tube banks immediately above the furnace may receive up to two-thirds of their total heat pick-up in the form of radiation, the rest being convection. As the gas temperature reduces in its passage through the boiler, the proportion of radiative heat pick-up reduces until, at the economiser, virtually all heat transfer from the gas is by convection.

For a gas flow normal to banks of plain tubes, the gas side convective heat transfer coefficient has been shown experimentally to agree with an expression of the form:

$$Nu = A\, Re^{B}\, Pr^{C}$$

where

Nu = Nusselt number $= \frac{h_c d}{k}$

Re = Reynolds number $= \frac{\varrho u d}{\mu}$

Pr = Prandl number $= \frac{C_p u}{k}$

h_c = convective heat transfer coefficient

d = tube outside diameter

k = gas thermal conductivity

ϱ = gas density

u = gas maximum velocity

μ = gas dynamic viscosity

C_p = gas specific heat at constant pressure

The constant A is chosen to suit the tube arrangement, in-line or staggered, and the constants B and C are generally taken as 0.6 and 0.33 respectively. All transport properties of the gas are taken at the mean film temperature.

The convection heat transfer coefficient has also been shown to depend upon the arrangement of tubes within the bank, the depth of the bank and the angle of gas flow relative to the tubes. Experimental factors describing the behaviour of each of the above effects have been evolved, and are generally included in the equation as multipliers to the constant A.

The total effective gas side film coefficient is now expressed as the algebraic sum of the component non-luminous radiation and convection heat transfer coefficients.

3.3.3 Water and steamside heat transfer

In addition to a gas side film coefficient, heat transfer is also dependent upon the resistance of the steam water film on the inside of the tubes.

In the vast majority of boiler applications the flow inside the heated boiler tubes is fully turbulent, although at the low steam flows encountered at start-up and low load operation transient flow may occur. The heat transfer coefficient (h_c) at the inside surface of such tubes may also be calculated using the equation.

In the application to heat transfer inside tubes, the constants A, B and C are given values appropriate to the condition of the steam/water within the tubes, whether water, wet vapour or superheated steam. The values of the constants are given in standard texts on heat transfer, and may be chosen from validated work applicable to the temperature and pressure range desired by the design engineer.

In general the values for the equation constants A, B and C are taken as 0.0133, 0.84 and 0.33 for superheated steam, and 0.023, 0.8 and 0.4 for water.

3.3.4 Application to design

Having established the resistance to heat transfer of the gas and water/steam films on the outer and inner surface of the tubes, the engineer is now in a position to design the physical heat exchanger for the desired application in the required position in the boiler. This involves the selection of a suitable tube material and wall thickness, based on pressure and temperature of application, and corrosion resistance.

The overall heat transfer coefficient is then established, taking due consideration of the tube wall thermal conductivity and thickness. The additional resistance to heat transfer of dirt layers on the tube walls must also be considered. Dirt resistance is normally based on experience or experimental data and may be included in the overall heat transfer coefficient, or as an effectiveness factor applied to the calculated heat exchanger area.

When calculating the size of heat exchanger for a particular duty, the factors usually known by the design engineer are the amount of heat transfer required, and the temperatures and flow rates of the gas and steam. The overall heat transfer coefficient is then calculated as described.

The transfer of heat through the various film and metal layers from the gas to the steam is described by the Fourier Law [4] of heat conduction:

$$Q = U A \Delta T_{ln}$$

where

Q = heat transferred

U = overall heat transfer coefficient

A = heat transfer area

ΔT_{ln} = logarithmic mean temperature difference

The logarithmic mean temperature difference is defined as:

$$\Delta T_{ln} = \frac{\Delta t_o - \Delta t_i}{Ln(\Delta t_o / \Delta t_i)}$$

where Δt_o and Δt_i are the terminal temperature differences between the gas and steam at outlet and inlet of the heat exchanger respectively.

Most heat exchange surfaces within boilers are considered to be either parallel flow or counter flow. Although the tube arrangements are generally of a serpentine or pendant nature, and the gas passes over them at right angles, the effects of this are negligible and pure parallel or counter flow can be assumed without significant error being incurred.

The heat transfer coefficient is unaffected by the direction of gas/steam flow within the heat exchanger and, for a constant heat transfer duty, it can be seen from the Fourier Law that the required surface area is inversely proportional to the logarithmic mean temperature difference:

$$A \alpha \frac{1}{\Delta T_{ln}}$$

If typical gas and steam temperatures existing in a modern boiler final superheater are considered, as shown in Fig 2.5, the logarithmic mean temperature difference is seen to be 1.2% lower for parallel flow than for counter flow. This indicates that a larger heat exchanger would be required for a parallel flow arrangement.

Nevertheless, a parallel flow heat exchanger was installed for this application. The reason was that, while the heat exchanger has to be larger than for counter flow, there is the benefit that the tube metal temperatures are lower than for the counter flow arrangement. It can be seen from Fig 2.5 that the tubes carrying the steam at the highest temperature are located in the cooler gas temperature zone. The temperature difference between the steam and gas at this point is reduced, resulting in a smaller temperature drop across the internal steam film and tube wall. This then keeps the tube wall metal temperature to the lowest practicable level.

3.4 Draught loss in tube banks

From the analysis described in Section 3.3.2, of this chapter, the convection heat transfer coefficient for gas flow normal to plain tubes can be shown to be directly proportional to the gas velocity raised to the power 0.6. This means that a doubling of the gas velocity will cause a 50% increase in convection heat transfer coefficient. This gain will, however, be offset by a corresponding four-fold increase in the draught loss of the gases passing over the tubes.

The engineer, in designing boiler tube banks, has to balance the cost of heating surface against power penalty for the draught loss. This power is supplied by the boiler draught plant, consisting of forced draught and induced draught fans.

In general terms the draught loss across a tube bank is defined as a function of the tube geometry, gas physical properties, gas velocity and surface condition of the tubes.

Boiler tube geometry is defined from heat transfer considerations. In a predominently radiative zone, the tubes are generally arranged in widely spaced platens. As convection becomes the predominant mode of heat transfer, in the top pass and rear pass the heating elements are more closely spaced, to increase gas velocity and hence increase convective heat transfer. In the economiser section where heat transfer by radiation is very small, tube spacing may be very close, inline or finned, in an attempt to improve the convective heat transfer of a considerable amount of low grade heat.

The layout of tube geometry throughout the boiler, as described above, is also beneficial from the point of view of draught loss. The wider spaced platens,

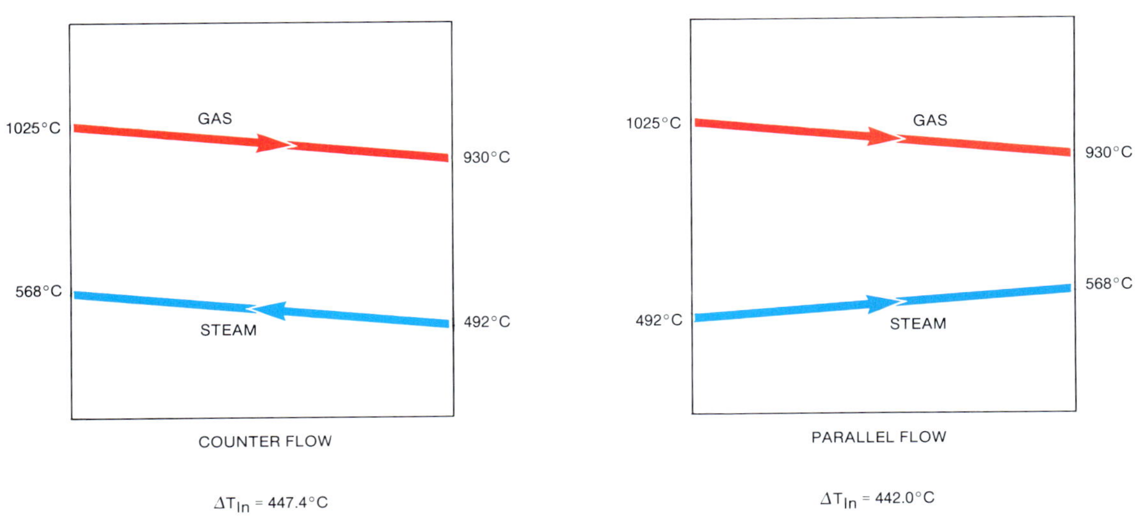

FIG. 2.5 Superheater gas and steam temperatures

situated in the zones of highest gas temperature and hence lowest specific volume, present the smallest resistance to flow.

At the colder end of the boiler, the economiser, the need for closely spaced tubes due to heat transfer considerations need not involve excessive draught losses since the bulk gas velocity will be reduced due to its reduced temperature. Care must however be taken at the design stage to avoid reducing the free area for gas flow too much by employing excessively close-packed tubes, as this would considerably increase the draught loss across the economiser, as well as lead to higher erosion rates and/or dust blockages.

The surface condition of the tubes may adversely affect the draught loss, particularly in the high temperature zones above the furnace. Operational experience has indicated that considerable amounts of slag may accumulate on platen elements, to such an extent that the gas free flow area may be considerably reduced, thus raising the draught loss. This increased draught loss would then have to be accommodated by adjustments to the induced draught plant, with a consequent penalty in reduced operating margins or system efficiency.

The configuration of boiler geometry ensures that gas flow over tubes is always fully turbulant, except possibly at very low loads. Equations used in the calculation of draught losses of gas flow normal to tubes are generally available in standard texts and usually take the form:

$$\Delta P = \frac{Ku^2}{2g}$$

where u = gas velocity at minimum section.

K is a function of tube bank depth and geometry and Reynolds number of the gas. This function may be obtained in various forms from empirical equations based on test and experimental data.

3.5 Water and steamside pressure loss

Pressure losses on the water and steamside of a pipework system within a boiler are caused by one or more of the following:

- Surface roughness — this is a characteristic of the tube material and of the tube fabrication process. The surface roughness is usually expressed in the form of the ratio of the average height of irregularities in the tube surface to the inside diameter. This ratio is called the relative roughness of the tube.
- Artificially introduced irregularities — such as internal rifling of the tube bore.
- Geometrical variations in tube runs such as bends, tee junctions, etc.
- Irregularities introduced during the fabrication process such as weld protrusions or tube deformation.
- Valves, flowmeters, etc.
- Differences in potential height between positions in a tube run.

Water/steamside flow in a boiler may be assumed to be turbulent at all but the lowest loads. Turbulent flow exists where the Reynolds number (Re) is in excess of about 2500, Reynolds numbers existing in boiler tubes at full load may well exceed 10^6.

$$Re = \frac{\varrho u d}{\mu}$$

where ϱ = fluid density

u = velocity

d = internal diameter

μ = dynamic viscosity

The head loss inside a tube due to surface roughness may be described by the Darcy equation:

$$h = \left(\frac{4fl}{d}\right)\left(\frac{u^2}{2g}\right)$$

where h = head loss

f = friction factor

l = tube length

d = internal diameter

u = fluid velocity

g = acceleration due to gravity

The friction factor is given by the Colebrook-White transition formula, usually represented graphically by the Moody Chart. This factor is a function of both the relative roughness and the Reynolds number of the fluid within the tube. The term $u^2/2g$ is called the *velocity head.*

The concept of velocity head is useful when analysing the losses due to geometrical changes in tube runs and losses across valves, etc. Standard data exist to describe the number of velocity heads (K) lost as fluid passes through various pipework configurations and types of valves. The overall head loss due to pipe roughness and geometrical changes can therefore be expressed as:

$$h = \frac{u^2}{2g}\left(\frac{4fl}{d} + K\right)$$

where K = total number of velocity heads lost due to bends, valves, etc.

Other correlations exist for the derivation of friction factor for two-phase flow.

It can be seen from the foregoing that the pressure or head loss is a function of the square of the velocity. The velocity itself is an inverse function of the square of the tube internal diameter (d). The pressure loss is therefore seen to be an inverse function of the fifth power of the tube diameter. This indicates the critical importance of knowing the tube bore accurately since any errors in tube bore will be magnified to the fifth power in pressure loss calculation.

A further source of possible error in pressure loss calculation stems from the indeterminate nature of the effect of butt weld protrusions during fabrication. A weld protrusion has a similar effect to an orifice, but since the exact dimensions of weld protrusions are not readily definable, the friction head loss cannot be calculated accurately. This source of error may be further compounded by the possible occurrence of several welds in series in a particular tube run.

The velocity of steam through tubes is important both for its effect on the pressure drop and on the conductivity of the steam film on the inside of the tube, which in turn affects the tube metal temperature. It has been shown in Section 3.3.3 of this chapter that the film conductivity is proportional to $u^{0.84}$, while the pressure drop from the foregoing description is seen to be proportional to u^2. By optimising the number and diameter of tubes, a steam velocity can be obtained which would cause a reduction in tube metal temperature, with a saving in material costs, but at a penalty from higher pressure loss. An economic balance must be obtained for each boiler section, particularly in the reheater, as heavy performance penalties can be incurred due to high pressure losses.

4 Furnace circuit

4.1 Boiler drum

The drum of a recirculating boiler receives the steam/water mixture from the evaporator tubes and separates this mixture into a water-free steam flow to pass to the superheater and steam-free water to pass down the downcomers. In order that these and other functions can be carried out various internal fittings are installed within the boiler drum (see Fig 2.6).

Water leaving the economiser is supplied to the drum via a feedpipe along its length.

The feed is distributed either by a number of pipes teed-off above each downcomer or by a series of holes along its length. Care is taken to ensure that the feed does not interfere with the separation process taking place in the water pool between the cyclones on either side of the drum.

A small pipe with a number of small holes along its length is used to distribute chemicals into the drum. The chemicals are fed in by a small high pressure pump, not continuously but rather as occasion demands, this being determined by monitoring the composition of the boiler water.

Boiler water sampling on some stations is achieved using a multi-tube sampling arrangement mounted along the length of the boiler drum so as to obtain a representative sample. Whilst this may be adequate, an arrangement by which samples are taken from a number of downcomers and subsequently combined by means of a manifold system is much preferred. This ensures that the sample taken for analysis is more representative of the water reaching the evaporative surfaces of the boiler.

The girth baffle acts as a plenum chamber in ensuring that the steam/water mixture from the risers is evenly distributed to the cyclones. The ratings of drums in terms of evaporation per unit area of water surface are now too great to rely on gravity separation, so that cyclone separators are used. The cyclone separator may have a vertical or horizontal axis.

In Fig 2.6 a vertical design is shown, the steam/water mixture enters tangentially near the top of the cyclone. The resulting vortex separates out the steam from the water. Steam passes out of the top of the cyclone into the primary scrubbers where most of the remaining small percentage of water is removed, and the water passes out of the bottom of the cyclone into the water pool where the final separation of the steam takes place. The water level outside the cyclone has an important influence on avoidance of carry-under; the bottom of the cyclone must discharge below the water level if steam re-entrainment is not to take place.

Problems can arise if the downcomer or pump geometry is such that there is significant axial flow along the drum; this causes a variation of water level along the drum which reduces the efficiency of some of the cyclones.

Vortex inhibitors usually in the form of a cruciform are located at the downcomer inlet. They prevent the formation of the *bathtub* vortex at this location.

Steam leaving the cyclones passes through the primary scrubbers situated on the top of each of the cyclones, through the secondary scrubbers just prior to the exit of the drum and into the saturated steam off-takes. The scrubbers consist of a series of corrugated plates which forces the steam into a number of sharp changes in direction. In this way the steam is completely dried.

4.2 Circulating pumps

Boiler circulating pumps are of the submerged type, with squirrel cage induction motor and pump installed in the same pressure vessel. A heat barrier separates the pump from the motor, the temperature of the motor being limited to approximately 100°C, while the pump operates with water at temperatures up to

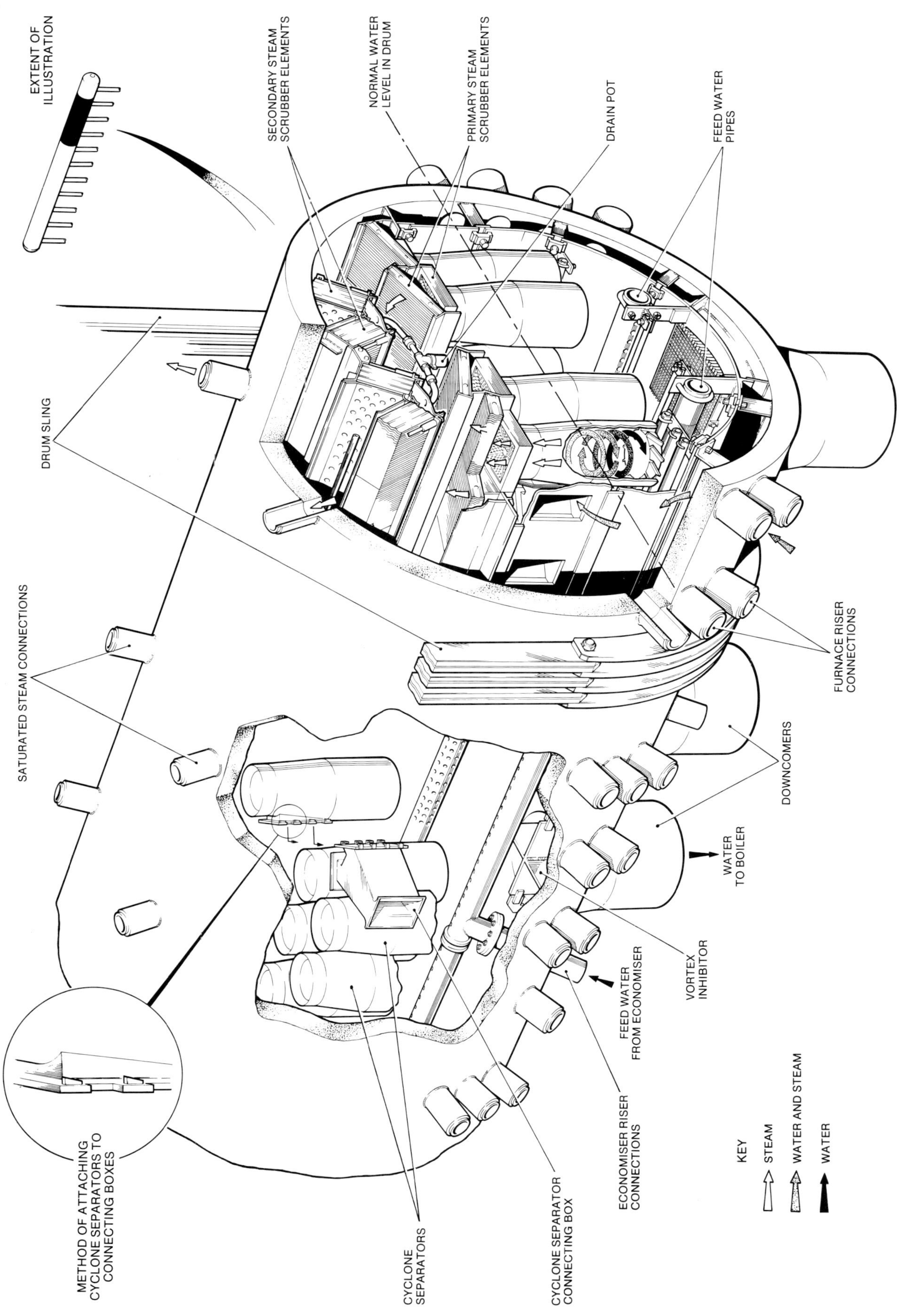

FIG. 2.6 Boiler drum

330°C. A high integrity cooling system has to be provided as, even if the boiler is shut down, the water in the pump remains at high temperature for long periods.

As the pumps are supported directly from the inlet manifold, no provision for boiler thermal expansion is required. In CEGB boilers, the unit is installed with the motor below the pump to inhibit natural circulation from the *hot* pump to the *cold* motor and avoid air pockets in the motor. With this arrangement, provision has to be made for ensuring that particulate matter in the boiler water is not allowed to accumulate within the motor.

5 Water and steamside chemistry

5.1 Drum boilers

In a drum type boiler any impurities entering with the feedwater, concentrate in the boiler water and will normally be either carried over with the steam or removed via the boiler blowdown.

Feedwater and boiler water are chemically controlled to prevent corrosion of boiler evaporator surfaces, superheaters, reheaters and turbines. The main risk to the boiler evaporative surfaces is from the concentration of acidic or alkaline impurities, either at regions of dryout or under porous oxide deposits. To minimise impurity ingress, make-up water to the system is demineralised and condenser design and materials are selected to reduce the risk of cooling water inleakage.

The feedwater system is operated at elevated pH to minimise the corrosion of condensers, feedheaters and connecting pipework and thus prevent the forward transport of particulate corrosion products into the boiler. The feedwater is de-aerated and hydrazine-dosed to remove oxygen down to a level of less than 0.005 mg/kg and to maintain a magnetite surface layer.

In normal operation, the boiler surfaces are also protected by an adherent film of magnetite, Fe_3O_4. If this film is removed by chemical attack or physical damage, corrosion of the underlying metal can take place. The main risk is from acid chlorides which can hydrolyse to form hydrochloric acid:

$$MgCl_2 + 2H_2O \rightleftharpoons Mg(OH)_2 + 2HCl$$

5.1.1 Boiler water chemical control

The control of chloride ingress is of major importance. Although feedwater ammonia and hydrazine dosing maintains alkaline conditions in boiler water, the ability of volatile alkali treatment (all volatile treatment (AVT)) to counteract acid chloride formation is limited. To provide greater protection against acid chloride ingress, additions of solid alkali are essential. Both sodium hydroxide (caustic soda) and sodium phosphate may be used, but it is normal CEGB practice to use low level caustic dosing. Sodium phosphate has only a limited solubility at high temperatures, which leads to problems of chemical control.

In boilers where high heat fluxes occur, e.g., oil-fired boilers, there is a greater risk of dryout, leading to concentration of impurities and there have been instances of corrosion due to caustic soda. In this case, the allowable caustic dose level is restricted and the tolerable chloride level is reduced.

The boiler water control limits for 160 bar pressure boilers, i.e., turbine inlet pressure, where the drum pressure is above 180 bar are as follows:

System	Boiler chloride mg/kg as NaCl	Sodium hydroxide mg/kg
AVT	<0.2	—
Coal-fired	<2.0	1.5 × NaCl min 2.0
High heat flux oil-fired	<0.5	1.5 × NaCl min 0.5

Boiler blowdown systems are designed for a continuous flow of about 1% of the maximum boiler steam flow, thus using caustic dosing, the allowable feedwater chloride level, neglecting carryover, is 0.02 mg/kg as NaCl or 0.012 mg/kg as chloride. It can be seen that the allowable limit for AVT control is only 0.0012 mg/kg as chloride, which corresponds to an extremely small seawater leak and explains why caustic dosing is always used for seawater cooled stations.

5.1.2 Steam purity

It is important that the steam leaving the boiler drum should be free from impurities which may deposit in the superheater, reheater or turbine. Such impurities may arise from the dissolution of boiler water impurities into the steam phase, notably silica and sodium hydroxide, and from mechanical carryover. Mechanical carryover occurs when there is incomplete separation of the two phases in the drum and droplets of boiler water are carried over with the steam.

It is accepted that silica in steam will not cause problems in the turbine if the concentration does not exceed 0.02 mg/kg. The distribution ratio between water and steam at about 180 bar drum pressure is approximately 10, thus the boiler water silica level must be maintained below 0.2 mg/kg. Boiler water silica is controlled by limiting the make-up water silica level to less than 0.02 mg/kg and by the operation of boiler blowdown.

The acceptable limit for sodium in steam is 0.01 mg/kg, but is reduced to 0.005 mg/kg where austenitic superheaters are used because of the risk of stress corrosion. The partition coefficient for sodium species in steam is less than for silica and the limits are usually achieved by operating within the guidelines for boiler water chloride and sodium hydroxide.

Boiler chemical conditions are maintained by high pressure dosing pumps which inject slug doses of dilute sodium hydroxide into the boiler drum, and by low pressure dosing pumps which inject dilute ammonia and hydrazine solution into the feed system. Extensive water and steam sampling and monitoring equipment is provided. It is normal practice to locate the final sample cooling systems and continuous analysers in centralised monitoring rooms.

5.2 Once-through boilers

The essential difference between subcritical once-through and recirculating drum type boilers is that all water entering the boiler tubes is evaporated in a number of parallel single-pass tubes.

All particulate matter and dissolved salts which enter with the feedwater and do not pass out with the steam flow will accumulate in the boiler tubes and may lead to corrosion, flow instabilities or overheating. In contrast to the recirculating drum boiler, there is no facility for removing dissolved salts by periodic blowdown. Control is therefore exercised by maintaining stringent limits on the ingress of impurities in feedwater.

Feedwater quality is maintained by the inclusion of full-flow condensate polishing plant in which the total boiler feed is demineralised and filtered in deep bed ion-exchange units.

Subcritical, once-through boilers are used in the later nuclear stations because their design requires a smaller number of pressure vessel penetrations than with drum boilers. The feedwater quality limits for boilers used with advanced gas cooled reactors (AGRs) are extremely stringent and sodium, chloride and sulphate levels of less than 0.002 mg/kg are specified. High feedwater pH levels or combined ammonia/oxygen dosing is used to control the dissolution of iron oxide and the rate of metal loss by erosion/corrosion.

In supercritical boilers there is no separation point between the water and steam phases, and impurities entering with the feedwater pass through with the steam. Boiler corrosion resulting from concentration of impurities should not be a major risk but there is still a risk of deposition in turbines and reheaters. The high solubility of copper in supercritical fluid presents a particular problem.

Feedwater chemical control is achieved as for subcritical boilers by the application of full-flow condensate polishing and ammonia or ammonia plus oxygen dosing.

6 Steamside and waterside corrosion

6.1 Steamside corrosion

Gaseous water (steam) flowing through superheater and reheater tubes reacts with the alloy constituents of the tube material (Fe, Cr, etc.) as shown in the equation:

$$4H_2O + 3Fe \rightarrow Fe_3O_4 + 4H_2$$

The magnetite (Fe_3O_4) scale thus formed normally adheres to the tube surface and limits the extent of further reaction. In this case the oxidation rate is diffusion-controlled and is parabolic in nature.

Most of the superheater and reheater tube materials are chromium containing alloys of iron, and possibly nickel. If chromium-rich oxides can be formed a substantial benefit in oxidation resistance is obtained as these oxides are thin and highly protective. Low alloy steels generally form iron-rich spinel oxides at significant rates at the temperatures of interest and may be subject to problems such as spalling.

As the oxide scales get thicker they become more brittle and when a critical thickness is reached the protective behaviour is often lost as they become more susceptible to spalling. Spallation almost always occurs as a result of cooling strains due to differential thermal expansion between the oxide and the metal. The proper consideration of the operational strains and the oxidation kinetics in most cases ensures that the critical thickness for spallation is not reached within the design life of a component.

A major difference between the spalling behaviour of austenitic and ferritic steels, is that for austenitic steels only the outer layer of the duplex oxide becomes detached leaving the inner rate-controlling layer behind. Subsequent metal loss rates for austenitic steels are not therefore increased by spalling, although they are for ferritic steels which lose the oxide at the metal/oxide interface.

Figure 2.7 gives a schematic representation of types of oxide on Type 316 austenitic steel. As can be seen, parabolic rate constants can vary by some two orders of magnitude depending on whether duplex iron oxides or chromium-rich sesquioxides form. It has also been established that silicon and molybdenum additions improve the steam oxidation resistance of austenitic steels. Appropriate surface finishing, such as cold working, promotes the formation of uniform sesquioxide films by refining the surface grain structure, which encourages chromium diffusion and availability at the surface. These highly protective and thin scales avoid the spalling problems associated with thicker scales.

Corrosion attack in ferritic steels is generally more uniform with iron-rich duplex scales being prevalent. Figure 2.8 indicates that alloys with higher chromium content are more corrosion resistant, particularly 12% Cr steel. There is also evidence that addition of silicon of about 0.5% enhances the oxidation resistance of 12% Cr steel.

Relatively low chromium ferritic steels operating in conditions where creep distention takes place can exhibit accelerated (linear) corrosion rates. The suggested mechanism for this is the transfer of excessive strains

Before

During

After

FIG. 1.3 Typical off-load water cleaning of platen superheater elements

FIG. 1.8 Corrosion of furnace wall tubes at Drakelow C Power Station

Fig. 1.9 Failed furnace wall tubes due to fuel ash corrosion at Drax Power Station

Fig. 3.1 660 MW boiler economiser being erected

FIG. 4.25 Twin seal flap isolating damper

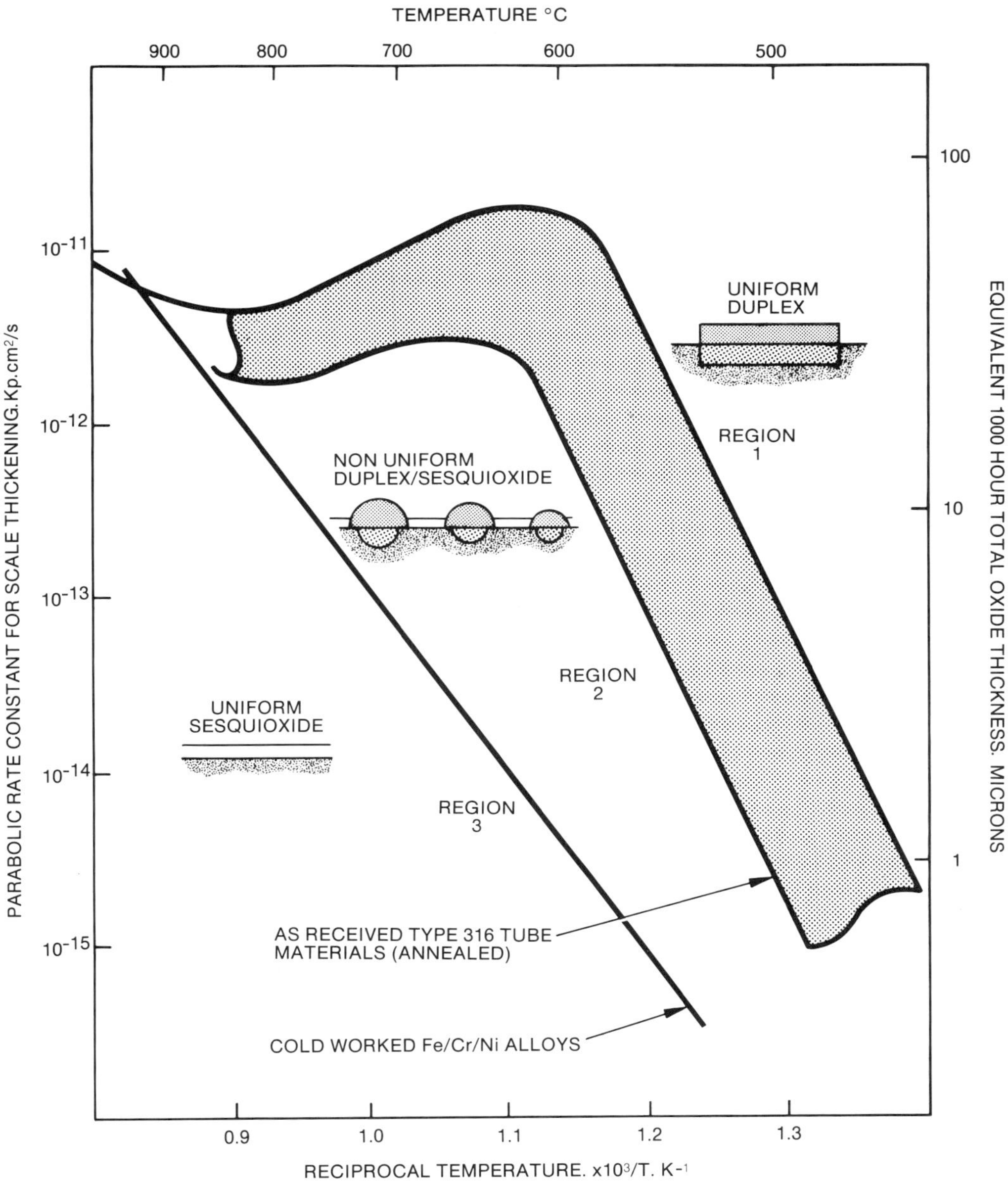

FIG. 2.7 The range of oxidation behaviour of austenitic steels

to the oxide, as the result of creep, which fails periodically, hence leading to laminated scale formation. This results in additional metal loss equal to the radial enlargement of the tube. As the oxide thermal conductivity is very much lower than the tube metal, formation of thick oxide scales elevate the tube metal temperatures which in turn lead to increased creep strain rate and higher oxidation rates and so on, creating a runaway situation. While this is not a common phenomenon, there have been a few instances where operation in off-design conditions has led to its occurrence. Figure 2.9 shows a typical laminated oxide layer formed on a ferritic reheater tube. Both the thick protective oxide scales and the laminated scales can spall as mentioned earlier. Figure 2.10 shows the appearance of a tube after some 70 000 hours of operation where spalling had taken place, and Fig 2.11 is a section through the spalled oxide scale.

Spalling of steam-grown oxides can cause tube blockages and solid particle erosion. In CEGB plants, tube blockages due to spalled oxide have not been a major problem, nor indeed the solid particle erosion of the type commonly encountered in the US and elsewhere. This suggests that the choice of design and operating parameters for UK plant has in most cases been appropriate for the materials used, and the material choice and manufacture has also contributed to the general lack of spallation and its consequences. Further contributory factors are the reduction and/or elimination of tight radius bends, use of tubes with bores greater

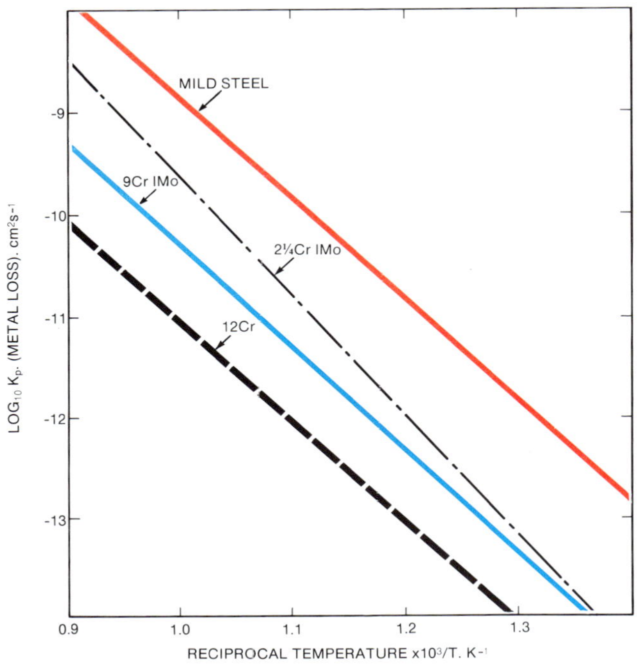

FIG. 2.8 Metal loss rates for parabolic oxidation of ferritic steels

than 25 mm, the equalisation of heat pick-ups between the tubes of a multitube bank, and possibly the use of appropriate steam strainers before the turbines.

6.2 Stress corrosion

All available evidence indicates that stress corrosion cracking can not be attributed to a single mechanism but rather to a multitude of mechanisms, in many cases interacting with each other either sequentially or in combination. It is nevertheless clear, that fissure is promoted in pre-existing compositional or structural features in the material within an environment conducive to the initiation and extension of cracking. Both corrosion and stress appear to be requisites for the development and propagation of cracking whatever the particular mechanism or mechanisms.

Stress corrosion attack on high strength austenitic alloys, such as the 300 series of steels, can lead to particularly rapid failures. These steels are generally grain boundary sensitive, are highly stressed and, if operated in a corrosive environment, are prone to intergranular and sometimes transgranular cracking.

In CEGB plants, with superheater and reheater outlet temperatures of 568°C, the use of high strength austenitic steels is common. The tube failures due to intergranular attack or stress corrosion cracking have been confined to very few isolated instances where the presence of a corrodent, e.g., concentrated chloride or sulphate containing water, was allowed to remain in the non-drainable austenitic tube banks. As this type of failure cannot occur, even in the sensitised tubes, in the absence of a corrodent, susceptible tubes must be kept dry until they reach a state of immunity. A typical diagram shown in Fig 2.12, which is drawn for type 316 steel, indicates that at an operating temperature of about 620°C the tubes will become immune at approximately 30 000 hours. It is therefore clear that when prolonged boiler outages occur, the still sensitised tube which can collect condensate in the non-drainable areas must be cleared of condensate to substantially eliminate stress corrosion failures when the boiler returns to service. This is particularly important in reheaters as, because of their thinner tubes, the crack growth path is shorter.

It has also been the CEGB practice for many years to ensure that austentic tubing is placed in a position in a boiler where the steam is superheated to a minimum of 40°C above the saturation temperature at the prevailing pressure under all operating conditions. This avoids water droplets containing concentrated solutions of corrodents coming into contact with austenitic tubes. Adherence to this criterion has resulted in there being no tube failures in CEGB boiler plants by this mechanism.

6.3 Off-load corrosion in reheater tubes

This type of corrosion results from the formation off-load of water pools containing ionic material such as sodium and/or calcium sulphate. In the presence of air, this acts as a differential oxygen cell in the same manner as rusting. The problem is typified by standing water lines and white deposits in non-draining tube sections. Pitting rates of up to at least 6×10^{-4} mm/h can occur, so a 2.64 mm wall tube could be penetrated in one 26-week outage. However it is thought more likely that pits reinitiate and propagate over a number of outages. A spate of tube leaks normally occurs immediately after return to service. Once these have been repaired, further failures prior to the next outage are uncommon.

The major source of the sulphate is believed to be attemperator spray water and, because the solubility of sodium sulphate is near minimum at reheater pressure, sulphates deposit in the low temperature sections of the reheater. When the unit is shut down, condensing steam then dissolves the deposits and corrosive pools are formed.

This condensation process can be avoided or minimised by two techniques:

(a) After closure of the turbine stop valves, condenser vacuum is maintained for a further 15–20 minutes. This, together with residual heat in the boiler, minimises the water vapour remaining in the reheater.

(b) Displacing the remaining steam after shutdown by compressed air. This process has to be carried out before condensation occurs in pendant sections, as the resulting U-seal in any tube diverts air flow

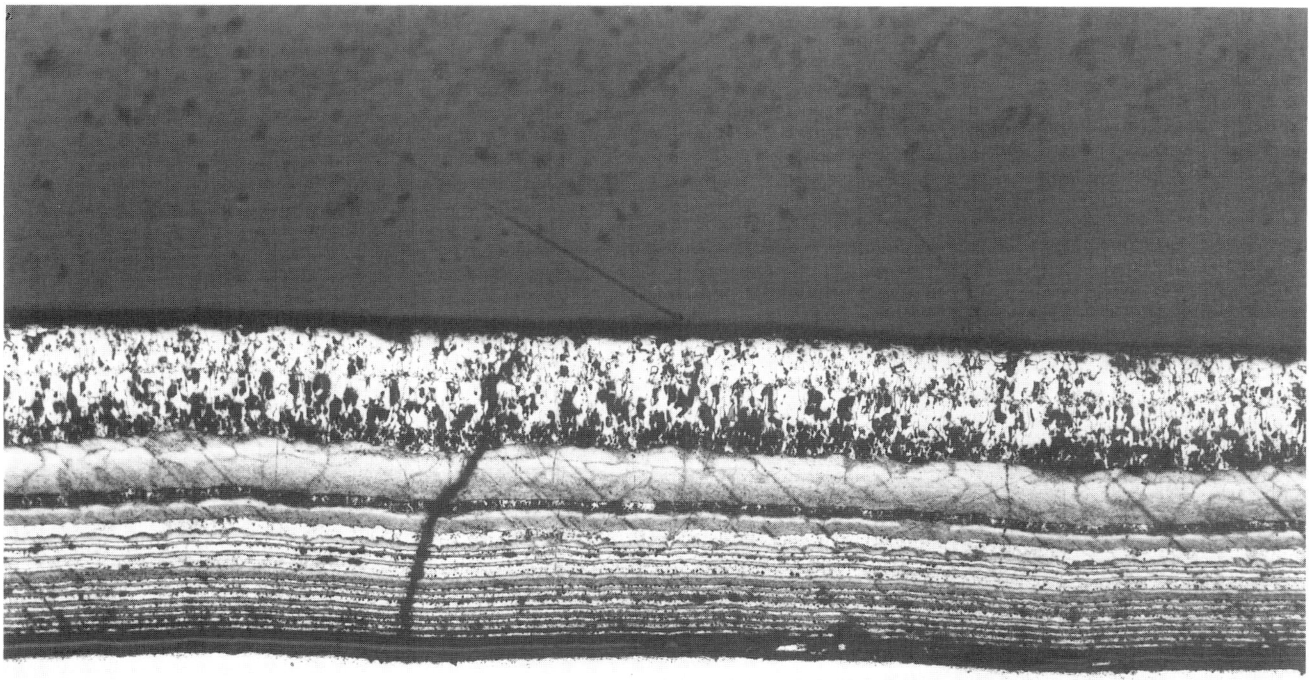

FIG. 2.9 Section through laminated oxide formed on a reheater tube – ×105 magnification

to clear tubes. If condensation has occurred this can only be cleared by refiring.

6.4 On-load corrosion in evaporator tubes

On-load corrosion in evaporator tubes is usually localised causing thinning of the tube wall to an extent where stress rupture occurs. When corrosion has been accompanied by hydrogen embrittlement, it can result in pieces of tube being blown out.

On-load corrosion occurs at flow discontinuities or where the evaporation process is disturbed by heat flux effects (departure from nucleate boiling). To avoid corrosion, significant weld protrusions must be avoided in high heat flux zones, the tubes must be free of inside surface laminations and mill scale, and other foreign matter must be removed by acid cleaning before commissioning. Departure from nucleate boiling can only be avoided by careful design of the circulation system (see Section 3.2.3 of this chapter). Subsequent operational procedures must be aimed at preventing metal oxides (arising from feed line corrosion) entering the boiler, where they could contribute to the corrosion process by depositing in areas of high heat transfer, thereby providing a boiler water chemistry which combats corrosion.

6.5 Off-load corrosion in economisers

During long term shutdown periods, economisers must be protected against corrosion. A particularly effective method is to drain the economiser when the boiler is taken off load and refill with a nitrate/borax solution. The treated water can be left in the economiser for the full period of storage or may be drained to allow maintenance and repairs to be carried out. In areas where pools of water remain, the nitrate/borax solution prevents corrosion even under aerated conditions.

Alternative treatment such as hydrazine/ammonia requires nitrogen sealing to avoid aeration and if repairs are being carried out, this is often not practicable.

FIG. 2.10 Ferritic reheater tube showing spalling after 70 000 hours operation

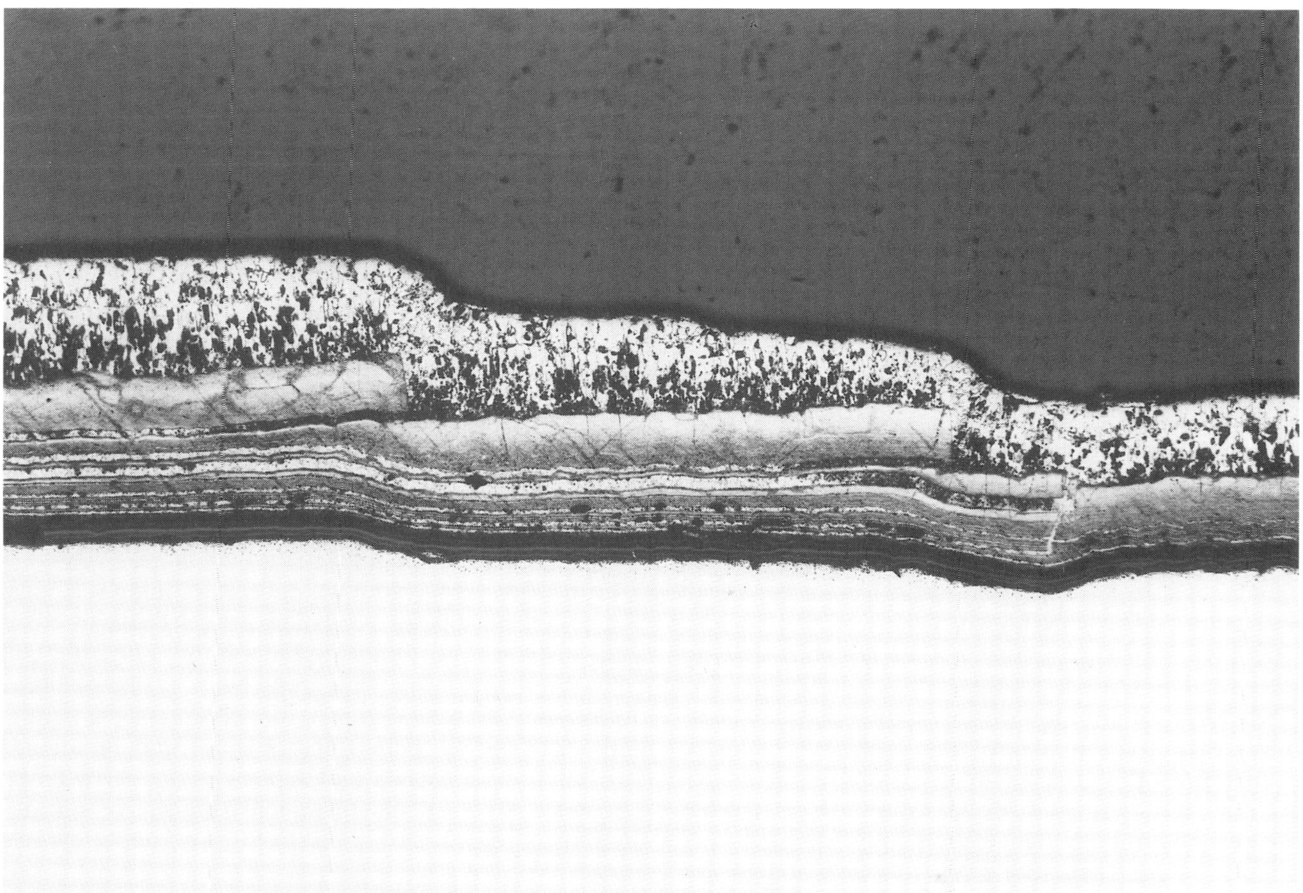

FIG. 2.11 Section through spalled area of oxide – ×105 magnification

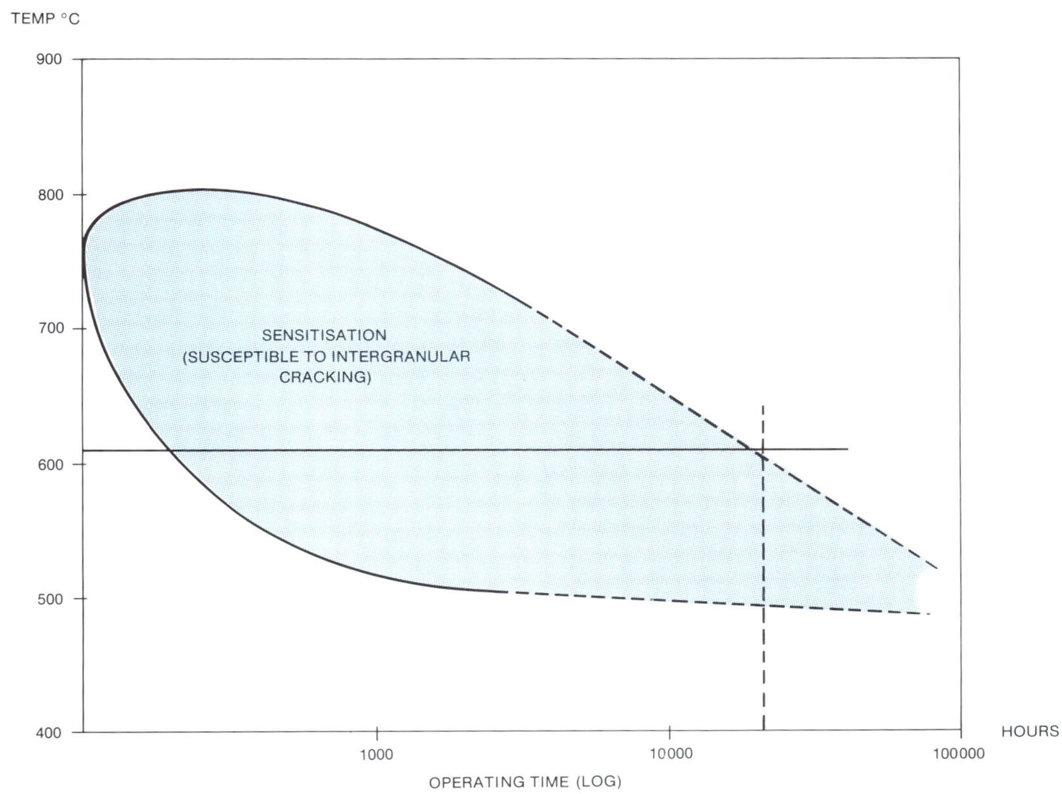

FIG. 2.12 Time/temperature sensitisation diagram for solution treated Type 316 austenitic steel reheater tubes

7 Steam temperature control

7.1 General

It is a requirement of all boilers designed for use in CEGB power stations that they should deliver steam at a controlled temperature from both the superheater and reheater at loads from full load down to 70% BCMR (boiler continuous maximum rating). This requirement ensures that the efficiency of the turbine is maintained over this load range and that the metal temperatures within the turbine and of boiler tubes themselves are kept within design levels. The latter point is essential in order to design the boiler plant to provide an economic installation.

7.2 Superheat temperature control

Variations in superheat steam temperature at constant load are usually attributable to varying degrees of ash deposition on boiler surfaces. The boiler designer must take into account the likely rates of surface fouling in order to determine the heat transfer surface area of the various superheater banks. He must also allow for variations in effectiveness of sootblowing. Outside this, he has at his disposal, a number of methods of controlling steam temperatures.

By incorporating into the boiler design a mixture of radiant and convective superheat surface, the overall characteristic of the superheater is such that the temperatures do not vary significantly with change of load. In a radiant superheater section, the steam temperature falls as load increases whereas the temperature would rise with load in a convection superheater section.

Modern designs of boilers include sufficient superheat surface for operation within the control range and the steam temperature control means employed is used largely to regulate reheat steam temperatures. In such situations, any increase in superheat duty brought about by the adjustment and control of reheat steam temperatures is removed in superheater spray attemperator stages. Where possible, the water for the superheater attemperators is taken from a position after the HP feedheaters where the use of this water does not affect the unit cycle efficiency. The quantity of spray water is controlled by the required temperature downstream of the spray attemperation points. It is important to site the temperature measuring device in the pipework after the attemperator in a suitable position to avoid contact with any water droplets in the steam flow. Such contact would lead to a false reading of mixed temperature and consequently maladjustment of the spray water valves.

7.3 Reheat temperature control

As discussed in the previous section, modern designs of boilers contain specific provisions for the control of reheat steam temperatures which also, inherently, affect the steam temperature within superheater sections.

There are various methods available for control of reheat steam temperatures, some are variations on a theme. The most common methods are now considered separately.

7.3.1 Divided rear pass

This method is sometimes referred to as the *gas bypass* method. Essentially, the boiler is designed with two parallel gas flowpaths at the rear of the boiler, prior to the air preheater. One gas pass contains convective superheater surface whilst the other contains convective reheater surface. A gas-tight baffle separates the two flowpaths. A common economiser is usually situated across both the two flowpaths. The quantity of gas flowing down each of the two parallel flowpaths is regulated by gas dampers usually situated below the economiser surfaces. The reheat steam temperature control system is linked to control the damper position regulating the gas flow down the reheat gas pass. A shortfall in reheat steam temperature would actuate the damper control to allow more gas down the reheat gas pass at the expense of the gas flow down the parallel superheater pass. Thus, as reheat steam temperature increased, the superheater temperature would decrease. This would be automatically corrected for in operation by a reduction in superheater attemperator spray water flow in order to maintain the superheater steam temperature at its design level.

The main disadvantages with this type of control are as follows:

- Boiler mass: The provision of two parallel gas flowpaths forming the rear pass of the boiler greatly increases the overall mass of the boiler requiring heavier supporting steelwork and making the boiler a more expensive capital outlay.
- Control damper characteristic: The gas flow versus damper position characteristic is non-linear. During the initial damper opening periods, there is a rapid increase in gas flow which increases with damper position to about 60% open, beyond which little increase in gas flow results from increase in damper position.
- Control damper seizure: The dampers are situated in relatively cool gas temperature zones but become prone to dust deposition in operation. Such dust deposits hamper, and sometimes prevent, the dampers being moved and hence the control of steam temperatures is impaired.

7.3.2 Gas recirculation and air injection

In a divided rear pass design of boiler, the control of reheat steam temperature is achieved by varying the quantity (mass flow) of combustion gas over the primary reheater surface at the expense of the superheater gas flow. The total gas flow leaving such a boiler is constant for a given fuel flow. This is not the case for boilers using gas recirculation or air injection as the control means.

Gas recirculation is a method by which a quantity of combustion gas is tapped from the boiler outlet ducting at a suitable position after the economiser surface and reintroduced into the boiler furnace. The gas recirculated should not interfere with the combustion conditions within the furnace and serves to alter the heat absorption within the boiler unit.

The introduction of a quantity of cool combustion gas into the furnace reduces the furnace absorption due to a reduction in the mean gas temperature within the furnace. As the furnace absorption reduces, for a given fuel flow, the gas temperature leaving the furnace and entering the superheater and reheater sections of the boiler is little affected. Thus the heat available for the convection sections of the boiler has been increased by the introduction of the recycled gas. The major portion of heat absorbed in the superheater, reheater, and the economiser of a boiler unit is by convection heat transfer which depends upon gas mass flow and gas temperature, both of which are affected by gas recirculation.

The variation in gas mass flow leaving the furnace brought about by the variation in gas recirculation quantity will also affect the heat transfer to superheater elements as well as reheater elements. The boiler designer would allow for this by providing a superheater control system linked into the spray attemperators as discussed previously. The reheat steam temperature would control the quantity of the gas being recirculated.

In oil-fired boilers, the gas tapping point is usually taken immediately after the economiser, ducted to a gas recirculation (GR) fan provided specifically for the purpose and from there into the furnace hopper. Dampers either side of the GR fan are used to regulate the GR flow.

In coal-fired boilers, the abrasiveness of flyash particles makes the tapping of gas before the precipitator plant unattractive. Such a system would require specific dust removal equipment before any GR fans in order to protect them from dust erosion damage, etc. This problem can be overcome by taking the gas tapping after the induced draft (ID) fans. As well as recirculating clean gas back to the boiler, this system has the added advantage that the ID fans can be sized to produce the head necessary to recirculate the gas without the need for additional GR fans. This saving in capital cost is offset slightly by the capital cost of the gas recirculation ducting between the ID fan outlet and the boiler furnace.

Air injection works in the same way as gas recirculation except that instead of using combustion gas to control temperatures, additional capacity is introduced into the forced draught (FD) fans to allow some of the heated combustion air to be diverted into the furnace hopper area. Air injection avoids the need for the ducting between the furnace and ID fan outlet but, compared to gas recirculation, has two major drawbacks which makes this method unattractive for UK plant. Firstly the boiler efficiency is impaired by a greater stack loss (gas weight). The weight of gas entering the stack is lower for a gas recirculation system (as a quantity of gas would have been tapped off before this point) than for an air injection system where the air and gas weights from inlet to outlet are (ignoring tramp air, etc.) a constant. Secondly, unlike a gas recirculation system, both the FD and ID fans for an air injection system require to be sized to handle the largest quantity of air/gas flow required. In a gas recycling system, only the ID fan is required to handle additional flow.

7.3.3 Tilting burners

Gas recirculation, as stated previously, alters steam temperatures by reducing furnace heat absorption and by increasing the mass flow of gas across convective heat transfer surfaces.

Another method available to vary the heat absorption in the furnace is by the use of movable (tilting) burners. This method is applicable to tangentially-fired furnaces where the burner boxes *inject* a fuel and air supply tangentially into a central fireball. The adjustment of the burner tilt alters the position of the fireball within the furnace and hence alters the furnace absorption. A clean furnace would require burner tilts to be positive (upwards) to avoid excessive evaporative duty and steam temperature shortfalls. Similarly, a dirty furnace would require negative (downwards) tilts to ensure the correct evaporation. The gas temperature leaving the furnace for a given fuel flow would be directly related to the furnace absorption and hence the burner tilt angle. The burner tilt control can therefore be linked into the reheat steam temperature control system.

As the reheater surface in UK boilers is situated in relatively cool gas regions, in order to avoid metal temperature problems during start-up when there is no steam flow through the reheater, the main effect of the variation in burner tilt is absorbed by superheat surfaces situated immediately beyond the furnace. A large variation in burner tilt is therefore required for a moderate change in reheat steam temperature. The usual variation is $\pm 30°$ about the horizontal. The buoyancy of the hot furnace gases tend to make tilts below $-15°$ less effective. Another disadvantage of this type of control is that the burner boxes can become prone to seizure and thus unable to control steam temperatures.

7.3.4 Heat exchangers

Non-contact attemperators are provided on some boilers in order to reduce the reheat steam temperature by attemperation without the efficiency penalty associated with spray (contact) attemperators. The usual arrangement for this type of system is to have a number of shell and tube heat exchangers in parallel, each with a bypass line. Inside each heat exchanger, the reheat steam is directed across tubes containing feedwater in order to achieve the temperature reduction.

The steam flow through each attemperator is regulated by control valves to give the required steam temperatures at the reheater outlet, the remainder passing through the bypass lines. The hotter steam flowing within the bypass lines has the effect of slightly increasing the mixed steam temperature at the position where the two streams join. The control system regulating the steam flow within each attemperator will take this into account by siting the control temperature measuring point downstream of the mixing position.

7.3.5 Equalisation of outlet temperature

The equalisation of reheater outlet temperatures across the width of the boiler can be achieved by the use of valves installed in the cold reheat steam pipework local to the reheater inlet headers. These valves, usually butterfly valves, can be used to control the quantity of steam flowing in a given section of the reheater and hence achieve a more uniform final reheater outlet temperature profile. These valves only serve to equalise the steam temperature leaving the reheater and, unlike the methods previously mentioned, do not serve to control the level of reheat steam temperatures. Furthermore, their use is limited in order to avoid high steamside pressure drop within the reheater.

7.3.6 Emergency sprays

The use of spray attemperators to control reheat steam temperatures incurs a cycle efficiency penalty as the heat added in order to evaporate the spray water is not available for use within the HP turbine. For this reason, the use of spray attemperators to control reheat steam temperatures is reserved for emergency use only when, for whatever reason, the control method employed is at its limit and rising steam temperatures are leading to excessive boiler and turbine metal temperatures. The reheat emergency sprays are normally located prior to the reheater inlet connections.

8 Pressure parts design

8.1 Calculation of metal temperatures

For design purposes, it is necessary to know the tube mid-wall metal temperature. This is essential for the selection of the appropriate tube material and wall thickness, where design parameters are temperature dependent.

The difference in temperature between the gas on the outside of boiler tubes and the steam flowing within them is made up from the combined temperature differences across several components as follows:

- Gas side film.
- Slag/fouling on tube outside wall.
- Tube wall.
- Oxide layer on tube inside wall.
- Steam/water side film.

This composite temperature difference is shown diagrammatically in Fig 2.13.

In order to commence a metal temperature calculation it is necessary to determine the heat flux falling on the tube and the steam/water temperature within the tube. The heat flux is derived from a study of the combined effects of incident radiation and convection at the point on the tube under consideration (together with any design factors for upset conditions, etc). This is normally expressed as a heat flux per unit length of tube (kW/m).

Internal steam temperature is normally taken as a starting point as it can be more accurately determined than the gas temperature at the same point in the tube. Both heat flux and steam temperature may be derived using the computational techniques described in Section 9 of this chapter.

In most applications of the calculation method, the internal surface of the tube is assumed to be clean. There is, therefore, only a steam/water film between the internal bulk temperature and the inside wall temperature. The internal film heat transfer co-efficient is calculated as shown in Section 3.3.3 of this chapter, where the fluid properties should be taken at the mean film temperature. This means that an initial estimate of inside wall temperature must be made in order to derive the mean film temperature. This initial estimate may be of the order of 50°C more than the steam temperature, its value is not of critical importance as the calculation is of an iterative nature.

$$\text{Mean film temperature } t_f = \frac{t_w + t_s}{2}$$

where t_w = estimate of wall temperature

t_s = bulk steam temperature

The temperature difference across the steam/water film is then obtained by dividing the heat flux, referred to the inside wall, by the film coefficient.

$$\text{Inside wall heat flux (kW/m}^2) = \frac{\text{outside wall heat flux (kW/m)}}{\text{tube ID (m)}}$$

$$t_f = \frac{\text{inside wall heat flux (kW/m}^2)}{\text{Steam film heat transfer coefficient (kW/m}^2\text{K)}}$$

The inside wall temperature t_w is determined from the equation:

$$t_w = t_s + t_f$$

The calculation is then repeated using this temperature to obtain a second estimate of mean film temperature, for use in calculating a new value for film coefficient.

This iterative process is continued until the calculated inside wall temperature shows no significant change from its previous value.

The tube wall temperature difference is determined from the product of the heat flux, this time referred to the mid-wall diameter, and the wall thickness, divided

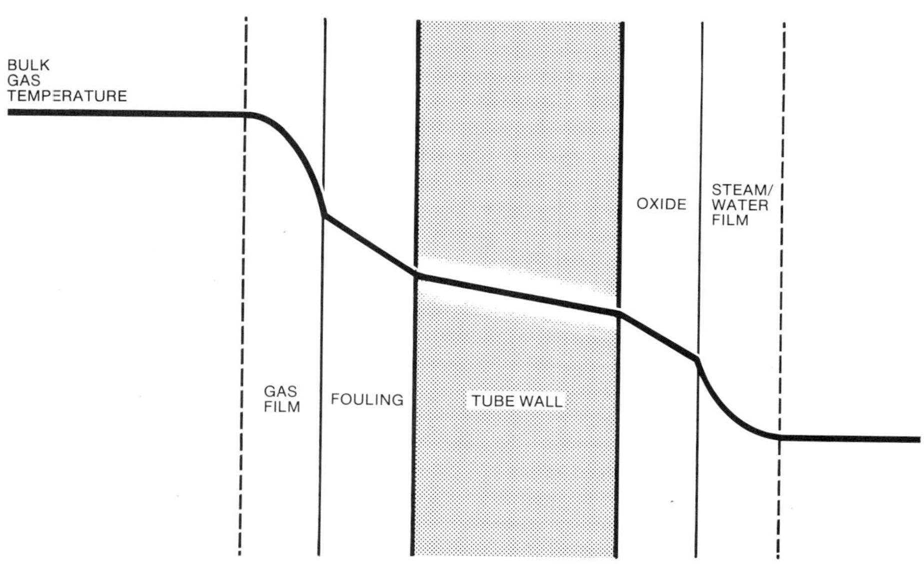

FIG. 2.13 Composite temperature drop from gas to steam/water through a boiler tube wall

by the thermal conductivity of the tube material, as follows:

$$\text{Mid-wall heat flux (kW/m}^2) = \frac{\text{outside heat flux (kW/m)}}{\text{mean diameter (m)}}$$

$$t_{metal} = \frac{\text{mid-wall heat flux (kW/m}^2) \times \text{tube thickness (m)}}{\text{metal thermal conductivity (kW/mK)}}$$

Again, an iterative process should be used, whereby an estimate of mid-wall temperature is made, from which the tube thermal conductivity may be determined. Each successive calculation therefore gives a more accurate value of tube wall temperature difference.

The inside wall temperature and the tube wall temperature difference having been calculated, it is then straightforward to determine the outside and mid-wall temperatures.

$$\text{Outside wall temperature} = t_w + t_{metal}$$

$$\text{Mid wall temperature} = \frac{t_w + t_{metal}}{2}$$

This process may be repeated at any position in the tube bank where heat flux and steam temperature are available. In this way the most onerous metal temperatures, significant to design purposes, can be evaluated.

The foregoing analysis is only valid for thin-walled tubes, where wall thickness is small in comparison to tube diameter. For thick-walled tubes an analysis involving a logarithmic mean diameter in preference to an arithmetic mean diameter would be required.

8.2 Boiler code requirements

The design, manufacture and quality of boiler pressure parts for the CEGB is currently in accordance with BS1113 [5] which in turn refers to other standards for specific requirements — mainly BS806 [6] and BS5500 [7].

BS1113 specifies requirements for design and manufacture including materials and design stresses, scantlings, workmanship, inspection, testing, documentation and marking, valves, gauges and fittings, as discussed in the following sections.

8.2.1 Materials

BS1113 includes a table of materials used in the manufacture of pressure parts complying with appropriate British Standards. In addition rules are given for the composition, manufacture, heat treatment and testing of materials not included in the table.

8.2.2 Design stresses

In discussing design stresses it must be kept in mind that the purpose of the design code is to apply appropriate material parameters and design methods which result in an economic, safe and reliable construction for the required duty. It follows that code requirements should be specified to take into account the type and quality of construction, the methods of analysis, the range of operating conditions and applied loadings, and the possible modes of failure. Therefore the design methods have been derived and allowable design stress limitations specified and modified as and when necessary to achieve this result. This approach has then been proved generally satisfactory by experience over many years.

A table of design stresses for approved material is given. For design in the proof range, stresses are related to temperature, and for the creep rupture range, stresses are related to time and temperature. Briefly they are the minimum of:

$$\frac{Re_{(T)}}{1.5}, \frac{Rm}{2.7} \text{ or } \frac{S_{Rt}}{1.3} \quad \text{for carbon, carbon manganese and low alloy steels}$$

and

$$\frac{Re_{(T)}}{1.35}, \frac{Rm}{2.7} \text{ or } \frac{S_{Rt}}{1.3} \quad \text{for austenitic steels}$$

where

Rm = the minimum tensile strength at room temperature.

$Re_{(T)}$ = the minimum 0.2% proof stress for carbon, carbon manganese and low alloy steels or the 1.0% proof stress for austenitic steels.

S_{Rt} = the mean value of the stress to produce rupture in time t and temperature T.

Figure 2.14 shows a graph of material proof stresses and creep rupture stresses for 160 000 hours rupture life for 2.25% Cr, 1% Mo steel. Curves are also shown for these stresses and minimum room temperature tensile stress divided by appropriate factors. Curve A to B will give design stresses within the proof range while B to C gives design stresses within the creep range. The actual values plotted are as given in BS1113 for design stress and BS3604 [8] for proof and creep rupture stresses.

The methods used for the derivation of design stress are given so that materials other than those tabulated can be used. Further details are given in Sections 8.3, 8.4 and 8.5 of this chapter.

The foregoing stress factors on Re and S_{Rt} refer only where the components are not subject to fireside corrosion. The method adopted when an allowance for corrosion is appropriate is described fully in Section 8.5 of this chapter.

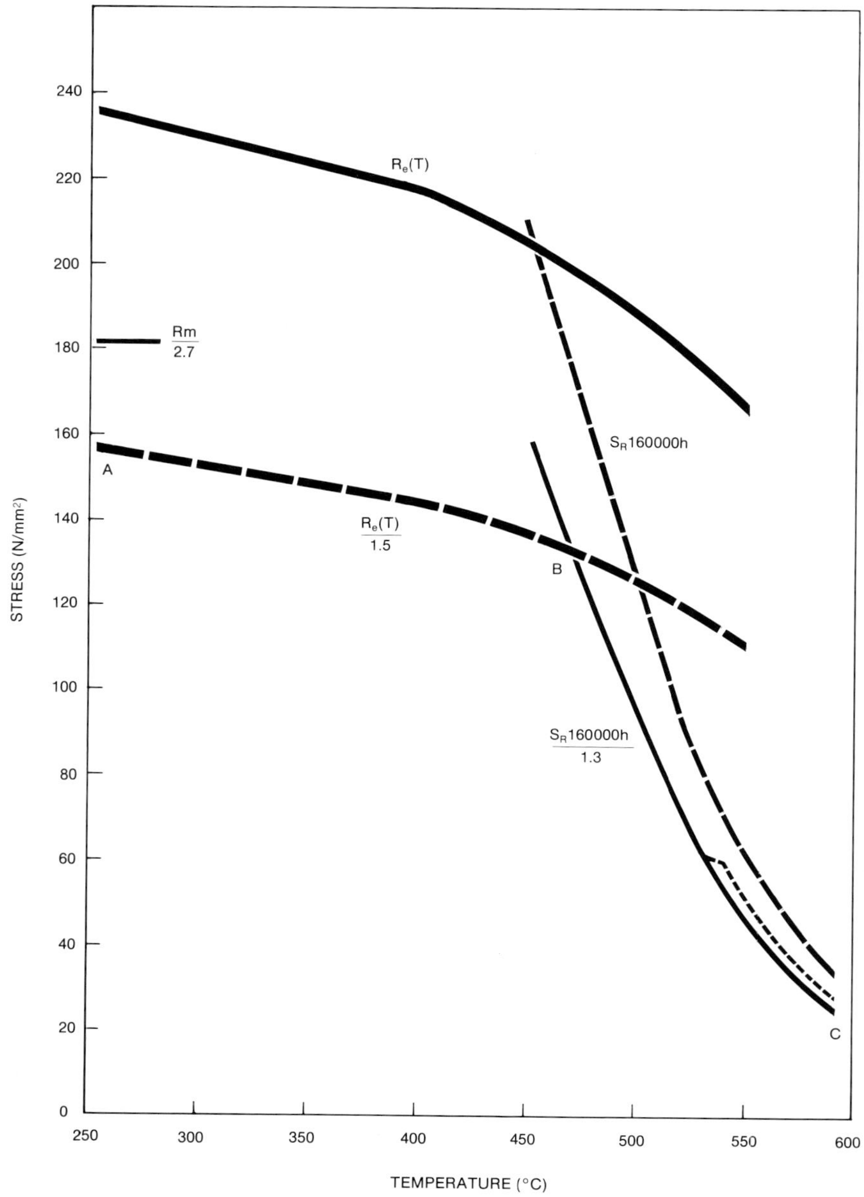

FIG. 2.14 0.2% proof, 160 000 hours creep rupture and design stresses for 2.25% Cr 1% Mo steel tubes

8.2.3 Scantlings

Rules and design equations are given to calculate the required thickness for the commonly used features of boilers including drums, headers, tubes and pipes. Most boiler pressure components have the common feature of being basically cylindrical in shape. For a simple thin cylinder:

$$\text{Circumferential (hoop) stress} = \frac{Pd}{2t}$$

$$\text{Longitudinal stress} = \frac{Pd}{4t}$$

where P = internal pressure

d = internal diameter

t = thickness

In a high pressure boiler the pressure parts are relatively thick and it is not a valid assumption that they are 'thin' for calculating stresses. It has been shown that there is a good correlation between test results from straight-pull test specimens and tubes under pressure if the diameter used in the formula is the mean diameter, that is the arithmetic mean of inside and outside diameters.

$$\text{For pipes and tubes } t = \frac{PD}{2f + P}$$

where D = outside diameter

f = allowable circumferential stress

t = thickness

P = internal pressure

For headers and drums the presence of tube stub holes is taken into account by a ligament efficiency η when the thickness formula becomes:

$$t = \frac{PD}{2f\eta + P}$$

Various formulae for ligament efficiency are given depending on the layout of the tube holes, i.e., either square pattern, diagonal, regular or irregular, etc., and are all based on the ratio of the remaining ligament area in the perforated zone divided by the total area prior to perforation.

Further rules are provided to limit combined stresses in headers and drums due to the sum of internal pressure, self weight and contents loading. For local loads at supports, etc., reference is made to the methods of BS5500 Appendix G.

Formulae are given in BS1113 for other commonly used features in pressure parts design and include reinforcement requirements for nozzles and branches in cylindrical vessels, headers of rectangular section, ends of drums and headers, structural attachments to tubes, and pressure parts of irregular shapes.

The flexibility of integral piping must be such that it does not impose high loads on the connected component or generate within itself unacceptable stress due to construction mismatch, expansion stress and internal pressure. Reference is made to BS806 for methods of calculating and assessing stress levels.

8.2.4 Manufacture and workmanship

BS1113 describes the type of material used for various components, also methods of fabrication, tolerances, welding requirements and heat treatments for thermal cutting, forming, welding and stress relief. Thus drums are usually fabricated from rolled plates by fusion welding, and headers constructed from pipes and forgings. For the CEGB, nozzles, branch and stub attachment is by welding, normally using a set-on weld preparation.

When tubes and pipes are bent, either hot or cold, thinning occurs at the extrados of the bend and the originally-circular section tends to become oval. Limits on thinning and ovality are given for tube bends and ovality for pipe bends. To allow for thinning on pipe bends a thickness allowance is included in the scantlings.

General requirements are specified to ensure good welding practice although for detailed specification reference is made to other British Standards such as BS2633 [9] for pipe and tube welding.

Special attention is drawn to austenitic-to-ferritic transition joints which are required to be made under workshop conditions and always located at butt welds clear of any geometric discontinuity.

Heat treatments for the different stages of manufacture are specified, where required, for different material classification groups. Heating and cooling rates, temperature ranges and time at temperatures are all tabulated as appropriate to the stage in manufacture, material and component thickness.

Subsequent to the occurrence of a number of failures of cold bent pipes in carbon and carbon manganese material, requirements for the heat treatment after bending have been tightened and related to bend radius and ovality. The tighter the bend the greater the cold work and consequent material hardness, the greater the ovality the higher is the stress concentration in a pressurised pipe bend. The specified stress relief or normalising treatment reduces residual bending stress and softens the material structure.

8.2.5 Inspection and testing

Inspection during construction is required by an Inspecting Authority which has the responsibility for making sufficient inspection to ensure that materials, design, construction and testing complies in all respects with the standard. This includes the approval of welding procedure tests and welders' certificates.

For major seams in the boiler drum, weld test plates are required which are welded as a continuation of the seams. These are subject to tensile tests, macro examination, bend tests and Charpy tests.

To ensure their integrity, completed welds on pressure components are subjected to some percentage of non-destructive testing by either surface or volumetric examination, or both. Surface examination is by visual, magnetic particle, or dye penetrant inspection. Volumetric examination is by either radiography or ultrasonic testing. The code specifies the percentage and the method of inspection to be applied, depending on the component duty, size, material and suitability for inspection. Generally, main weld seams and branch welds are subject to 100% crack detection and ultrasonic examination, other welds being subjected to 10%.

In addition to specifying inspection requirements the code sets acceptance standards for weld appearance and finish, surface and sub-surface defects.

To demonstrate the strength and integrity of individual components and of completed boilers, and to establish that no major errors have occurred, the boiler has to undergo hydraulic testing. Completed boilers are tested to either:

(a) 1.5 times the maximum permissible working pressure or 1.25 times the highest calculation pressure, whichever is greater; or

(b) 1.25 times the maximum permissible working pressure or 1.1 times the highest calculation pressure, provided all components have been tested before assembly in the manufacturers' works in compliance with (a).

8.2.6 Documentation and marking

The manufacturer is required to allow the inspector full access to all drawings and documents necessary to check all aspects of the design. The completed boiler is required to be accompanied by a dossier which incudes a certificate of conformity with the code, all drawings, materials certificates, weld procedures, non-destructive test procedures and hydraulic test certificates. Each boiler has to be marked to show the details of the manufacturer, maximum working pressure and other significant aspects.

8.2.7 Valves, gauges and fittings

BS1113 describes the requirements for safety valves, main stop valves and boiler feedwater valves; also for water level gauges and alarms, and boiler blowdown and drain mountings.

8.3 Design on proof stress criteria

As previously described in Section 8.2.2 of this chapter and with reference to Fig 2.14, the design stresses in the proof range are derived from part A to B of the curve. In this case, for the particular material used as an example, for a design temperature below 470°C (the crossover temperature) the design stress is time independent. Curves similar to Fig 2.14 could be drawn for all the different materials used for boiler pressure parts, although the detail shape and crossover temperature would be different.

Figure 2.15 shows typical data used to define the minimum proof stresses shown in material standards and in the form of design stress in BS1113. The data shown represents 111 cases within the appropriate material composition requirements.

For a component designed at a temperature below the crossover temperature, creep failure cannot nevertheless be precluded because creep occurs at all temperatures, the rate being dependent on the temperature and stress level. Thus if the stress level is raised by, for example, a structural discontinuity, creep failure could occur.

Code scantling calculation methods allow stresses to be greater than design at geometric discontinuities, e.g., at branch openings in drums or headers where a stress concentration factor of up to 2.25 is allowed for. Additionally no allowances are included for thermal stresses, weld residual stresses, or stress induced by boiler erection.

Two recent examples where failures have occurred due to creep crack growth at a temperature within the proof design criteria are:

(a) Cold bent furnace supply pipes where stresses have been increased by excessive section ovality, cold pull used to correct misalignment during erection, residual bending stresses, and local stress raisers such as pipe manufacturing defects or defects due to mechanical damage.

(b) Furnace membrane tube failures where the basic pressure stress has been enhanced by thermal stress, welding residual stresses and local geometric com-

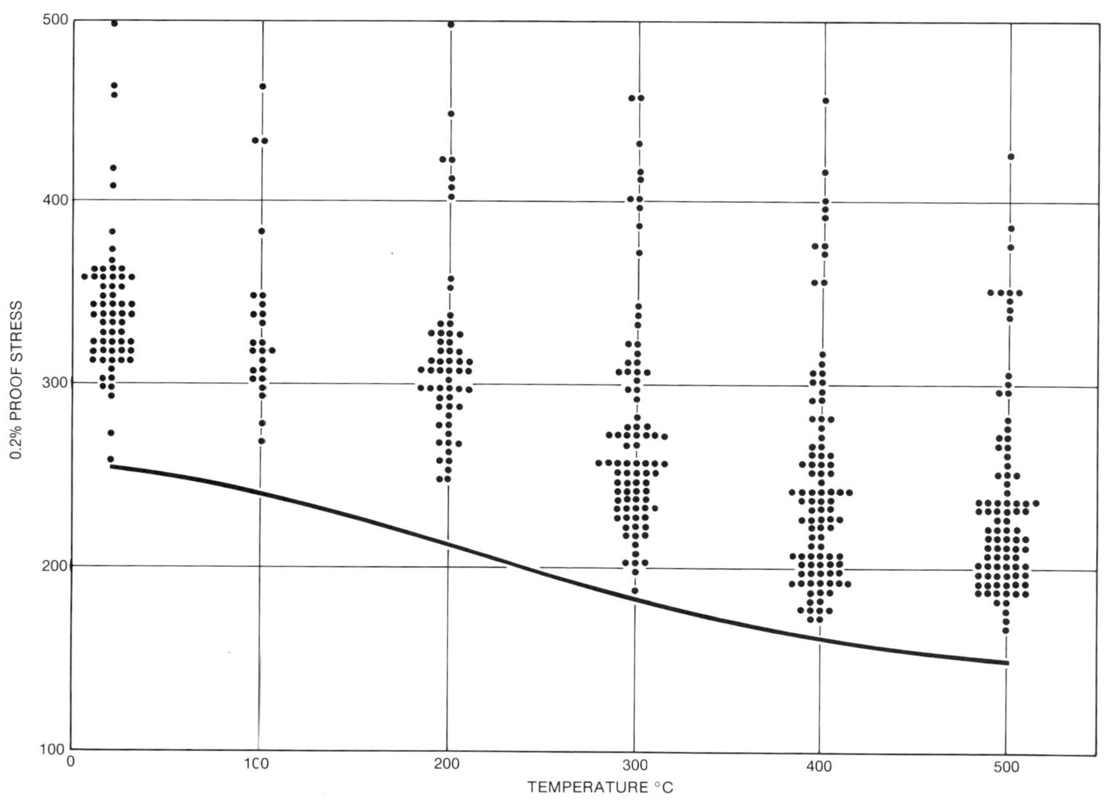

FIG. 2.15 Comparison of data submitted with minimum 0.2% proof stress derived at a tensile strength of 440 N/mm^2

plexities causing stress concentrations at fin-to-tube welds.

Fatigue due to cyclic stress loading is a further cause of failure despite all component scantling design being within the code requirements. Again the location of failure is usually at local stress concentrations. While components designed to the code are generally proof against low cycle/high strain fatigue damage due to the margins included in the design methods, loadings not specifically covered in the code such as thermal stresses due to temperature differentials, or stresses due to mechanical load, weight or restraint loads can lead to failure.

Recently, modifications have been necessary to components in economiser inlet circuits, particularly the inlet header arrangement and recirculation pipework. These have been made in order to reduce temperature differentials and their consequential effects. On some of these components both top and bottom and through-thickness (thermal shock) temperature differentials have been measured and shown, by analysis, to generate unacceptably high stresses which have caused severe cracking in some instances.

8.4 Design on creep rupture criteria

Again, referring to Section 8.2.2 of this chapter, and particularly Fig 2.14, design stresses in the creep range are those (for the 2.25% Cr 1% Mo steel example) above 470°C and are shown in curve B to C.

For recent CEGB boiler contracts all pressure parts operating at temperatures higher than the crossover temperature, i.e., where the design stress is time dependent, are designed for a specified minimum life. This includes the next generation of boilers being planned at present.

For 2.25% Cr 1% and 0.5% Mo 0.25% Va materials there is a provision in the code for design stresses above 540°C to be increased by 10% and this is used for CEGB boiler designs. The code requirement allows this on condition that components are subject to inspection after two thirds of the design life has been used. This increase in design stresses is shown in Fig 2.14 as a dotted line.

Typical source data for creep rupture stresses are shown in Fig 2.16 for 1% Cr $\frac{1}{2}$ % Mo material tested at 550°C. Similar curves are available, covering a range of temperatures for every relevant material. It will be noticed that although many relatively short term data are available, longer term data are sparse and some extrapolation is required. The complete average rupture data for 1% Cr $\frac{1}{2}$ % Mo are shown in Fig 2.17. The creep rupture stresses derived from the source data are the average rupture stresses. The data exhibit considerable scatter (about ±20%) about this average. The factor of 1.3 used to derive the design stress, therefore, gives a small margin on minimum creep rupture stress. When the data are not well founded, such as for a relatively new material, further factors can be introduced until confidence in the longer term values is gained.

Creep damage prior to failure is usually characterised by cracking along the interconnecting grain boundaries which ultimately can be detected visually. As with fatigue, normally the first signs of potential problems will be seen at stress concentrating design features.

In common with components designed in the proof range, components designed for creep rupture do not necessarily fail through a creep mechanism.

Fatigue or a combination of creep and fatigue can lead to premature failure. Cracking has been experi-

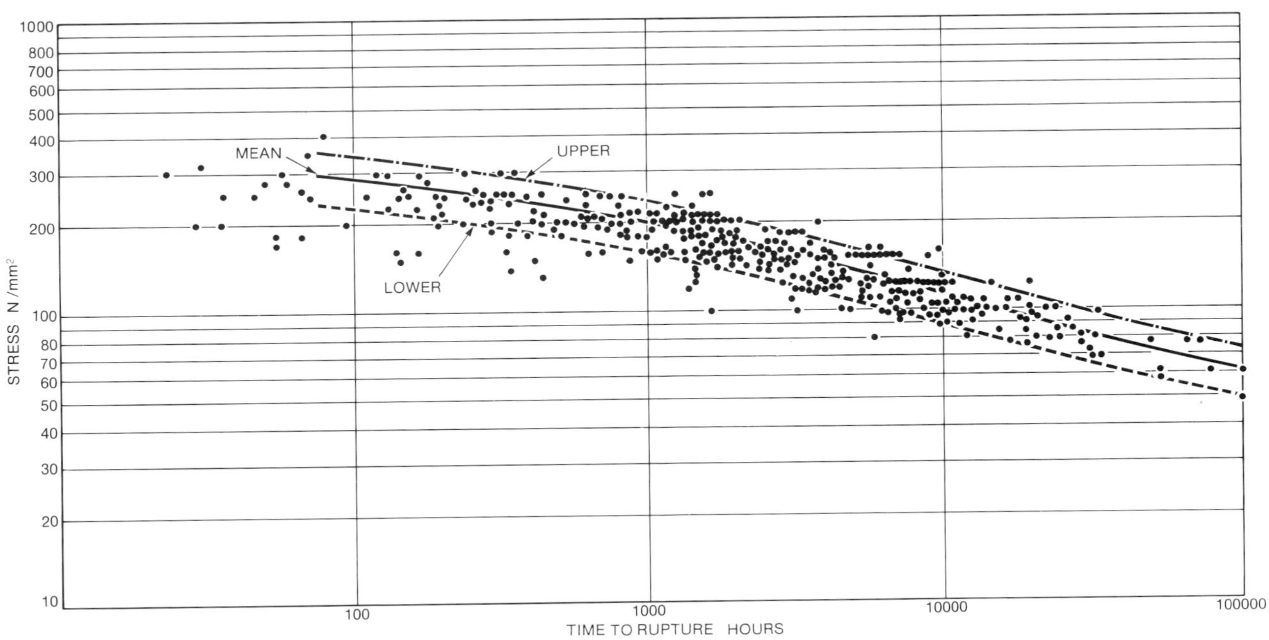

FIG. 2.16 Stress-rupture data for 1% Cr $\frac{1}{2}$ % Mo steel at 550°C

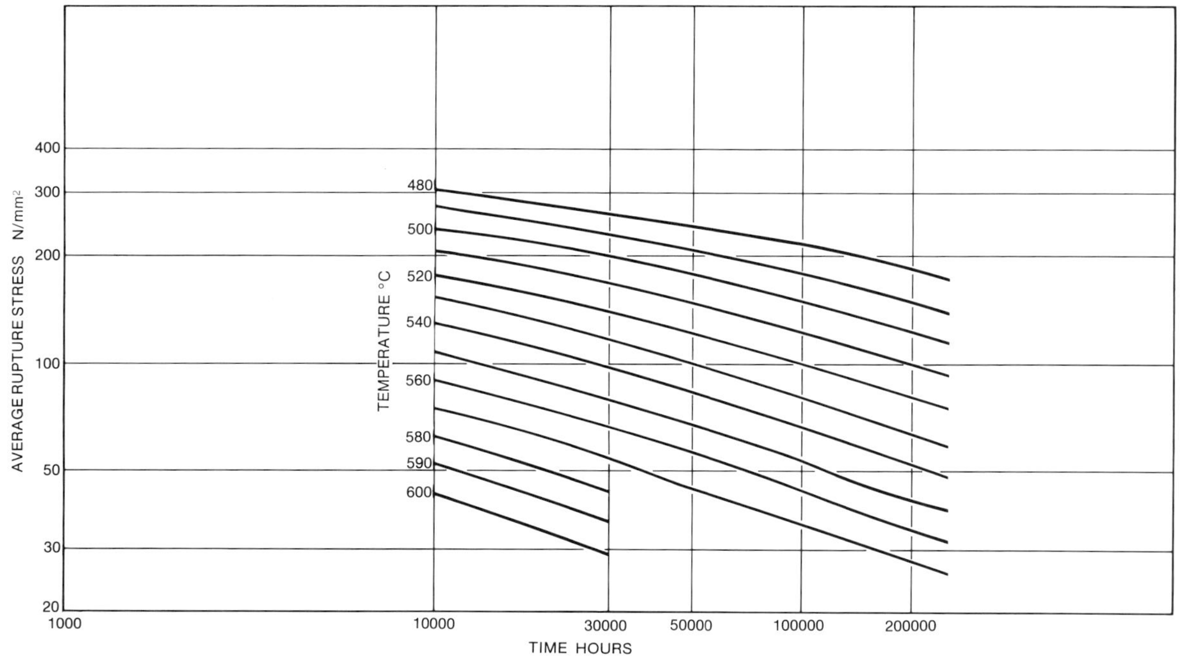

FIG. 2.17 Average rupture data for 1% Cr $\frac{1}{2}$ % Mo steel

enced in some headers made of 2.25% Cr 1% Mo material operating at creep temperatures but also subject to cyclic transient thermal stress loadings. Common features of the headers are thick sections (about 100 mm) and small ligaments due to tubes being pitched at 22.25° angle. Although the scantlings comply with code requirements the reason for cracking observed has been identified as high strain thermal fatigue.

8.5 Combined corrosion and creep

This section applies to the design of tubes subject to flue gas corrosion. The procedure involves the determination, initially, of the metal and gas temperatures. It is often necessary to carry out complex and detailed thermal performance calculations to ascertain the gas and metal temperatures, as referred to elsewhere in this chapter.

The thermal analysis is in two main parts, firstly the overall boiler performance assessment, and secondly the detailed steam and gas flow and temperature maldistribution assessment for each individual heating surface/tube bank. Sections 3 and 9 of this chapter describe the methods and the computing systems, respectively, available to carry out these analyses.

Having thus determined the appropriate gas and metal temperatures, together with the knowledge of fuel properties/corrosion propensity and tube position, the scantling design can proceed. The fuel ash corrosion aspects are covered in Chapter 1, Section 2.2. At this stage it must be borne in mind that the mid-wall temperature is used for the evaluation of the required stress data and tube outer surface (gas side) temperature for the evaluation of the corrosion rate.

Life fraction rule is used to determine the life of a given tube. The interpolated stress rupture data is divided by the appropriate design factor, at the prevailing metal temperature, to calculate the creep rupture life fraction used up for a time interval ≤ 1000 h. The tube waste during this interval is calculated, using the corrosion rate data, and subtracted from tube outside diameter (OD) at the end of each interval. At each interval the tube surface temperature and mid-wall temperatures are recalculated, as tube thickness is reduced due to wastage, using constant OD heat flux but unchanged steam flow and temperature. The calculations are repeated until one of the following criteria are met, at which time the tube is deemed to have failed:

(a) When total life fraction exhausted reaches unity, or;

(b) When the hoop stress becomes equal to the minimum proof stress divided by an appropriate design factor.

The whole calculation procedure is repeated for different tube thicknesses until the required life is achieved. Section 9.5 of this chapter describes the computational model in greater detail.

8.6 Coextruded tubing

This type of tube is referred to in Chapter 1, Section 2.2.3. Design using coextruded tubing is fully covered in GDCD Standard 243 Part 1 [10].

9 Application of computers to boiler design

Validated computational programs have been developed and used within the CEGB to aid the design engineer in many fields of power plant assessment. The main objectives of computer modelling as applied to boiler plant are as follows:

- To assess technical specifications and tenders.
- To predict the performance of proposed modifications to boiler components and evaluate their effects on the boiler as a whole.
- To investigate part load performance of boiler plant without the need for access to actual power plant, which necessarily must operate in a strict merit regime.
- To investigate the sensitivity of plant to deviations from design conditions, such as a variation in fuel specification or the simulation of fouled boiler tubes.

9.1 Furnace performance

In Section 3.1 of this chapter the method of calculating furnace heat flux distribution is outlined, together with the information required from laboratory tests and plant trials as input to the program. The program provides information necessary for design of the water/steam circulation system and the gas temperature and heat flux at the furnace exit plane, required for determining the performance of superheat surface installed in the furnace. Programs of this type normally have the facility for examining part load as well as full load conditions and can allow prediction of performance with various asymmetric firing arrangements.

9.2 Total boiler/station system performance

9.2.1 Computer modelling system

A system widely used at present in the CEGB utilises a modular concept, in which the steady state performance of each power station component is described by a series of non-linear algebraic equations. Each of these series of equations is known as a module. The modules are held in a library from which they can be accessed by the programmer/designer and arranged mathematically to describe a wide range of power plant systems, the completed system being known as a *model*.

The modules are identified by type names and are referenced by a name which must be unique to the model, thus allowing the same module to be used several times within one model structure.

In addition to modules which describe plant items, further modules are available which describe the behaviour of fluid streams. Mixing of gases or steam/water flows can be described using such modules. Modules exist to enable the construction of models of whole power stations, including turbine and feedheating plant. The module library is being continuously updated and expanded to accommodate new developments in the mathematical treatment of existing plant items, and to include the more novel methods of power generation such as gasifiers and pressurised fluidised bed combustors.

9.2.2 Construction of a boiler model

The first step in the construction of a boiler computer model is the production of a block diagram showing the gas and steam/water flowpaths between the boiler components. A block diagram of the form shown in Fig 2.18 is prepared from the diagrammatic arrangement shown in Fig 2.1.

The block diagram shows that some of the flowpaths can be specified from outside the model. Such quantities as coal and air flow to the furnace, and feedwater inlet conditions are usually obtained from plant measurements, and are specified as boundary conditions to the model. Other data such as intermodular steam and gas temperatures and flowrates, together with system outlet conditions are entered as first approximations based on the experience of the programmer/designer. The final values for these parameters are obtained as the model is run and iterations performed to within a user specified residual error criterion.

The individual modules are connected together mathematically by equations specifying the variables to be used as inputs and outputs. The use of the same variable name as both the output from one module and the input to another would indicate that this quantity is then routed mathematically between the two modules in the direction specified.

The module which describes the boiler furnace is supplied with a fuel firing rate and combustion air conditions in the variable input section of the model. A further section is provided in which all salient physical characteristics of the furnace are entered. The dimensions of the furnace combustion chambers, emissivity of the flame, fuel calorific value and various fouling factors are included in this section.

On leaving the furnace, the hot gasses pass through banks of steam cooled tubes. Immediately above the furnace the heat transfer mode is predominantly radiation, with a smaller convective component. The modules describing tube banks in this area are necessarily designed to accommodate the high proportion of radiant heat pick-up. The input gas conditions to the platen superheater are supplied by the output from the furnace module, and the input steam conditions are provided by the primary superheater module. Again, the physical characteristics of the platen, tube dimensions and pitch, tube material, emissivity and the dimensions of the cavity containing the platens, are included. These are used within the program to evaluate the thermal performance of the platen.

All the superheater and reheater sections in the predominantly convective zones of the boiler are described by a further module, one for each section,

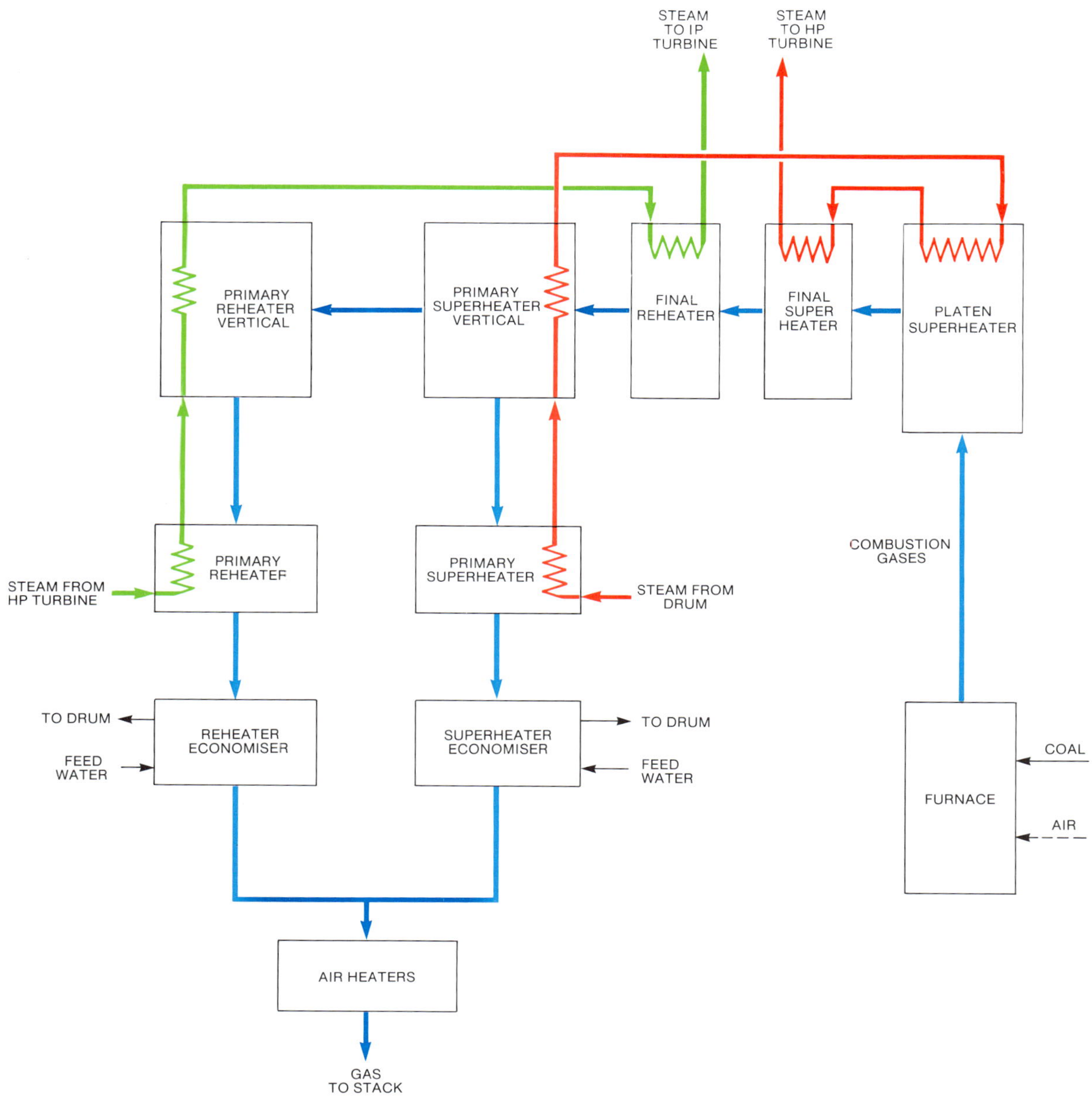

FIG. 2.18 Block diagram showing boiler elements and flowpaths for computer model

which is designed to represent the more convective nature of the heat transfer. This module is also used to represent the economiser. Inputs and outputs are provided either from boundaries or other modules, and all physical data on each tube bank are included in each module, together with heat transfer effectiveness factors.

The airheaters are normally modelled in order to provide a true representation of furnace air inlet temperature and stack gas temperature. Inlet air temperature to the airheaters is provided as a boundary, and the inlet gas temperature is obtained from the output of the economiser. The airheater module again contains all salient physical parameters, including the dimensions and arrangement of the heating matrix, and details of cross leakage from the air to the gas.

In addition to the equations which are used to connect the modules together, the user is at liberty to add further equations in the form of algebraics. These equations can be used to describe such functions as splitting and mixing of flows or the calculation of particular quantities not given in the standard printed output for any particular module. The use of algebraics may, in certain configurations, reduce the number of modules required. This can reduce the complexity of the total system and hence reduce computing time and cost.

Additional computer packages such as gas mixture properties and steam tables are available and are accessed as required from within the modular system.

9.2.3 Uses of the boiler model

The uses to which the completed boiler model may be put are defined in the objectives outlined in the introduction of this section. Dependent upon the requirements of the user, the model must be tuned to either a set of design or operational data. The tuning process is accomplished by identifying from design or test data the correct steam conditions at the output from each module, and adjusting the heat transfer effectiveness factors to enable the computer to reproduce these conditions. The effectiveness factors are a function of the resistance to heat transfer caused by fouling of the heat exchanger tubes. The values of these, taken initially in the setting up of the model, are based on previous experience.

When the model is to be used in the assessment of a design submission, a range of expected effectiveness factors will be known. If the factors required by the model in order to reproduce the performance claimed are seen to lie outside this range, then further investigation of the design submission will be required. In cases where the model is required to assess the performance of a modification to the plant, a mimic of the pre-modification system is required. The effectiveness factors introduced into the model to reproduce the test results are a representation of the state of the plant at the time the data were recorded. The model is now in a condition which will represent accurately the boiler system. Operating techniques can be varied or modifications made, as if the operator were working on the actual boiler.

9.3 Pressure loss, flow and temperature distribution analysis

At the inlet to a platen or pendant the steam is nominally in a homogeneous condition, after being thoroughly mixed in large diameter pipework and headers. At the exit from the platen or pendant considerable tube-to-tube temperature variations can occur before the steam is again mixed in the outlet header. Such temperature variations can indicate that while the mixed steam temperature may be within acceptable limits, steam and hence metal temperatures in some individual tubes may exceed the mechanical design limits for the tube material.

The causes of uneven temperature distribution within a platen are as follows:

- The tubes near the outside of a platen are considerably longer than those near the inside. They have, therefore, more surface area within the gas path and hence their heat pick-up is greater.

- The tubes at the inside and outside of the platen have more of their surface exposed to incident thermal radiation, as they are not shielded on both sides by adjacent tubes.

- Some tubes are wrapper tubes, installed for tube alignment within the platen. These are necessarily longer and therefore have more heating surface.

- Tube geometry and tube length cause changes in resistance to steam flow. Tubes carrying a reduced flow of steam will run at a higher temperature.

- Incident thermal radiation decays exponentially into the tube bank. Tubes nearer to the radiation source receive more radiant heat pick-up.

- Differential fouling of tubes can cause uneven heat pick-up in both radiative and convective modes.

The arrangement of tubes within a typical superheater platen is shown in Fig 2.19. The differences in length

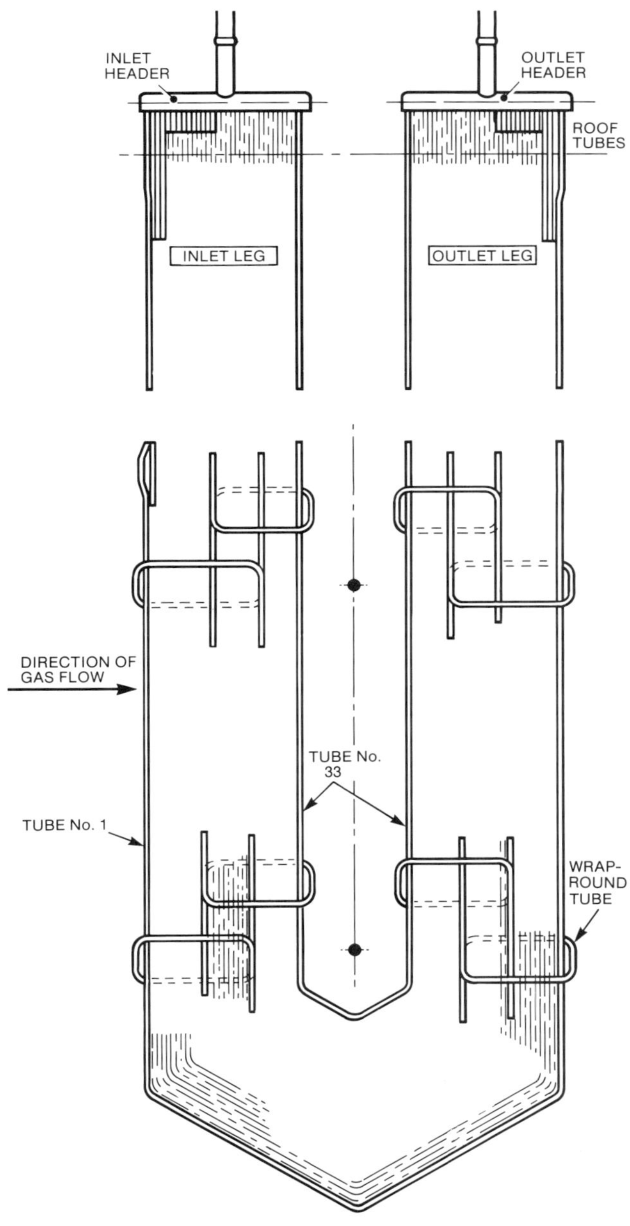

FIG. 2.19 Typical arrangement of platen superheaters

of individual tubes caused by the inclusion of wrapper loops can be clearly seen. The increase in tube length from the inner to outer tube (tubes 33–1) is also illustrated.

Individual tubes which exhibit excessive metal temperature give cause for concern in that a reduction in creep life and an increase in corrosion rate at high temperatures can cause premature tube failure, and since there is no redundancy in pressure parts, a single failure will cause an expensive unit outage. The computer can be used to aid the design engineer in the solution of this problem. A mathematical model of the existing arrangement can be set up and various forms of proposed modification can be investigated.

The software available from which to construct the mathematical model is based on established theory of pressure losses and flow in networks. Considerable experience has been obtained in its use and confidence in the results obtained is based on validation with specially instrumented plant.

The network is constructed mathematically of nodes and branches. A node is an entry to or an exit from the network, or a junction of two or more branches. A branch connects two nodes and may consist of one or more sections. Each section of a branch is defined by the user at any position along the branch exhibiting a change in geometry, material or heat flux.

When defining the platen shown in Fig 2.19 as a mathematical network, the inlet header and outlet header would each be designated a node. Each of the 33 platen tubes would be branches joining the two nodes, and each branch would be divided into sections describing all geometrical physical and heat transfer data within the branch. The number of sections within a branch is dependent upon the complexity of the particular system and the degree of detail in which the analysis is required.

The network program produces a set of thermodynamic and flow data for each section within the system. The individual tube outlet temperatures for the example platen illustrated in Fig 2.19 are shown in Fig 2.20. The temperature distribution shown in Fig 2.20 is for the platen as installed. This shows the expected distribution and it is seen that there is a downward trend from tube 1 to tube 33. Tube 1 is high due to the proportion of its surface, as a leading tube, exposed to incident thermal radiation. It is also the longest tube in the platen and therefore receives the greatest heat pick-up. Moving through the platen, both the tube length and incident radiation fall, causing the downward trend in temperature distribution. The spikes in the chart are due to geometric variations such as wrapper tubes.

There are two options open to the design engineer in which to reduce the peaks and even-out the temperature distribution at the outlet from the platen. The tubes with the highest outlet temperatures can be re-routed such that they are shortened and therefore subjected to less heat pick-up, or restrictions to flow can be introduced into the cooler tubes in order to direct more steam flow to the high temperature tubes and hence reduce their outlet temperatures without changing heat pick-up. Both methods have been implemented in differing situations and have been validated by plant results. The shortening of boiler tubes does however have a slight effect on the overall perform-

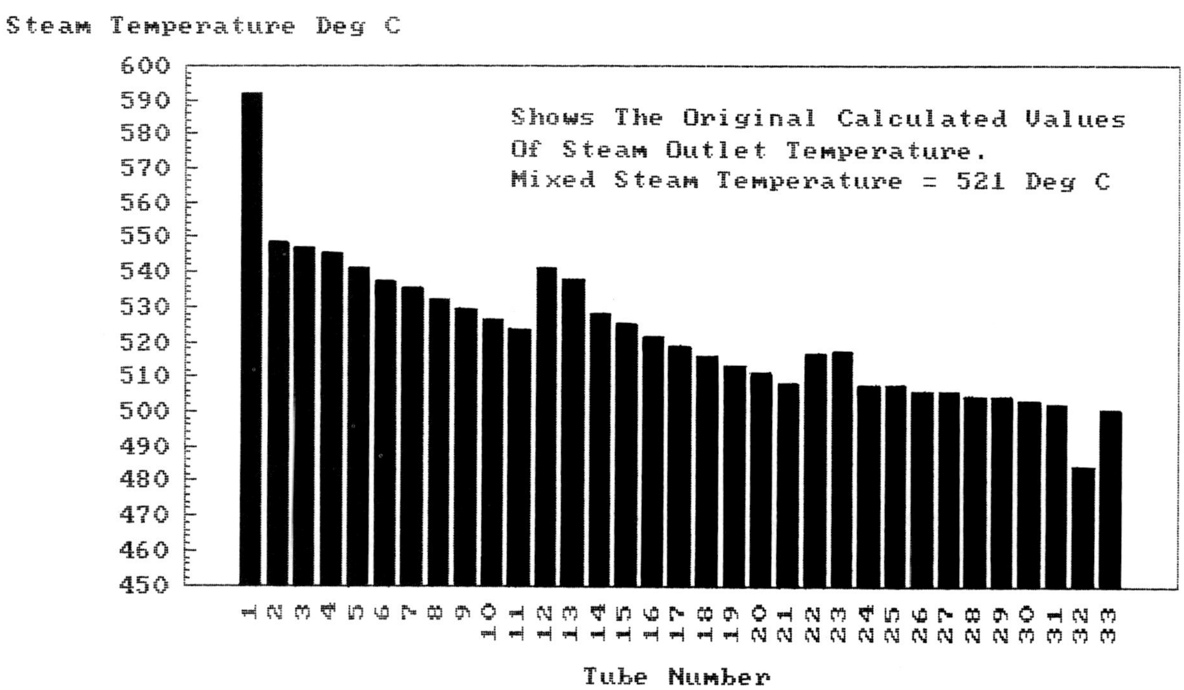

FIG. 2.20 Typical superheater platen outlet steam temperatures before modification

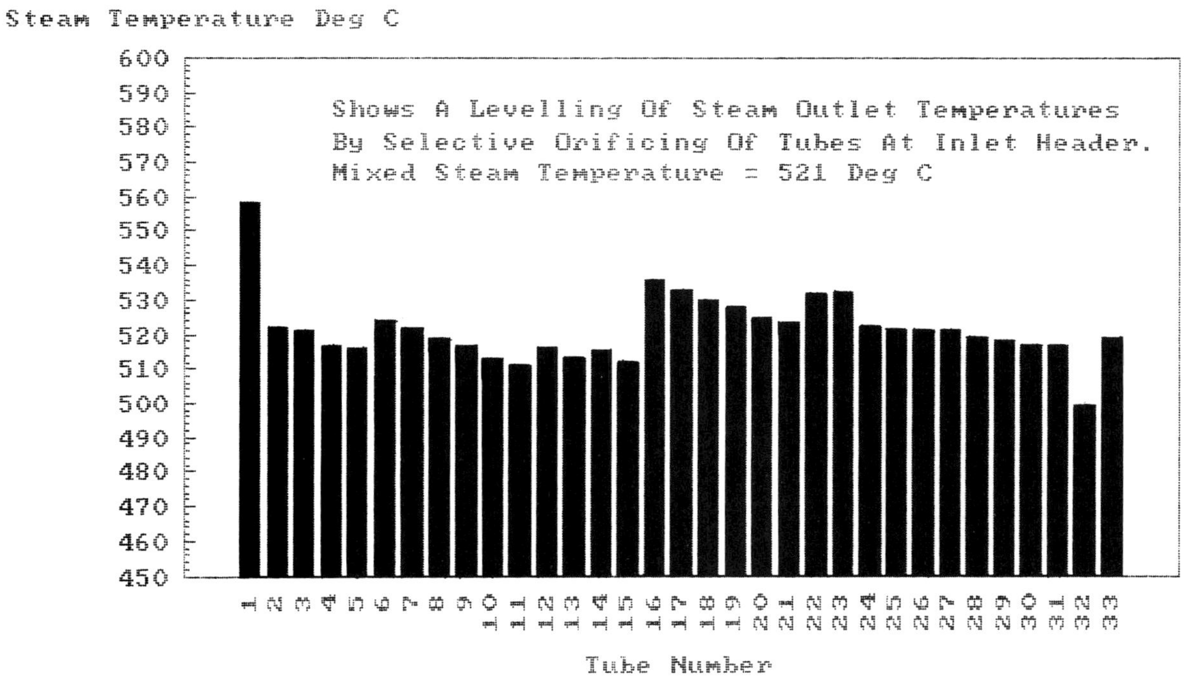

FIG. 2.21 Typical superheater platen outlet steam temperatures after modification

ance of the tube bank, but it is not normally significant when considering the performance of the boiler as a whole. The chart shown in Fig 2.21 shows the predicted results of selective orificing at the inlet header. Increased flow is directed to the outer tubes and a levelling of the temperature distribution is obtained.

Several sizes of orifice may be required, distributed along the inlet header, to bring the outlet temperature distribution to within acceptable limits. The ideal solution would produce a flat temperature distribution. Economic and practical considerations dictate, however, that a satisfactory solution has been achieved when the highest temperature peaks are brought down to an acceptable level.

9.4 Furnace circulation

The furnace tube circulation system of a drum boiler can be modelled using similar techniques to those described in Section 9.3 for the modelling of temperature distribution within the tubes of a platen. The flowpath of water is from the drum to the downcomers, then to the furnace inlet headers, through the furnace wall tubes to the outlet headers and back to the drum. This can be modelled as a network using nodes, branches and sections. For a forced circulation system, the circulating pumps can also be included in the model.

In setting up a boiler circulation model, the design engineer should decide on the detail of analysis required, since it is possible to describe the circuit in sufficient detail to evaluate flow and thermodynamic data in individual tubes anywhere within the system. If a general analysis only is required, perhaps of tube wall panel data rather than individual tube data, then considerable data preparation and computation time can be saved by setting up the model accordingly.

The preparation of a full scale furnace circulation model will involve the collection of geometrical data on every tube in the system. Tube wall roughness from which friction pressure loss is evaluated, is also required. The nodes of the system, or junctions of two or more branches, are defined first. These will include the boiler drum, all headers at the inlet and outlet of tube wall panels, and any other junctions of tubes such as take-off points for heated support slings. The branches joining the nodes are then subdivided into sections. Each section covers a portion of its branch exhibiting constant parameters such as tube geometry and heat flux. The description of furnace wall tubes in this way is flexible in as much as the programmer/designer is at liberty to choose how many discrete zones in which to divide each tube when allocating heat transfer information. Heat flux falling on the tubes is different dependent upon the position of the tube within the boiler, changes occurring as a continuous gradient rather than as discontinuous zones. The mathematical description of the change in heat flux must be in the form of discontinuous zones to conform to the concept of the division of branches into sections. Model accuracy therefore depends upon the number of zones, and hence sections, defined.

Greater accuracy is ensured if the division of heat transfer within a branch is represented as closely as possible to the continuous process existing in practice, suggesting a very high number of sections with small differences in heat flux. Experience and validated results have shown, however, that considerable simplification may be acceptable in most cases, and the division of heat fluxes into zones surrounding each burner level may be sufficient. Further heat flux calculations covering areas of the boiler walls not within the flame zone are also required.

The boiler is initially designed to enable the required steam production to be achieved while maintaining an adequate margin in order to prevent dryout in evaporator tubes. The expected ratio for a drum boiler is in the range 3:1 to 5:1 for the standard CEGB steam cycle. This value depends upon the relative effects of the thermal syphon, set up by heating of the wall tubes, and the pressure loss around the system.

Dryout in furnace tubes can result in premature failure due to excessive metal temperatures, or on-load corrosion. It may be caused by excessive heat flux levels in the furnace or, indirectly by the increase in steam carry-under into the downcomers from the drum. Excessive steam carry-under reduces the amount of subcooling of the water at inlet to the furnace, thus reducing the available margins against dryout.

A computer analysis of the circulation system can pinpoint areas of concern, since steam/water mixture quality is available at any point in any tube. The effect of proposed modifications and deviations from standard operating conditions can be evaluated with a high degree of confidence, based on validated analyses over many years.

In addition to the analysis of the circulation system itself, the model can be used to investigate the effect of modifications external to the system. Such an analysis has been carried out to examine the feasibility of installing a pumped recirculation loop from the furnace wall inlet headers to the inlet of the economiser. Such a loop was proposed to limit temperature differentials and thermal shock which arose in the economiser due to intermittent feeding during shutdown, and to reduce temperature differentials between various flowpaths in the evaporator circuit during the early stages of start-up. Achievable recirculation flow and its effect on the furnace circulation system were analysed using the computer model.

9.5 Tube life assessment

Proof stress or creep rupture criteria may be used in the design of boiler tubes, depending upon desired application. These are discussed in Sections 8.3 and 8.4 of this chapter and also BS1113.[5] They cover the design of tubes which are not subject to fuel ash corrosion. In the design of tubes which are subject to fuel ash corrosion the life fraction method is used as described here. This method of tube life assessment is a particularly useful application of computers to boiler design. The calculations involved are of an iterative nature, and often the results are required to show differences between large numerical values. If carried out manually, such calculations may be susceptable to cumulative errors which may severely affect the integrity of the results obtained.

In order to assess the life of a boiler tube the following data are required:

(a) Physical data:	tube material
	design pressure
	outside diameter
	wall thickness
(b) Thermodynamic data:	steam temperature
	outer surface heat flux
	tube material thermal conductivity
	steam film heat transfer co-efficient

In addition to the above data, safety factors for creep stress and proof stress are normally incorporated. The final data required are the rates of corrosion of the tube material for tubes subject to fuel ash corrosion, expressed as a function of the outside wall temperature. The corrosion rate is derived from a validated correlation of the type described in detail in Chapter 1 Section 2.2.2, in which gas and metal temperatures, together with fuel properties, are used.

The first stage in the calculation of tube life involves the determination of the mid-wall and outside wall temperatures; this analysis is described in Section 8.1 of this chapter.

The fuel ash corrosion of boiler tubes is a continuous process, it can however be considered mathematically as happening in a series of steps of a finite duration. The duration of these time steps will influence the accuracy of the calculation, but practical considerations limit them to between 100 hours and 1000 hours.

Having decided upon a time period, an iterative calculation is carried out as follows.

The hoop stress in the tube is calculated from:

$$\text{Hoop stress } \sigma = \frac{P\,(d + t)}{2t}$$

where P = tube design pressure

d = tube internal diameter

t = tube thickness

This hoop stress is then compared with the appropriate allowable proof stress from standard data at the calculated mid-wall temperature. If the proof stress criterion is not breached the corrosion wastage over the considered time period is then evaluated, the outside wall temperature being used for this purpose.

The creep damage sustained by the tube over the considered time period is expressed as a life fraction:

$$\text{Life fraction} = \frac{\text{time period}}{L\ (\sigma T)}$$

where L (σT) is the creep rupture life derived from the design stress rupture curve appropriate to each time period interval.

A new set of tube temperatures and hoop stress is now calculated using the reduced wall thickness due to corrosion over the first time period, and proof stress comparison and life fraction calculations are repeated. End-of-life criteria are that the summation of the life fractions reaches unity, or that as a consequence of metal wastage the hoop stress reaches the allowable proof stress.

During the calculation of metal temperatures for tube life assessment, it is standard practice within the CEGB to assume that the heat flux falling on the outside of the tube remains constant even as the effective tube thickness is reduced by corrosion. This enables metal temperatures to be calculated which will conform more accurately to actual practice, since corrosion is not uniform around the tube diameter but tends to affect localised areas only. The minimum thickness is however required for stress calculations, together with the appropriate mid-wall temperature.

This process enables a rapid assessment of tube life to be performed for both tender assessment, where all salient parameters should be known, and design work in which many tube materials and wall thicknesses may be required to be analysed.

10 References

[1] Stefan-Boltzmann Law, McAdams W. H.: Heat Transmission: McGraw Hill

[2] Hottel, H. C.: McAdams W. H.: Heat Transmission: McGraw Hill

[3] Sharan, H. N.: Journal of the Institute of Fuel: July 1963

[4] Fourier, McAdams W. H.: Heat Transmission: McGraw Hill

[5] BS1113: Specification for Design and Manufacture of Water-tube Steam Generating Plant (inc. superheaters, reheaters and steel tube economisers): 1985

[6] BS806: Specification for Design and Construction of Ferrous Piping Installations for and in Connection with Land Boilers: 1986

[7] BS5500: Specification for Unfired Fusion Welded Pressure Vessels: 1985

[8] BS3604: Specification for Steel Pipes and Tubes for Pressure Purposes; Ferritic Alloy Steel with Specified Elevated Temperature Properties: 1978

[9] BS2633: Specification for Class 1 Arc Welding of Ferritic Steel Pipework for Carrying Fluids 1987

[10] GDCD Standard 243: Co-extruded Boiler Tubes Part 1: 1987 Design

CHAPTER 3

Ancillary plant and fittings

Contents

1 Introduction

In Chapters 1 and 2 of this volume, the design of the main components of a power station boiler are considered. In this chapter, equipment which makes a major contribution to the high efficiency achieved in modern boilers is described. The principal items are economisers and airheaters which both provide heat recovery and also, in the case of the airheaters, the provision of air at temperatures which ensure efficient combustion. Boiler plant cleaning methods and equipment, and boiler fittings and mountings are described in the two final sections.

2 Economisers

2.1 Function of an economiser

The function of an economiser in a steam generating unit is to absorb heat from the flue gases and add this as sensible heat to the feed water before it enters the evaporator circuit of the boiler.

The economiser is located after the superheater and reheater banks and before the airheater, see Chapter 2 Fig 2.1. Its thermal duty is defined by the gas temperature drop from inlet to outlet. Gas inlet temperature is dependent on the furnace, superheater and reheater duties, and gas outlet temperature on the requirements of the airheater. On a pulverised fuel boiler, high primary and secondary air temperatures are required for drying and efficient combustion of the coal and this sets the airheater duty. The airheater outlet temperature is specified to be above acid dewpoint to avoid deposition and corrosion and this, together with the required airheater duty, defines the economiser outlet temperature.

Economiser duty can then be calculated from gas weight and the inlet and outlet gas temperatures. Since the water inlet temperature and flow are also known, the water outlet temperature must be below the saturation temperature corresponding to evaporative circuit pressure, otherwise steaming will occur with consequent flow instability problems in the economiser and, for drum boilers, loss of drum performance.

In once-through boilers, it is essential to have single-phase flow at the inlet to the furnace cooling circuit.

The gas temperature in an economiser is sufficiently low for there to be no intertube radiation and all the heat transfer will be by convection. The tube arrangements have small transverse and longitudinal pitches with resulting high gas mass velocities, to obtain high convection coefficients.

2.2 Types of economiser

2.2.1 Plain tube

Plain tube economisers are constructed with several banks of tubes. The height of each bank is limited to no more than 2 m to enable full sootblower penetration. The tubes can either be in-line or staggered tube formation. Staggered formation induces more turbulence in the gases than the in-line construction. This gives a higher rate of heat transfer and requires less surface for a given duty, but has a higher draught loss. If there is a possibility of fouling occurring, the in-line arrangement is preferred because it is more easily cleaned and inspected.

The tube diameters are generally 38 mm or 52 mm OD and the horizontal pitch is chosen to give sufficient gap between the tubes to prevent bridging-over by deposits that may form on the tubes.

2.2.2 Welded fin tube

Cast iron was originally used in the manufacture of finned economisers because of its resistance to corrosion, but was only strong enough for pressures up to 20 bar. To overcome the pressure limitation, cast fins were welded on to mild steel tubing. This had the disadvantage that the heat transfer was dependent on having good thermal contact between the fins and tubing. The fins themselves were bulky items and the resulting gas pass area was comparatively small, which led to cleaning problems, this was further aggravated by difficulties in aligning the fins. With the present day higher feedwater temperatures, corrosion at the gas outlet has been greatly reduced and cast iron protection is no longer required. With the development of precision production welding it is possible to use and produce perfectly aligned mild steel fins.

In comparison with plain tube economisers, steel-finned economisers occupy less space for the same thermal performance and draught loss. The reduction in tube length for similar tube diameters and pitches is usually around 4 to 1. This results in a smaller casing, less structural steelwork to support the reduced weight, fewer bends and fewer welds, giving a saving in overall cost.

2.3 Erosion

Unless particular attention is paid to avoidance of high localised velocities in welded steel-finned economisers, high rates of erosion can occur with subsequent failure of the pressure parts. High localised velocities can occur if:

- There is bypassing of the finned section either because of the absence of end shields or the end shields provided have been damaged.
- Localised ash deposition in the bank; the result of poor sootblower coverage or in some cases cascading of dust from upstream hoppers. To avoid blockage of a typical 660 MW economiser, 16 sootblowers are fitted.

Design velocities in economisers with UK coals should be between 8 m/s and 10 m/s, this velocity being similar in both plain tube and steel-finned economisers. Above these velocities erosion becomes progressively greater, the erosion rate increasing to the power 3.3 of the velocity.

2.4 Construction

One of the reasons in the past for outages was weld failures, especially site welds. Now, with the object of carrying out as much work under ideal workshop conditions, blocks of elements are manufactured as complete units in the factory. Each block is complete with headers, supports and support bars welded into position before dispatch to site. On a typical 660 MW unit there would be 24 units per economiser each weighing 20 tonnes, see Fig 3.1. The units are then transported to site and lifted into position.

It is now normal practice to suspend the economiser on water cooled slings, pendant fashion. This allows the economiser to expand downwards with the boiler and does not require an expansion joint between the boiler and the economiser casings.

FIG. 3.1 660 MW boiler economiser being erected
(see also colour photograph between pp 66 and 67)

2.5 Two-shift considerations

When two-shifting there are occasions during the shutdown period when the boiler is intermittently fed to maintain drum water level. In this situation the feedwater is not heated by the feedheaters and will be at the de-aerator temperature of about 100°C and, because of internal circulation on the waterside, the whole economiser will have become heated up to near the drum fluid temperature of about 300°C. Two problems can arise and have to be allowed for in the economiser design:

(a) The thermal shock of the cold fluid in the hot economiser.

(b) Stratification; the cold fluid may only partly fill the horizontal headers or pipework causing hogging and, if the structural system is constrained, this will cause high stresses.

During an overnight shutdown, there may be three or four short periods when water is fed to the economiser and, during the life of a boiler, there can be upwards of 6000 overnight shutdowns which would result in some 24 000 cycles. In assisted circulation boilers, this problem may be readily overcome by recirculating water from the circulating pumps discharge into the feed pipework upstream of the economiser, the recirculating water being introduced through a thermal sleeve into a mixing device. The recirculation rate has to be sufficient to increase the temperature of the feedwater to a level which reduces thermal shock and stratification stresses to within the design capability of the economiser system.

Unless special recirculating pumps are provided on natural circulation boilers, the difference in temperature between feedwater and economiser cannot be readily reduced. Even without a recirculating pump, stratification stresses can be significantly reduced by avoiding long horizontal headers and, where headers are unavoidable, by supporting them at only two points. If it can be arranged that stratification stresses are either small or are not additive, the system can normally accept the thermal shocks associated with the feed cycles over the station life.

3 Airheaters

3.1 Introduction

Airheaters are used in all current CEGB central station boiler units to recover heat from flue gases at temperatures lower than is economically possible with economisers. On a 660 MW boiler unit up to 140 MW of heat is recovered. The minimum flue gas temperature in the economiser is controlled by the feedwater temperature, whereas in the airheater it is the avoidance of fouling of the airheater and corrosion of flues, etc., which limits the gas outlet temperature. Apart from the need to preheat air to dry coal and improve combustion efficiency, the benefit from preheating combustion air can be shown to be a 1% improvement in boiler efficiency for each 22°C rise in the combustion air temperature. The demand therefore is to reduce the gas outlet temperature whilst at the same time minimising possible damage to boiler equipment. There are a number of alternative schemes for air preheating and each is considered by the CEGB in overall economic terms. No absolute guide can be given without considering all aspects of the installation, but the following is intended to explain the design, operation and control of those systems available.

3.2 Airheater types

3.2.1 General

There are two basic flue gas airheater designs currently operating in the CEGB on quite different principles, namely:

- *Recuperative* — in which the flue gas glows on one side of a plate or tube and the combustion air on the other; heat transfer occurring across the boundary. These are of static construction with theoretically small nominal leakage of air-to-gas through expansion joints, access doors and casings. In older stations, up to 120 MW, tubular or plate recuperators were commonplace. These were large, for the thermal duty, difficult to clean and did not allow any replacement of damaged heating surface.
- *Regenerative* — where the gas flows through a closely packed matrix with consequent increase in matrix temperature and subsequently air is passed through the matrix to pick up the heat. Either the matrix or the air hoods are rotated to achieve this. There are consequent sealing problems at the moving surfaces caused by the differential pressure between the air and flue gas. The bulk of the airheaters are of the regenerative type and all are arranged in contraflow, in general with air up and gas down.

On all modern boilers the CEGB uses regenerators for the principal duty of combustion air preheating. Their advantages over recuperators being:

- A significant reduction in overall size and weight.
- Easy and economic replacement of the heating surface with separate cold and hot end packs.
- Low weight of metal permitting the economic use of alloy steel or enamelled elements in the low temperature sections.
- Minimum metal temperatures at the cold end of a regenerative heater are higher than on a recuperator operating under the same conditions. This metal temperature oscillates some 22°C above and below a level close to the mean of the air entering and gas exit temperatures, and is not depressed by deposits as occurs with plate and tubular types.
- Holing of element plates due to corrosion will not materially affect the performance of the heater until the plates actually disintegrate.

Nevertheless there are the following disadvantages which have in some instances caused significant reductions in unit output:

- Moving parts with attendant risk of outage.
- Leakage of air into gas and gas into air due to entrainment, and air-to-gas leakage through seals due to the air/gas pressure difference. The effect of leakage is important in that it affects the duties of the FD, ID and PA fans.
- A relatively thin layer of deposit on the elements can reduce the flow area quite appreciably and increase the fluid pressure losses.

3.2.2 Tubular recuperators

Designs suitable for PF or oil-fired plant usually consist of steel tubes of 37 mm to 63 mm diameter. If the gas flow is through the tubes the larger diameters are advisable to minimise choking. The tubes can be arranged horizontally or vertically, the advantage of the latter being that dust fallout is less likely to obscure heating surfaces. The airheater can be divided into a high temperature section with tubes up to 12 m long and a low temperature section with tubes up to 4.5 m long, thus simplifying repair of tubes susceptible to corrosion. Single or two passes on the gas side with multi-passes on the air side usually fits in with the overall plant layout. In an effort to reduce the corrosion problems of the low temperature section, glass tubes have been tried, but there remained the problem of the fragility of glass along with the difficulty of sealing the ends at the tubeplate.

In comparative terms the space requirements and weight of these heaters imposes an economic disadvantage as compared with rotary types. The volume occupied by the heat transfer surface is up to nine times greater in a recuperator compared to a regenerator, the latter being approximately half the overall weight. It should be noted that duct and flue arrangements reduce the overall volumetric ratio considerably. Their simple and robust construction and minimal leakage of combustion air allows their continued possible use for coal drying and transport requirements. Indeed in at least one UK boiler unit regenerative mill airheaters have been replaced by a tubular recuperator.

The CEGB has not precluded the use of recuperative airheaters in its specification for new plant. However, with the emergence of the Rothemuhle airheater as a competitor to the widely used Ljungstrom regenerator, the CEGB has not been offered a recuperator for some 20 years in new plant orders. In future enquiries the CEGB will continue to allow boilermakers to put forward alternative airheater schemes. On existing plant CEGB economic criteria does not allow expenditure on retrofitting of alternative airheaters unless an economic case can be based on overcoming shortfalls in output. Early difficulties with regenerators which caused output shortfalls on the 500 MW units have now largely been overcome.

3.2.3 Plate recuperators

These comprise parallel plates which provide alternate passages for gas and air. Economic gas spacings for PF-fired units are 12 mm to 16 mm with 12 mm air spacing. The plates are welded-up in batches to form easily erected blocks. Spacers are used to maintain the design gaps over the plate to prevent localised fouling and corrosion as replacement of damaged areas is difficult.

Good distribution of gas and air especially at the low temperature end of the heater is important to avoid local cool spots which can initiate corrosion. Splitter plates have to be fitted in the air inlet ducting if a change of direction occurs immediately before the heater. Generally the gas ductwork is arranged to avoid changes of direction upstream of the heaters but, in spite of this, low temperatures and velocities tend to occur near casings.

The minimum plate metal temperature can be improved to minimise corrosion by recirculating a percentage of the hot air from the outlet to the inlet. Air bypasses are normally used during starting up periods and low load to prevent unduly cool conditions in the heaters.

A recent development of the plate recuperator termed the Open Channel Air Preheater (OCAP) [1] has been put forward as an alternative to the regenerator when separate mill airheaters are provided. This preheater has low leakage and no welds within the fluid flow environment. Expansion is accommodated by flexible seals at the plate periphery. An added performance benefit arising from the low leakage is the ability to utilise more flue gas heat, as the gas has less dilution from leaking air. The heaters are normally arranged as a multi-pass system in crossed counterflow. The block arrangement of each pass allows the selection of appropriate plate materials for each pass.

Reference should be made to Table 3.1 which provides a comparison between rotary regenerative and OCAP preheaters for mill duties for a 900 MW unit. The CEGB has no operating experience with the OCAP and similar comments must be made regarding its use as those for the tubular recuperator.

3.2.4 Howden (Ljungstrom) regenerative airheaters

The modern Howden rotary regenerative airheater is a development of an original design patented by Frederic Ljungstrom in 1920. In contrast to the recuperator, heat is transferred not by passing through tube or plate walls but by absorption and release from the same surface. The entire mass of heat exchanger elements is rotated in a continuous cycle through alternate streams of combustion air and flue gas. Figure 3.2 shows a typical airheater for a 660 MW coal-fired unit. Each airheater weighs in excess of 500 t and is supported on steelwork from the boiler floor.

The mild steel rotor is the central part of the airheater and contains the heat transfer matrix. Radial plates extending from the hub divide the rotor into 24 sectors which in turn are subdivided at the hot and intermediate end element containers. At the cold end of the rotor, grids welded between the radial divisions allow the sideways removal of the cold end elements without disturbing the upper tiers. The cold end elements generally have a shorter life than upper tiers due to the combined effects of corrosion, fouling and sootblowing.

The weight of the rotor is carried on the underside by a spherical roller thrust bearing, whilst at the top, radial loads are resisted by a spherical roller guide bearing.

Enclosing the rotor is the casing which has internal insulation held by 'Cor-Ten' plates and supports the sootblowing and water washing devices. The four transition ducts convey the air and flue gas to and from the airheater. As the airheater is supported on cold steelwork from the floor then adequate allowance has to be made for the differential expansion between the airheater and the slung boiler unit. This is accomplished with duct expansion joints. The casing is attached to the main structure of the airheater.

TABLE 3.1

A comparison between rotary regenerative airheaters and OCAP plate recuperative airheaters for mill duties on a 900 MW unit

	ROTARY REGENERATOR	OCAP RECUPERATOR
Heating surface/heater	8715 m^2	11982 m^2
Total weight/heater	139 t	331 t
Number/boiler	2	2
Leakage	12% to 14%	nominally 2%
Undiluted gas outlet temperature	131°C	122°C
Fan shaft power per boiler FD/PA/ID	12664 kW	13172 kW

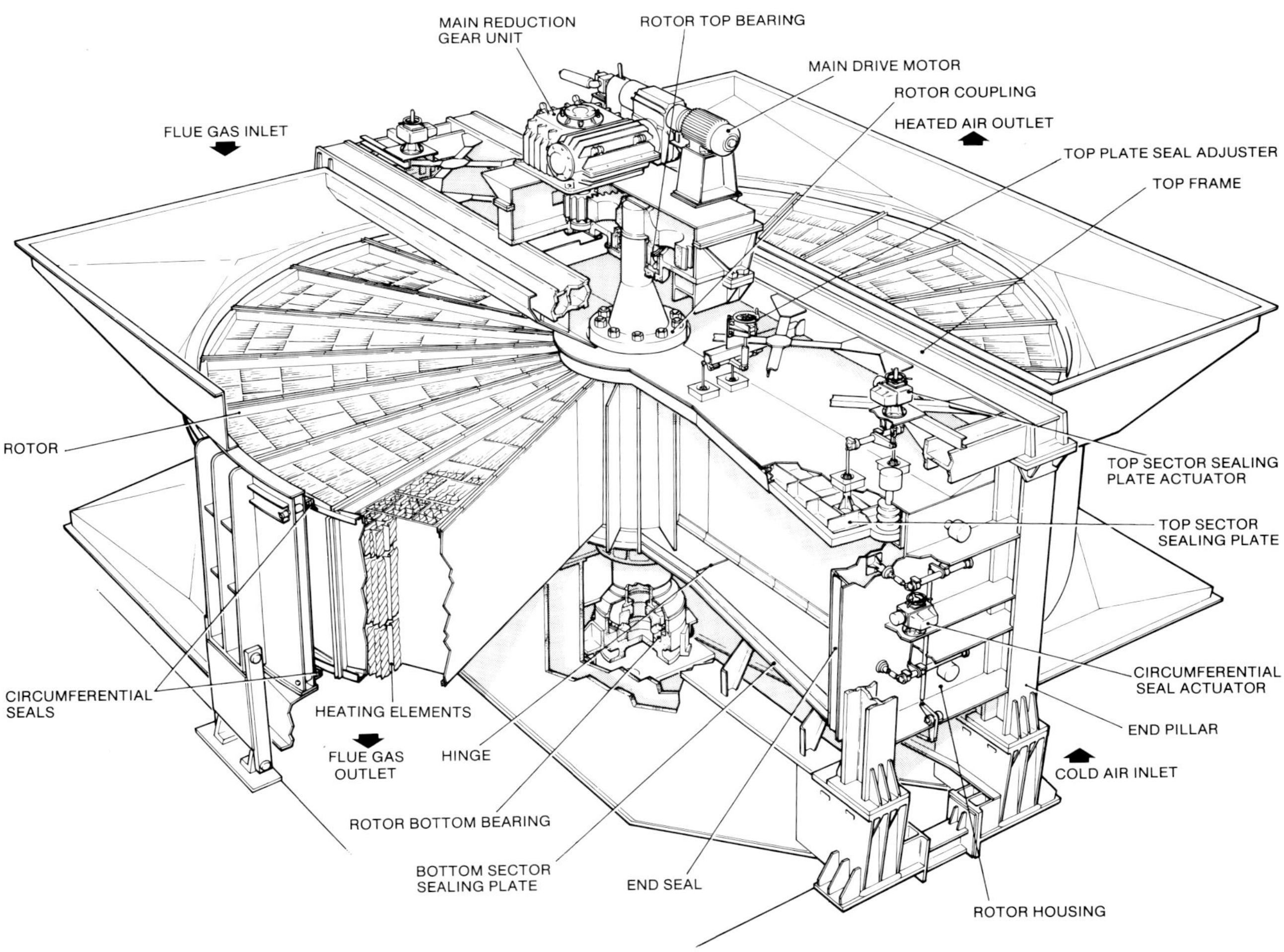

FIG. 3.2 Gas airheater for a 660 MW coal-fired boiler — Ljungstrom

The latter consists of deep support girders carrying the rotor thrust bearing and two end pillars which in turn support the top frame. Bolted and welded connections join these members to form a rigid frame which resists external loads which tend to disturb the vertical alignment of the rotor bearings.

In operation, the rotor assumes the distorted shape illustrated in Fig 3.3 with the casing and non-rotating seal surfaces remaining square, which would cause rubbing at the cold end and axial seals and an increase in gap at the hot end. In order to stop rubbing and minimise the air-to-gas leakage caused by the differential pressure between the air and gas, an adjustable sealing system is provided. The sealing system comprises stationary but adjustable surfaces mounted on the casing, consisting of sector sealing plates spanning the ends of the rotor, and axial sealing plates attached to the end pillars which seal against the periphery of the rotor. Sealing strips are attached to the edges of the rotor radial division plates along the top, bottom and down the outside edges of the rotor. In the unit illustrated, an automatically adjustable top sector plate is provided, in which inductive sensors are mounted at the periphery of the top sector plates which provide a signal proportional to the gap between the rotor and sector plate. The signal from the sensors is then used to modulate the upper sector plate to model the distortion of the rotor. The seal systems are shown diagrammatically in Fig 3.4. A particular operating condition which would cause severe rubbing and consequent damage to the seals is that termed *boxed-up*. In this condition the rotor assumes an even but elevated temperature. This is allowed for by providing remote retraction of the top sector plate and axial seal plates. On older units the gaps caused by rotor capping were catered for by off-load adjustment of the sealing strips with the top and bottom sector plates being fixed. Adjustment to the axial plate was available.

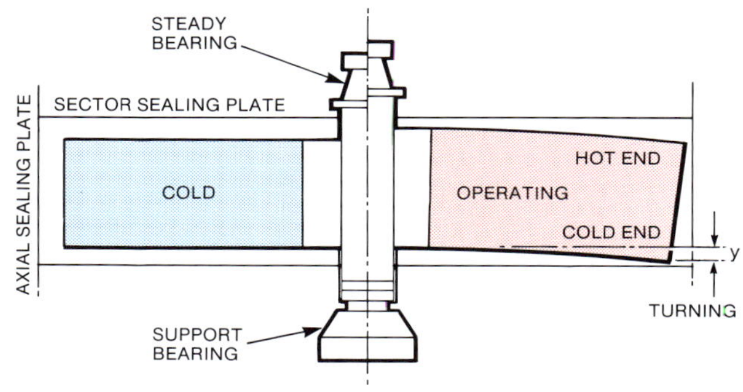

During normal boiler operation the top to bottom temperature causes the rotor to assume the shape shown above right. For a 14 m diameter rotor y will be approximately 35mm.

FIG. 3.3 Gas airheater distortion

Typical leakage levels range from 4% to 10% for main airheaters and from 10% to 20% for correctly adjusted mill airheaters. With the Howden regenerator, bypassing of the heat transfer surface by air and flue gas occurs down the sides of the rotor. To minimise this bypassing, circumferential sealing between the perimeter of the rotor and the casing is provided by strips attached to the casing which seal against

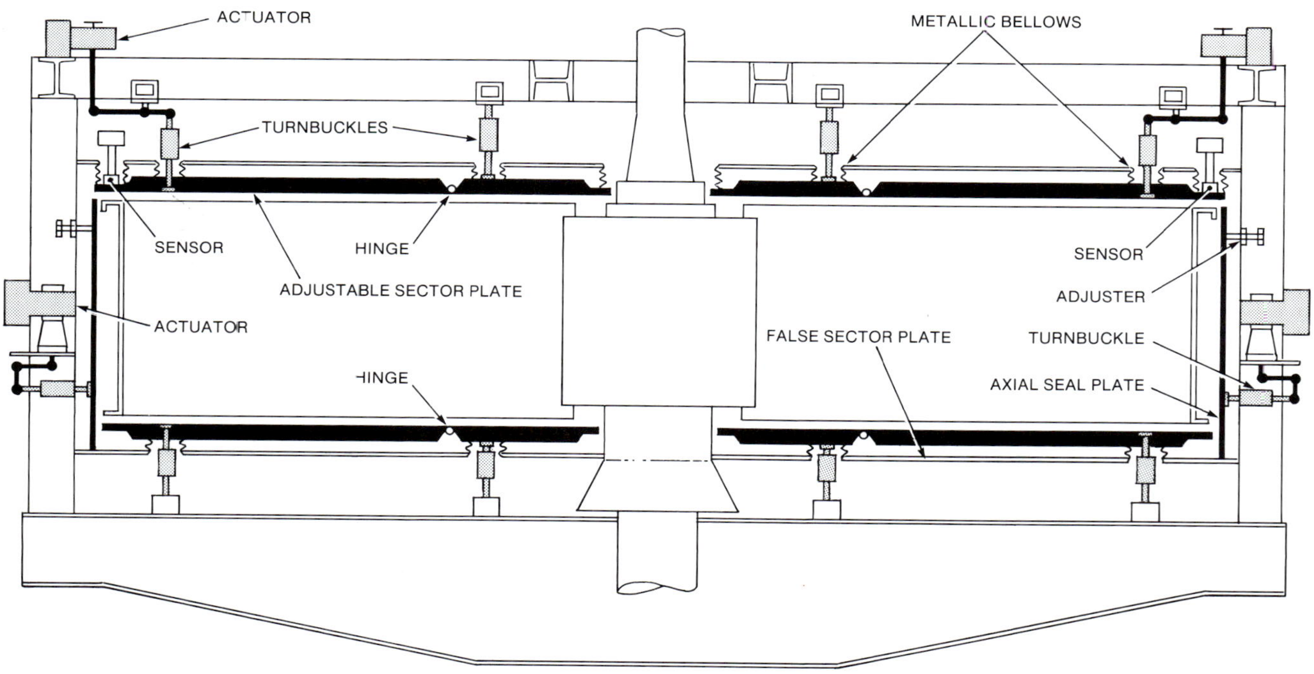

The figure shows the automatically adjustable airheater seal system for a 660MW coal-fired or oil-fired boiler. The top sector plate has an on-load automatic adjustment to cater for the downward movement of the rotor. The bottom sector plate and axial seal plate can be manually adjusted off-load.

In order to cater for the 'boxed-up' condition, in which the rotor is soaked at an even elevated temperature, the axial seal plate and top sector plate can be retracted by remote operation to a pre-set non-rubbing condition.

FIG. 3.4 Automatic adjustable seal system for gas airheater

the circumferential tyres on the top and bottom of the rotor. On later units adjustment of the seal can be made off-load but in earlier units, with the sealing strip attached to the rotor, this was not possible

The current design of airheater incorporates a bottom parallel girder support system which has allowed improvements to flow distribution to be made at inlet and outlet compared with the top supported A-frame system typical of the 500 MW units. This combined with improved flow distribution in the boiler rear pass established from model tests, and lower relative lengths of circumferential seals on larger airheaters, led to a marked improvement in performance in the airheaters supplied to units after the 500 MW unit station construction programme.

To minimise leakage across the rotor hub, seals are attached to the top and bottom sector plates and seal against the hub. To prevent dust build-up two small suction pipes are provided for the top hub seals which transmit the dust back into the gas ducting.

The rotor revolves at 0.75 r/min and is driven by means of a 15 kW electric motor through a double-reduction gear unit via a flexible coupling. The rotor can also be driven at slow speed for water washing the heating elements. In addition, a hand turning device for maintenance purposes is often provided.

Experience has shown that the automatic sealing system has not been without its problems and it is envisaged that future airheaters may revert to fixed sector plates which will also obviate the need for primary seals between the adjustable sector plate and the main structure, or for bellows for sector plate adjusters. The provision of separate air bypass ducts instead of the earlier integral bypass has significantly reduced leakage and improved performance.

The sector plates on the airheater illustrated in Fig 3.2 cover some 15° each side of the rotor, hence with 24 radial sectors, at any time only one radial seal can be guaranteed to be under the sector plate. It has been found that by doubling the number of radial sector plates whereby at least two seals are under the sector plate then significant reductions in leakage can be achieved. This has been of particular value where separate mill airheaters have been used (see Section 3.3.2 of this chapter). It should be noted that airheaters on the 500 MW and older units had 12 sectors and a 30° sector angle.

The heat transfer surface area of the airheater illustrated is in excess of 51 000 square metres and is formed in three tiers. The cold end tier is shallow (0.3 m) and is formed from 0.8 mm 'Cor-Ten' sheets which are of a plain form with notches formed in alternate plates to maintain spacing. The hot and intermediate elements are each 0.8 m deep and formed from 0.5 mm mild steel sheet packed in pairs with one sheet being undulated and the other notched and undulated. The undulations run at 30° to the notches and impart high turbulence to the fluid thereby improving the heat transfer. It has been past practice to provide additional depth in the rotor to accommodate further surface. This practice may be discontinued on new plant as the sootblower nozzles are too far away from the elements. All elements are packed in rigid containers to facilitate removal and handling. Further information on the heating element forms is given in Section 3.7 of this chapter.

3.2.5 Davidson (Rothemuhle) regenerative airheaters

These rotating head regeneration airheaters are manufactured by Davidson Ltd of Belfast under a licence agreement with Rothemuhle of Germany. This design was introduced into the CEGB somewhat later than the Ljungstrom type, the first being supplied for the early 500 MW units. The Rothemuhle airheater operates on the same principle as the Ljungstrom. The heat transfer surface is contained within a stator, and mounted above and below the heat transfer surface there are pairs of rotating hoods keyed to the same central shaft. In operation, hot flue gas from the economiser outlet enters the airheater at the top and passes downwards through the stator elements giving up heat. Cold air from the FD fans passes through the slowly rotating lower hood and through the stator where it absorbs the heat previously imparted to the stator by the flue gas. A system of sliding seals is incorporated in each air hood to minimise leakage of air to the flue gas. In Fig 3.5, details are shown of a typical modern regenerator for a 660 MW oil-fired unit. The airheaters are supported by steelwork from the boiler floor, necessitating the use of expansion joints between the ducts and the boiler unit. The airheater shown is 14 m in diameter.

The stator is of all-welded construction comprising an outer casing and inner hub and division plates as shown in Fig 3.6. It is split into seven rings by circumferential division plates and into a number of cells by radial division plates. On current designs the stator radial plates are continuous, unlike older units with staggered division plates. The former arrangement has caused seal jumping resulting from the successive build-up and breakdown of pressure under the seal shoes. It has therefore been necessary to stagger the sealing shoes to minimise this effect.

Each air hood is a fabricated steel assembly of circular shape to mate with the air duct and presenting two sectors at the stator face. Each air hood is supported by a frame and a hub which is keyed to the main shaft.

Referring to Fig 3.7, articulated sealing frames seal the air hoods against the stator faces. The sealing frames are supported by, and the seal clearance adjusted from, the spring pin units. The frames are connected to the air hood through double-layer expansion sleeves which are made in stainless steel. Replacement cast iron sealing shoes are attached to the frame. The collar

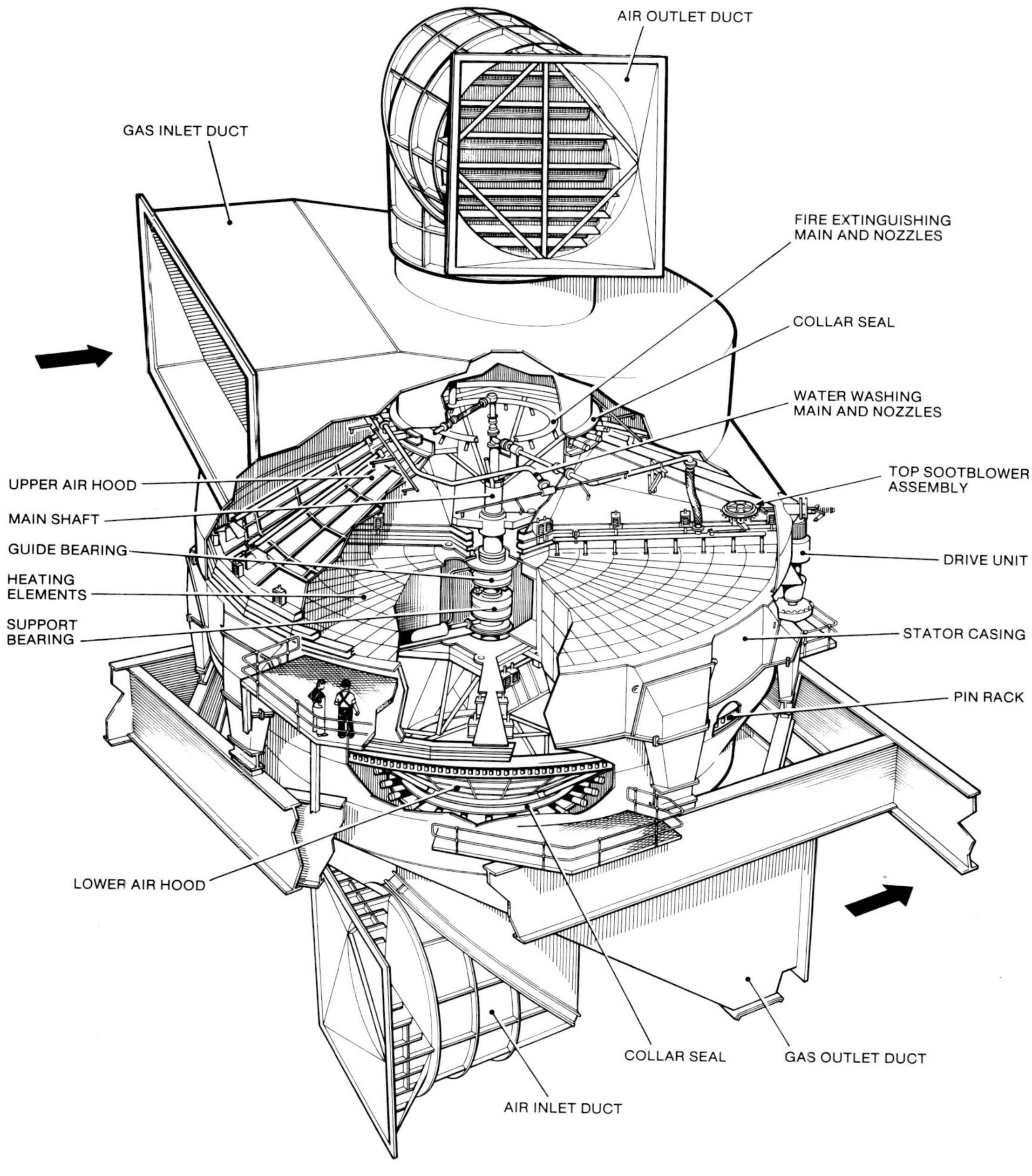

FIG. 3.5 Gas airheater — Rothemuhle

seal provides the seal between the stationary air ducts and the hoods.

In operation the stator distorts due to the top-to-bottom temperature differential. To allow for this a spring suspension system is incorporated as shown in Fig 3.8.

To ensure contact of the hot end seals at full load, the seals have to be set to provide a rubbing contact at lower loads and prolonged operation at part-loads results in unacceptable wear rates. In order to reduce this wear, thermal compensators utilising the difference in thermal expansion between mild and stainless steels can be located on the upper air hoods.

The airheater illustrated in Fig 3.5 has the support and guide bearings located in the stator hub region. This is contrary to previous practice on the 500 MW units where both top and bottom support arrangements were used with bearings remote from the airheater stator. This has led to wear problems caused by significant growth of drive shafts. Various techniques have been tried with success to minimise the consequent heavy wear; these do not necessarily reduce leakage levels but maintain post overhaul levels:

- Replacing the cold end stainless steel expansion sleeve with a more flexible 'Scandura' seal and replacing the spring pin assemblies with adjustable counterweights.

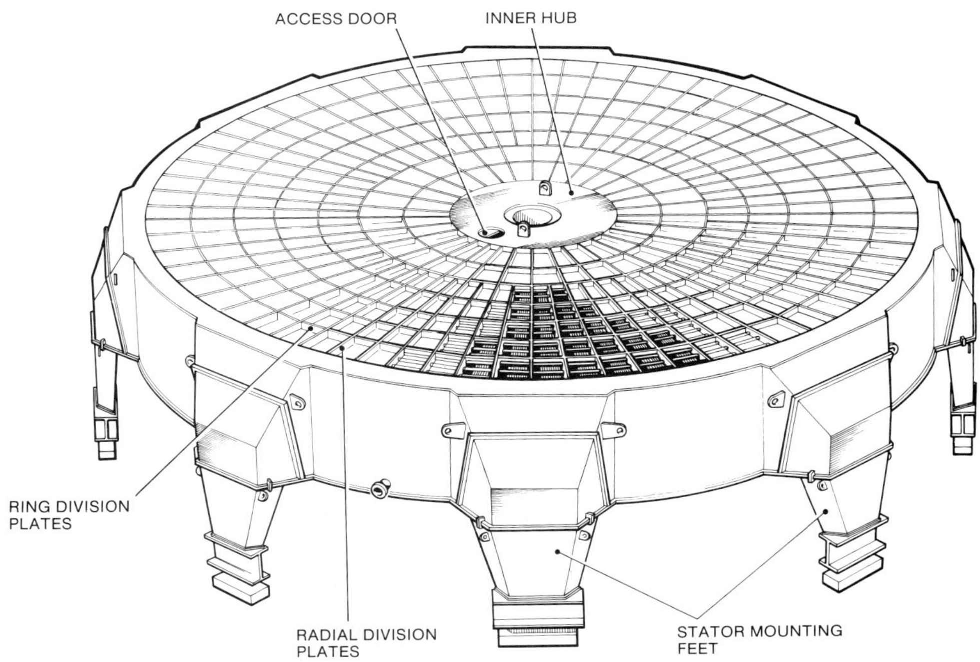

FIG. 3.6 Stator and heating elements — Rothemuhle

- On-load seal adjustment using mechanical actuators to adjust spring tensions.
- On-load seal adjustment using magnetic actuators controlled from inductive seal gap sensors.

It is a requirement of any airheater that seals should not require setting between overhaul periods. Much effort has gone into a revised design of seals to reduce leakage and some sealing surfaces are carefully machined. The combined effect is to reduce the leakage on large airheaters dealing with total combustion air or secondary air only to some 5%, in the presence of an air to gas differential of 40 mbar. On mill airheaters which deal with primary air only at 140 mbar differential pressure, leakage in a correctly set airheater can be as low as 10%.

The mainshaft bearings are cooled by a forced oil lubrication system, the oil being air-blast cooled.

The drive unit consists of an electric motor and gear unit mounted vertically on a fabricated frame attached to the outside of the stator. The output shaft of the gearbox is connected by a pin type flexible coupling to a short shaft, which runs in two bearings and carried a pinion at the lower end. The pinion meshes with a pin rack attached to the circumference of the lower air hood.

The sootblowers are mounted on the upper and lower hoods on the sector exposed to the gas flow (Section 3.5 of this chapter refers). The heat transfer surface for the airheater illustrated in Fig 3.5 consists of a number of 0.8 mm thick 'Cor-Ten' steel plates arranged in packs. Each pack contains alternate vertical and diagonal corrugated plates. The vertical corrugations are staggered, the whole producing turbulence and consequent high rates of heat transfer. Further details on element forms are given in Section 3.7 of this chapter.

3.2.6 Steam airheaters

Steam airheaters have been provided on oil-fired units to limit airheater cold end corrosion and fouling and to preheat combustion air until such time that the flue gas airheaters can be put into service. There are designs available from a number of manufacturers and they normally consist of banks of round or oval finned tubes as shown in Fig 3.9. The steam is normally slightly superheated at inlet and is condensed through the airheater banks. Arrangements need to be made to draw off any air that may be contained in the steam to prevent air locks. In Table 3.2, typical operating conditions are given for a 600 MW oil-fired unit. The condensate disposal unit consists of steam traps, strainers and bypass valves or, alternatively, a fully condensing system is arranged with a condensate tank and pumps for return of the condensate to the condenser flash vessel or auxiliary boiler. The steam airheater is located in the FD fan discharge.

Tubes have normally been arranged vertically although this is considered a more expensive arrangement. When arranged horizontally, sufficient fall on the tubes and enough intermediate supports should be provided to avoid local flooding of tubes.

A number of different fin attachment and tube types have been used by the CEGB:

(a) Circular tubes with tension wound fins — this arrangement allows thin fins to be used with up to

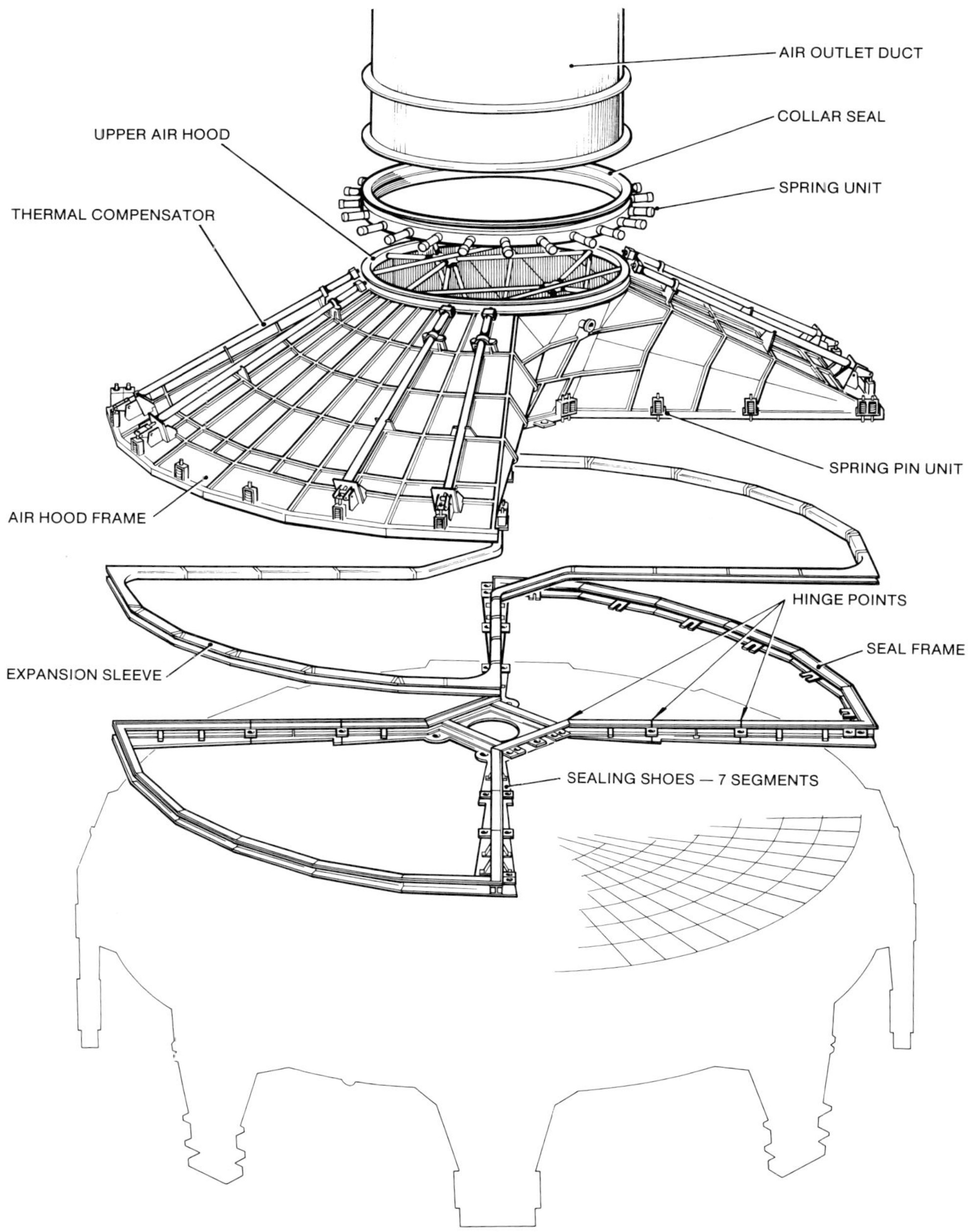

FIG. 3.7 Upper air hood and seals — Rothemuhle

11 fins per 25 mm tube length. Fin heights can be large but they are susceptible to damage.

(b) Oval tubes with rectangular plate fins — in this arrangement the oval tube gives a reduction in air flow resistance and provides up to 15% more surface on the air side compared with the round tube.

(c) Oval tubes with helically-wound fins — as for (b) but by helical-winding there are no corners to the fins and closer packing is possible.

Steam airheaters are used on dual-fired plant and on plant with open boiler houses. The steam airheaters can be supplied with steam from auxiliary boilers or, when sufficient load is generated, steam can be bled from the turbine.

3.2.7 Alternative static airheater systems

Although untried by the CEGB, nevertheless two alternatives have been put forward for both retrofit installations and new plant. These are the *liquid-coupled airheater* and *heat pipes and thermosyphons.*

Liquid-coupled airheater

The system consists of two or more extended surface heat exchangers, linked by pipework containing an

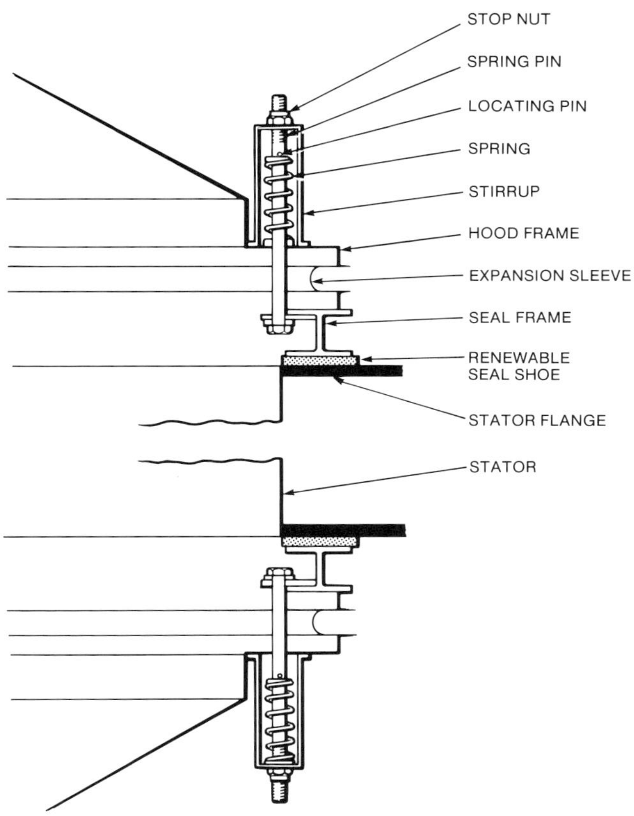

The suspension units are adjusted such that no spring pressure is applied to the sealing surfaces when the heater is at full boiler load. As the gas temperature decreases the suspension system tends to apply pressure through the hot end seal frame to the stator surface. Conversely the cold end seal frame tends to be pulled away from the stator.

FIG. 3.8 Gas airheater stator sealing arrangement

internal circulating fluid. One unit serves as the flue gas cooler and is mounted in the hot gas stream, with the other unit heating the incoming air. High fluid pressures are involved to avoid boiling and pressure parts standards would have to be adhered to. The advantages of these airheaters are the flexibility afforded to ductwork arrangement and zero air-to-gas leakage. The main disadvantages are cost as, in addition to high operating pressures, the required heat transfer surface must of necessity be twice that of an equivalent recuperator, and loss of the airheater should the pressure boundary rupture.

Heat pipes and thermosyphons

The heat pipe can be defined [2] as a device that transfers heat by a process of evaporation and condensation of a fluid circulating within a sealed cavity. For possible airheater installations an adiabatic section would be necessary in order to allow connection between the hot gas flue and air ducts. Furthermore, the surface would comprise circular finned tubes working in thermosyphons, which do not use an internal wick but rely on gravity for fluid circulation.

As with the liquid-coupled system, the CEGB has no experience with this concept for air preheating but is aware that interest has been shown elsewhere. However, only limited applications have been identified where equivalent heat loads to airheater duties are available. Doubts remain as to the cost effectiveness and operating life, but similar advantages of plant layout and zero leakage apply. The operating fluid is limited by considerations of reaction with its casing, its thermodynamic critical point and safety.

3.3 Arrangement of plant

3.3.1 Main airheaters only

A typical arrangement for a 500 MW coal-fired unit with three airheaters and eight mills, each with an associated PA fan, is shown in Fig 3.10(a). All the flue gas flows through the airheaters. The airheaters are sized to raise the air to a temperature which maintains mill exit temperature within close operating limits. If a drier fuel is used the hot air is tempered by cold air taken from the FD fan discharge and mixed at the PA fan inlet.

With wetter fuel, use can be made of gas which has been bypassed around the economiser and thus improve the temperature difference in the airheater and produce a higher air exit temperature.

These changes to accommodate the off-design moisture level result in an increase in stack loss.

Air bypass arrangements are made for start-up purposes but also to maintain the gas exit temperature and combined air inlet and gas outlet temperatures above the design levels.

3.3.2 Main and mill airheaters

With reference to Fig 3.10(b), the main airheaters deal with the secondary air requirements as do the FD fans and are sized accordingly. The primary air fans supply the air to the mill airheaters, with their suction taken from the FD suction duct. As with the main airheater scheme, tempering air and economiser bypasses can be fitted to cater for off-design fuel moisture.

Recent CEGB boilers have used the main airheater only scheme which has resulted in higher supply costs for the PA fans, higher operating powers and erosion from dust carried over by the airheaters. This course was chosen because of the poor leakage performance of the mill airheater schemes causing unit output shortfalls with both Ljungstrom and Rothmuhle air preheaters, as the mill airheaters have to deal with a differential pressure at least twice as large as the main airheaters. Over recent years much work has been done to improve the leakage performance and it is envisaged that the next generation of plant will use the main and mill scheme and thereby take ad-

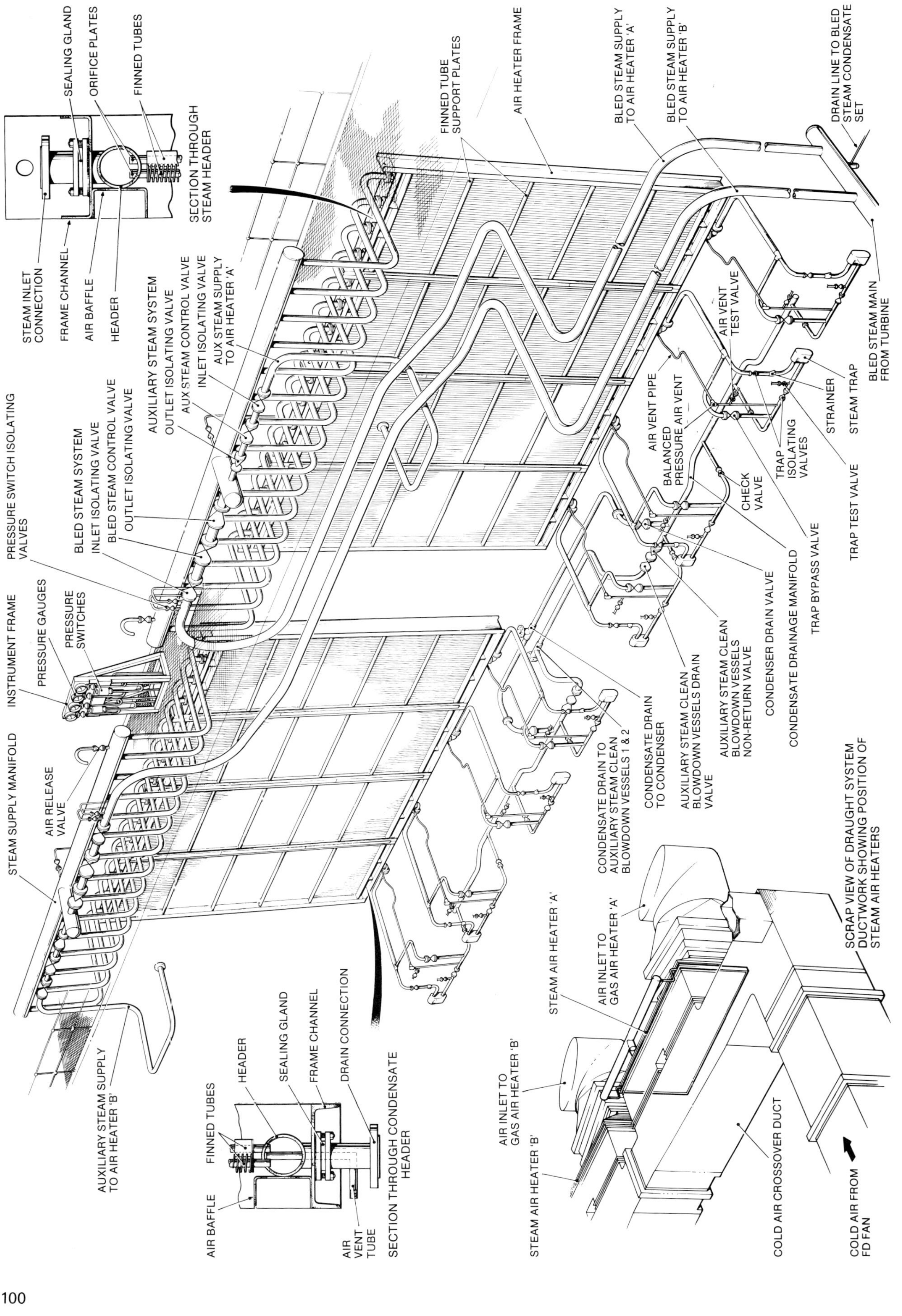

FIG. 3.9 Steam airheater

TABLE 3.2
Typical steam airheater operating conditions for a 660 MW oil-fired boiler unit at BCMR

Pressure at airheater inlet —	4.9 bar
Pressure at airheater outlet —	4.83 bar
Steam flow —	6.68 kg/s
Air inlet temperature —	32°C
Air outlet temperature —	60°C
Total heating surface —	4722 m^2
Pressure loss air side —	3.0 mbar

vantage of the supply cost savings and lower operating powers.

3.3.3 Combined mill and main airheaters

Regenerative airheaters of both the Ljungstrom and Rothemuhle designs are available with the main and mill sections incorporated in the same casing. Economies in size and ductwork layout are possible and also at least some of the primary air leakage is into the secondary air stream.

Difficulties are envisaged with control of mill airheater temperature to cater for operating with a range of coal moistures and allowing for temperature differentials between the primary and secondary sections.

3.3.4 Airheaters for oil-fired boilers and outdoor coal-fired boilers

Oil-fired units are not critical on the hot air temperature supplied to the burners, but even if the periodic replacement of at least the low temperature ends of the airheaters is envisaged, then some protection of the elements against corrosion is necessary by preheating the air prior to entry into the flue gas airheater. The same argument applies to outdoor boilers which do not draw their combustion air from the enclosed boilerhouse and as such the airheater air inlet temperature is only a few degrees above ambient.

Such preheating is economically carried out using a steam airheater situated between the FD fan and flue gas airheater, the steam being bled from a stage of the turbine.

Airheaters on oil-fired units are normally fitted with gas bypasses so that acid is not deposited on the plates during start-up.

3.4 Corrosion and fouling

Deposits on airheater surfaces are initiated through the combined effect of moisture, sulphur trioxide (SO_3) and ash at local metal temperatures below dewpoint. Apart from particular cases of hot-end blockage caused by the entrapment of large ash particles in regenerator surfaces, fouling generally occurs at the cold end of an airheater.

3.4.1 Coal-fired units

With PF firing corrosion is not as much a problem as fouling due to the absorption effect of the SO_3 by the alkaline ash, but fouling prohibits the adoption of very low gas exit temperatures. A practical guide to the minimum surface metal temperature is given by taking the mean of the cold air inlet and gas exit temperatures. In operation it is common practice to control the airheater by use of an air bypass, such that the combined air inlet and gas outlet temperature is above 152°C, with a minimum gas exit temperature of 120°C. The latter temperature has been chosen to limit corrosion. Studies are in hand to establish the overall economic case for reducing the minimum gas exit temperature of 120°C but caution is necessary as disastrous effects on flues, fans and chimneys may occur, and mill operating requirements may be infringed.

Severe problems were encountered with 500 MW boiler units through airheater fouling and in many cases unit output was affected. Normal cleaning methods with steam sootblowers or water washing did not contain the problem. The airheaters were fitted with double-undulated airheater plates which did not offer a free flow channel in the direction of the fluid flow. A modified design of element was optimised by the CEGB in which a parallel channel was provided, allowing more successful cleaning and fewer areas where ash can accumulate. Little fall in thermal performance was found and the airheaters have subsequently operated successfully. On the 660 MW units the double-undulated elements have been successfully used, but with a cold end consisting of alternating flat and notched surfaces.

3.4.2 Oil-fired units

Oil firing requires a higher minimum metal temperature than coal to avoid corrosion, typically between 105°C and 110°C. Operating at low excess air has reduced corrosion at the low temperature end of airheaters by reducing the formation of SO_3. It is current practice to have all the heating surface made from 0.8 – 1 mm 'Cor-Ten steel' which offers a small increase in life, the aim being to produce a life of at least 3 years for the lower tier of elements, which operate at metal temperatures of 150°C down to 105°C. The airheaters are operated so that the combined air inlet and gas outlet temperature does not fall below 210°C, thereby giving an approximate metal temperature of 105°C. The gas outlet temperature is kept in excess of 150°C in order that downstream equipment does not suffer from corrosion. It will be seen that the inlet has to be at least 60°C,

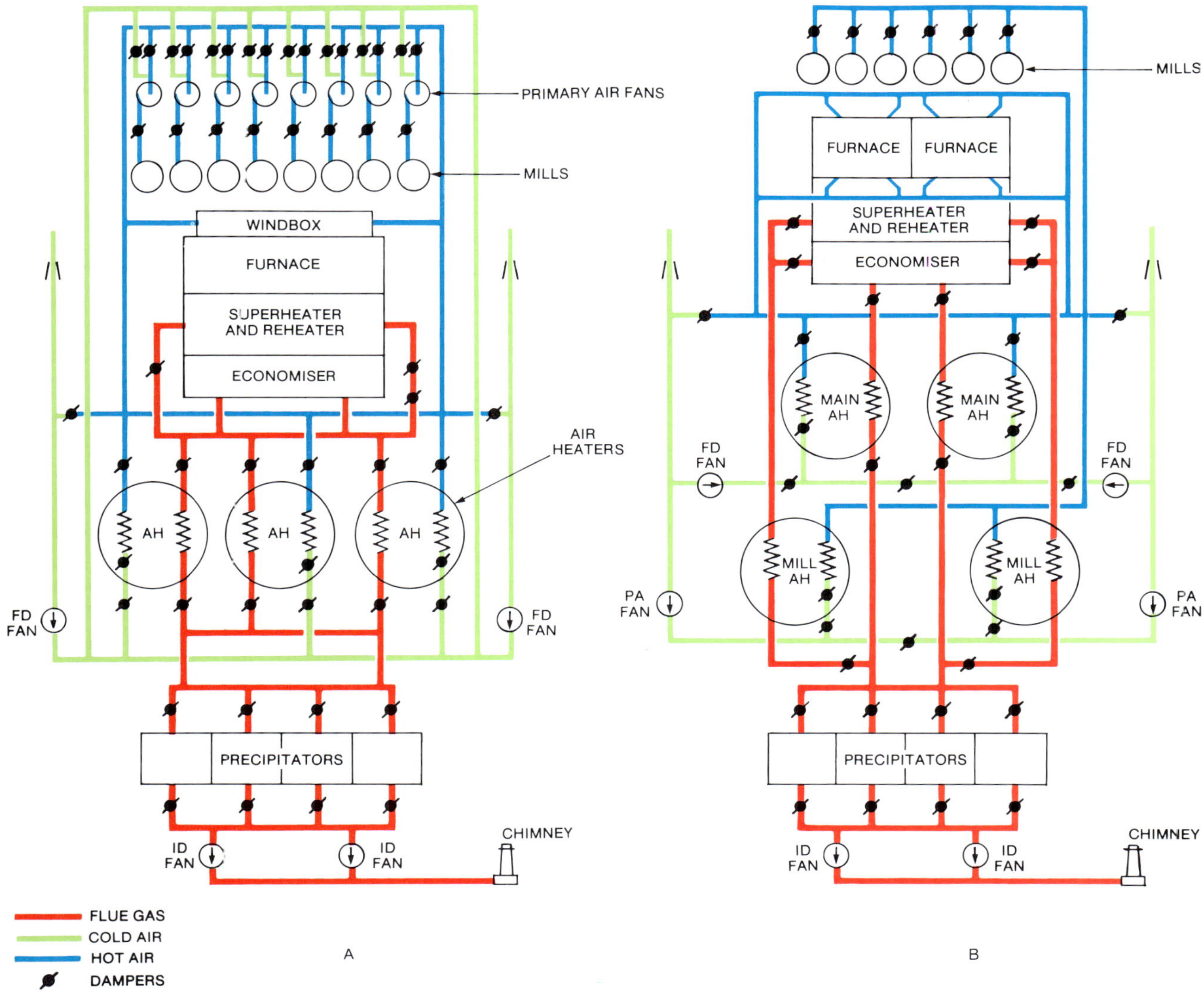

FIG. 3.10 Airheater layouts for coal-fired stations

and this is achieved using a bled steam airheater in the FD discharge duct.

A number of oil-fired units are operated at below 150°C gas outlet temperature. In these circumstances an additive, usually magnesium hydroxide powder, can be injected into the flue gas to neutralise the acid.

The cold end metal temperature can be adversely affected by maldistribution of air and gas flows, it therefore follows that proper design configurations of flues and ducts are essential together with efficient cleaning methods. Only the regenerative airheater has the facility for replacing cold end heating surfaces quickly and cheaply, and it can also continue to operate with fairly severe corrosion and fouling of the heat transfer surface.

The presence of a corrosive fluid can lead to damage to permanent access landings within the airheater and can present a safety hazard.

Removal of grits prior to the airheater considerably reduces the possibility of hot end plugging, which is believed to be a combustion/fuel related problem and is peculiar to particular sites.

3.5 Cleaning of airheaters

3.5.1 Tubular

Whether arranged with vertical or horizontal tubes the only proven practicable method of cleaning is by shot rain.

3.5.2 Plate

Localised cleaning of the gas exit ends can be accomplished using hand held air lances. More recent plate exchangers have arrangements which would allow the permanent installation of sootblowers. Shot cleaning methods are highly efficient. The CEGB has little up to date experience with plate or tubular recuperators.

3.5.3 Regenerative

For the Ljungstrom airheater, there have been a number of sootblowing systems including multiple fixed jets and swinging arm types. It is current practice for semi-retractable sootblowers to be incorporated as shown in Fig 3.11. The lances are situated in the top and bottom gas ducts with the bottom blowers normally blown first. For a 660 MW coal-fired boiler the operating time is 60 minutes per lance at an air pressure of 10 bar and flow of 80 m^3/min, the blower operating on the forward stroke only.

For the Rothemuhle airheater similar operating conditions would apply. However, two main designs of sootblower have been incorporated as shown in Fig 3.12 (reciprocating lance) and Fig 3.13 (escapement wheel), the reciprocating lance being the current preferred scheme. The sootblowers are fitted to the top and bottom trailing edges of the air hoods in the gas space, thereby the downward flow of the flue gas takes away the dislodged particles. This arrangement necessitates the use of a rotating union which has led to severe leakage of blowing medium.

The choice of steam or air as the blowing medium is covered in Section 4.4 of this chapter.

Thin layers of deposit on the heat transfer surface do not significantly affect heat transfer but increase pressure losses. Areas of complete blockage divert the air and gas to unblocked areas with a consequent increase in velocity and thus heat transfer. The increase in pressure difference across a regenerator from blockage will also cause greater bypassing and a consequent reduction in performance.

Off-load low pressure water washing at 5–7 bar is normally carried out on a regenerator prior to an outage with the elements still warm or when severe fouling is evident. During water washing the rotor speed is either reduced or stopped to minimise carryover of debris to the air ducts. Facilities must be provided for removing the mildly acidic wash water, the acidity of the outgoing water giving an indication of when to stop washing. The Ljungstrom airheater usually has a combined sootblowing and water wash lance, whereas on the Rothemuhle airheater fixed mains are installed above the stator in the gas flue. The mains are used sequentially to deluge the exposed stator face with the air hoods stationary, the hoods are then moved through 90° and the freshly exposed stator washed.

High pressure water washing (400 bar) typified by the Hydrotech process [3] has been successful in removing heavy fouling, particularly hot end (popcorn fouling) in regenerators. The Hydrotech process ensures that all areas receive the same volume of wash water by regulation of the rotor speed. The total volume of wash water is some 40% of that for a low pressure wash. The costs involved are a fraction of the cost of element replacement with the process carried out during an outage of two to three days and has been shown to be an economic way of increasing heating surface life.

3.6 Maintenance considerations

The thermal, pressure loss and leakage performance of regenerators can be significantly affected by the following and attention should be given to these during maintenance periods:

(a) Missing or damaged basket seals at the cold end of the airheater.

(b) Poorly fitting bypass dampers.

(c) Alignment of the hoods and rotor.

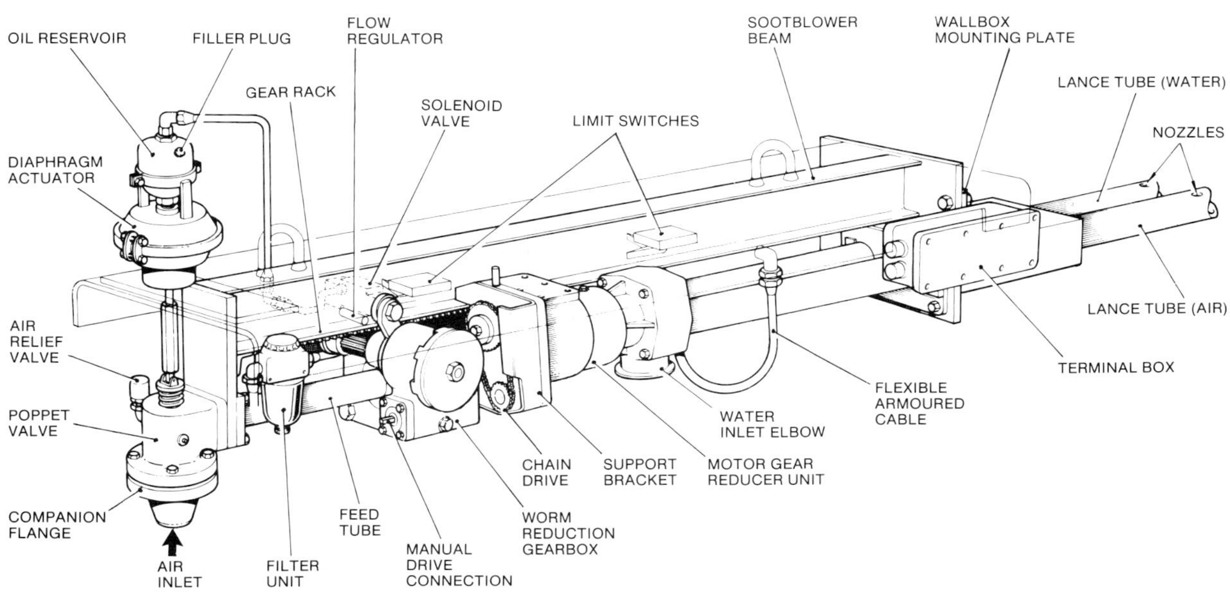

FIG. 3.11 Semi-retractable sootblower arrangement for Ljungstrom airheater

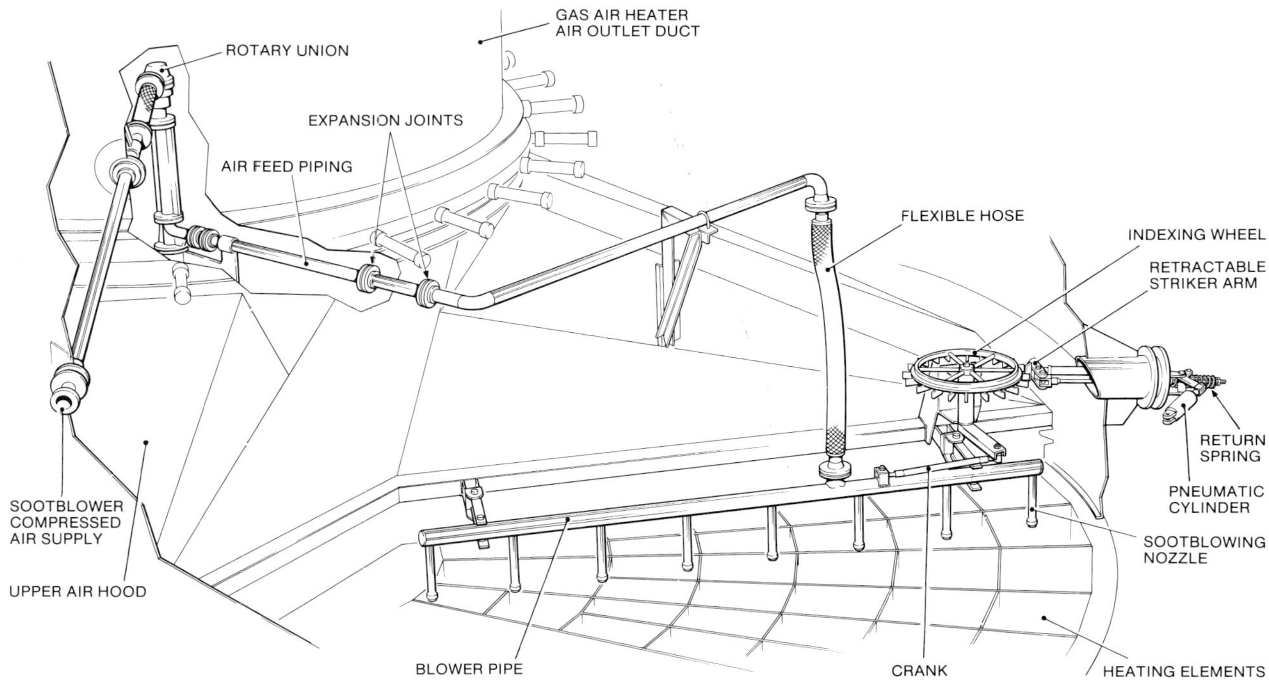

FIG. 3.12 Reciprocating lance sootblower arrangement for Rothemuhle airheater

(d) Inadequate sootblower lance travel and blowing medium conditions off-design.

(e) Poor setting of seal surfaces.

With regard to item (e), techniques are available for estimating rotor or stator distortions using capacitive or inductive sensor devices, thus allowing more accurate setting of seal surfaces.

3.7 Thermal design

The design of an airheater is based on thermal performance data derived from laboratory rig tests on clean elements or standard design parameters for plates and tubes. In addition, allowances are made for the following, all of which can contribute to a shortfall from the ideal:

- Flow maldistribution.
- Flow bypassing the heating surface.
- Fouling and corrosion.
- Heat losses from the structure.
- Changes in fluid properties.
- Distribution of cross-leakage.
- Manufacturing and erection tolerances.
- Pressure losses through transition ducts and flues.

In designing the airheater, each manufacturer makes allowances based on its experience, but it is advisable for the users to satisfy themselves as to the accuracy of the allowances or experience factors and the validity of the rig tests of the heat transfer surface by carrying out either themselves, or by an independent source, both laboratory rig trials and site tests.

The reference material[4] provides sufficient information on the fundamental thermal and hydrodynamic relationships used for the design of the heat exchangers, but the following relationships are established from the clean plate laboratory rig tests for regenerator heat transfer surfaces:

Stanton number $St = C\,Re^{-n}\,Pr^{-0.67}$

Fanning friction factor $f = D\,Re^{-p}$

where $$St = \frac{h}{G\,Cp}$$

$$Pr = \frac{Cp\,\mu}{k}$$

$$Re = \frac{G\,dh}{\mu}$$

The constants C and D and the indices n and p are established in the rig tests [5]. See also Section 7 of this chapter for nomenclature.

The thermal performance is normally expressed in terms of the convective heat transfer coefficients on the air and gas sides, suitably combined to give an ideal overall heat transfer coefficient.

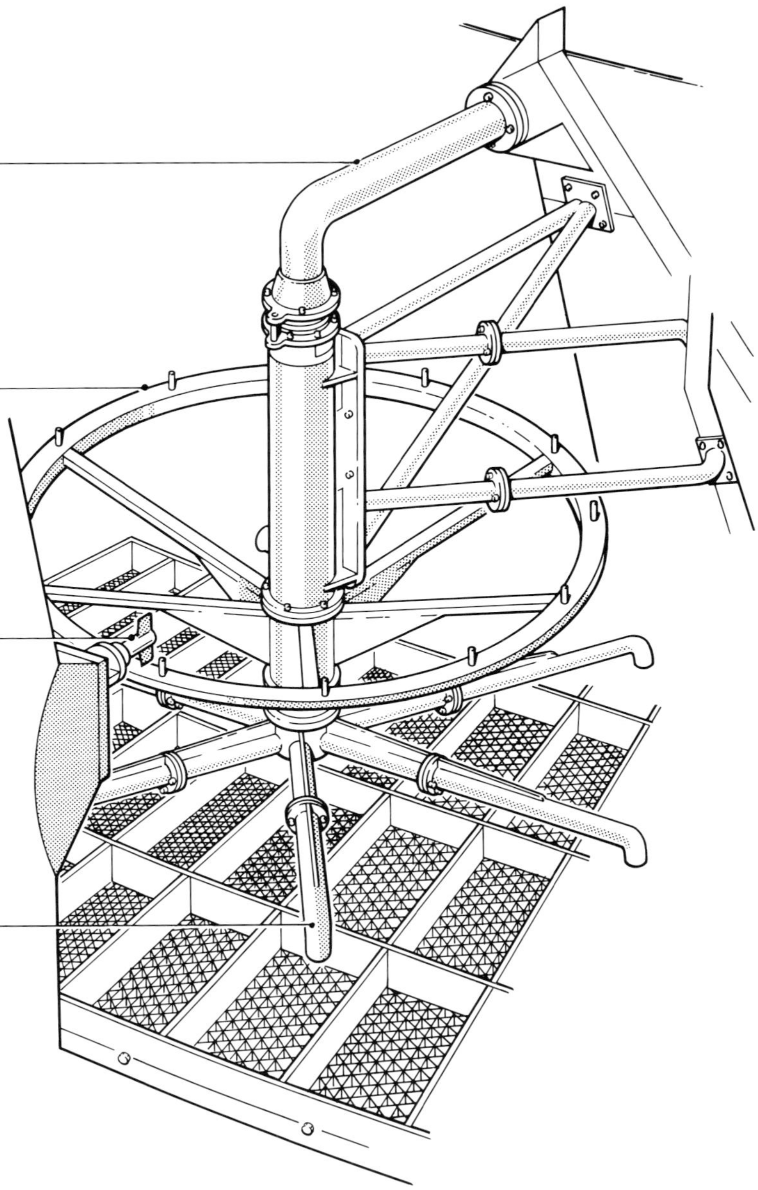

FIG. 3.13 Escapement wheel sootblower arrangement for Rothemuhle airheater

In the case of the regenerator, a further correction is made to cater for non-linearities in the time/temperature relationship at the inlet to the regenerator [6] which reduces the ideal overall heat transfer coefficient by 2–7% dependent upon size, rotor speed and heating surface.

Once the heat transfer surface characteristics and experience factors are known, not only can accurate assessments of the required size of an airheater be predicted and checked, but assessments can also be made of the effect of possible changes to the heat transfer surface [7]. In both cases the use of computer techniques considerably speeds up the process of selection and prediction [8, 9].

For regenerative airheaters, a wide range of airheater element profiles are available from different suppliers and numerous combinations of hot end and cold end elements have been tried in CEGB power stations. The type installed in any particular airheater is the necessary solution to operational problems or efforts at improving boiler efficiency. Typical examples of current types are shown in Fig 3.14.

The choice of element type should be based, not only on the predicted performance, but also on carefully controlled site trials or previous satisfactory operation with fuels of similar characteristics.

3.8 Airheater fire detection

Fires do not occur in clean airheaters. The most serious risk is during periods of low load or poor combustion

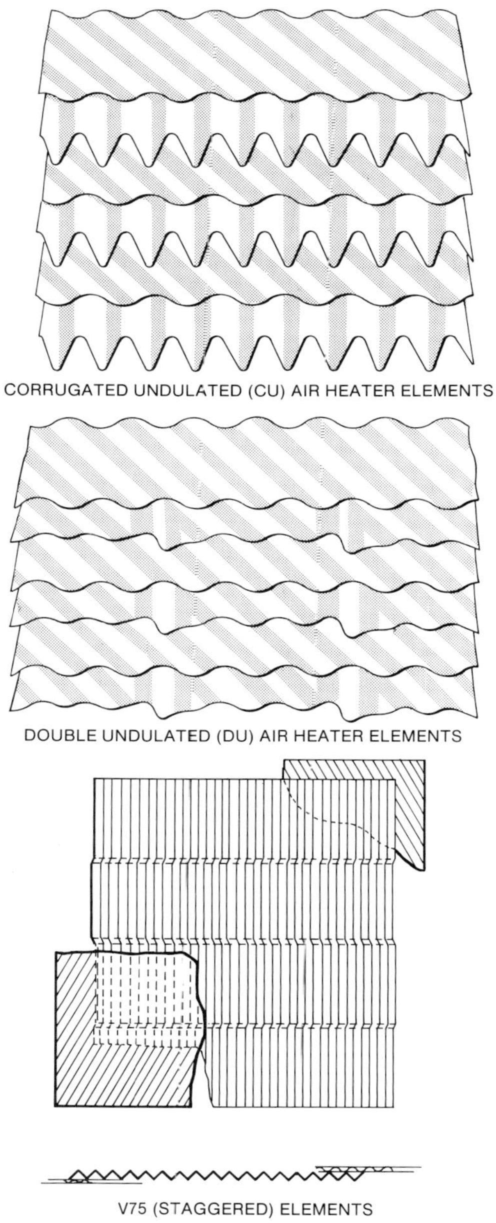

FIG. 3.14 Typical regenerative airheater elements

conditions particularly when oil firing or commissioning. When finely-divided carbon deposits are left on the heat transfer surface with the boiler in a boxed-up condition a local fire can occur, but there is normally a considerable delay in the fire becoming established and spreading.

Fire detection equipment using multiple thermocouples on the latest airheaters ensures that the whole of the heat transfer surface can be monitored at all times with alarms in the Central Control Room.

Set point reference temperature systems are not recommended as the reference temperature of necessity must be higher than the fluid temperature at normal operating conditions, hence under boxed-up conditions the alarm would not be initiated until the fire is well established. Thermocouple probes should be environmentally tested to ensure their integrity in the abrasive and acidic flue gas.

3.9 Regenerative heat exchangers in the flue gas desulphurisation process

Reference should be made to Chapter 4 Section 12 for a full explanation of the flue gas desulphurisation (FGD) system. However, in the wet processes, cooling of the flue gas occurs which would result in drooping of a wet chimney plume. Current practice is therefore to reheat the gas stream. The reheating can be achieved with direct injection of steam into the flue gas or combustors which on small plant may be acceptable. However, it is now common practice to adopt a gas/gas reheat system which transfers energy from the incoming untreated flue gas to the cleaned gas flow. A rotary regenerator is normally used, similar in concept to an airheater, which offers a compact heat exchanger with a low cost replaceable heating surface. Plastic materials are required in the low temperature end of the heat transfer matrix, with an enamelled steel hot end.

The two disadvantages of the regenerator are the leakage of untreated gas to treated gas through the seals and the need to run the flues close together in counterflow. The latter can be a major factor in retrofitting FGD to existing plant. In an effort to overcome these difficulties, booster fans can be positioned in the treated cool gas stream whereby treated gas leaks into the untreated gas stream, but the fans suffer from severe corrosion. Liquid-coupled reheaters would overcome both the disadvantages but are larger and require a pressure boundary.

4 On-load cleaning of boilers

4.1 Introduction

Sootblowers are provided to remove combustion deposits from the boiler surfaces and ensure effective heat transfer to the steam. Steam is used almost exclusively as a blowing medium in Europe, but many air installations are in use in the USA and the CEGB has pursued a policy of air sootblowing on new boilers since 1964.

4.2 Types of sootblower

There are several types of sootblower:

- Furnace wall sootblowers using steam or air.
- Furnace wall sootblowers using water.
- Long travel sootblowers for superheater, reheater and economiser passes.

- Multi-nozzle, semi-retractable lance sootblowers for economisers.
- Rake sootblowers for finned tube economisers.
- Sonic sootblowers.

An example of the furnace wall sootblower using steam or air is shown in Fig 3.15. When required, these sootblowers rotate forwards into the furnace, and the air or steam is turned on as the sootblowers pass through the wall tubes. The total travel is about 200–250 mm. Dependent on the design and the number of nozzles in the lance, the lance turns between an arc of 120° and several complete turns until a limit switch reverses the drive motor and retracts the lance. The cleaning radius is 1.5–2 m.

Furnace wall sootblowers using water, of which there are comparatively few in service at present, are used to clean tenacious or molten slag from furnace walls by means of a comparatively low velocity jet. In this case a variable speed motor rotates the lance forward to the initial blowing position, at which point the water is automatically turned on. The variable speed motor ensures that there is a uniform dwell time on the wall tubes and optimises the use of water. At full travel the water is cut off and the lance retracts at full speed. Total travel is approx 0.75 m. The water is supplied to the lance by a flexible tube which eliminates the need for a feed tube. The cleaning radius is up to 3 m. Water pressure varies from 10 bar to 20 bar. Water lancing is restricted to removing deposits in the furnace and surfaces containing steam at or below saturation state. The use of water lancing to remove tenacious deposits from furnace wall tubes would be expected to produce thermal fatigue cracks in the tubes, and this is the case. However, CEGB experience is that cracks up to 0.8 mm can occur but that the rate of crack growth will slow down as the water cleaning cycle continues. At present it is considered that cracking in combination with tube wastage by corrosion will give only a small reduction in tube life.

Long travel sootblowers are used to clean the superheater, reheater and sometimes the economiser passes, see Fig 3.16. When the blower is initiated, the lance moves forward and rotates simultaneously so that the opposed lance jets describe a helical motion as shown in Fig 3.17. When the jets have cleared the boiler wall, the steam or air valve is opened by links connected to the travelling part of the blower. The blowing medium passes through the inlet valve, the feed tube, the lance and the nozzles. The nozzles describe the helical path in the forward direction and when the lance is fully extended it turns through 90°, thus tracing a helical path in the reverse direction which bisects the helical path in the forward direction. Hence

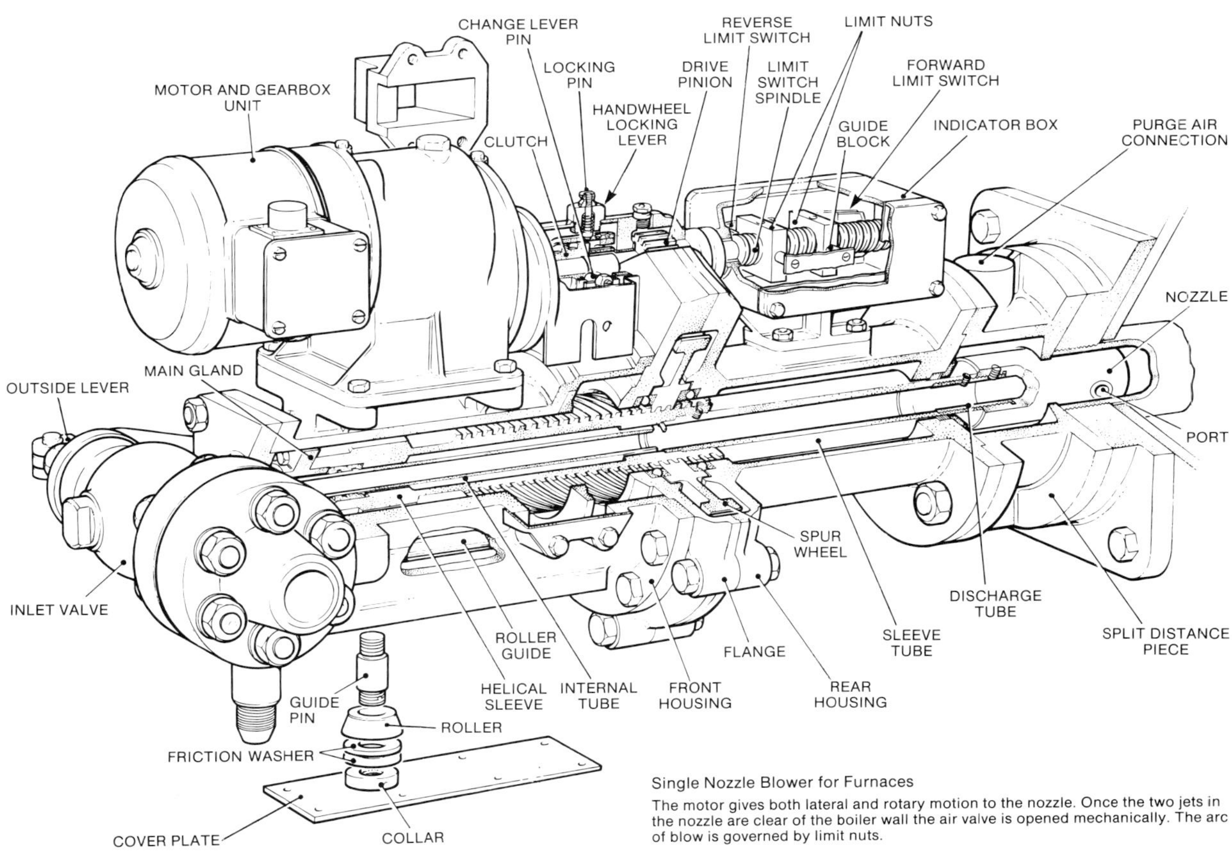

FIG. 3.15 Furnace wall retractable sootblower

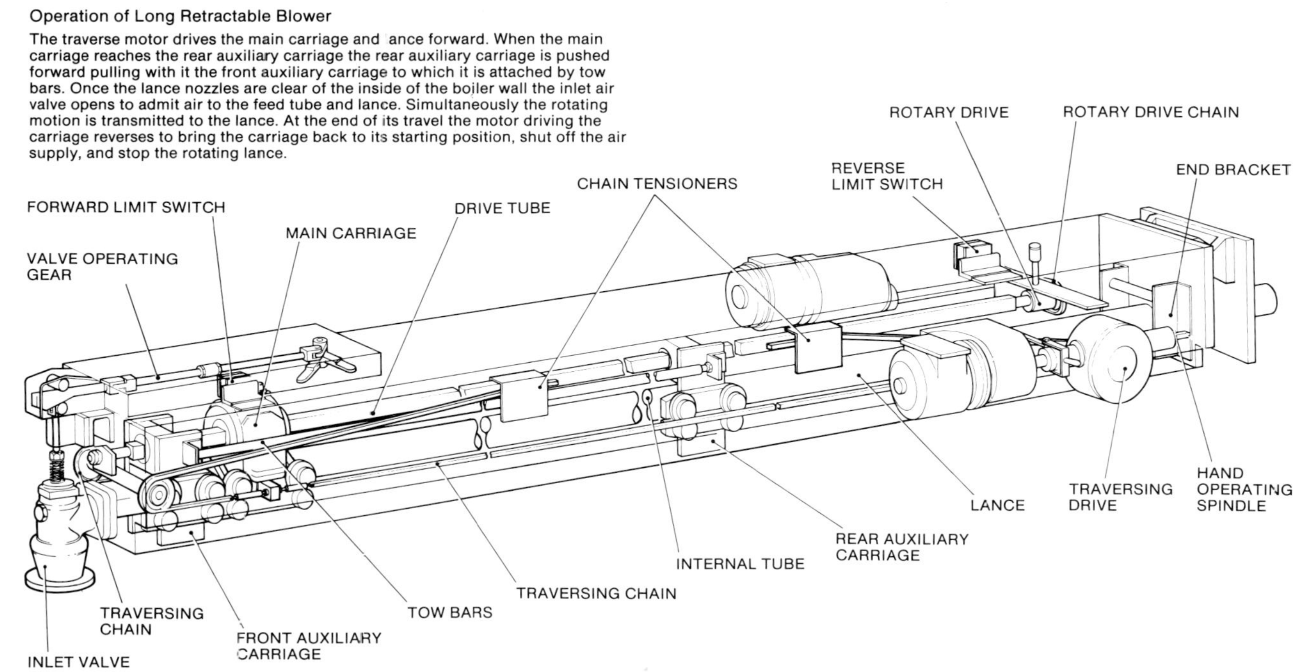

FIG. 3.16 Long retractable sootblower

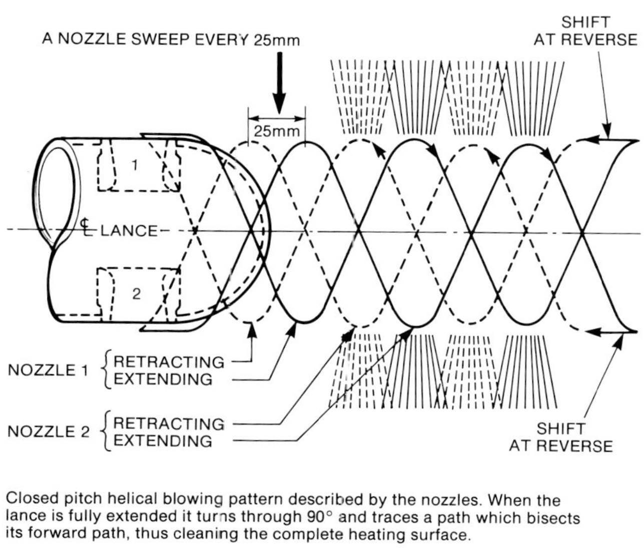

FIG. 3.17 Sootblower closed-pitch helical blowing pattern

the complete heating surface covered by the blower is cleaned. The speed of travel of the lance varies between 2 m/min and 6 m/min dependent on the surface to be cleaned. In general the hottest zones require the most cleaning. Improvements in design and materials now allow the use of unsupported lance lengths up to 16.75 m. This in turn permits the boiler designer to increase boiler widths to 33.5 m. The increased lance length in the hottest part of the boiler requires considerably more blowing medium, particularly air, than is necessary for cleaning, just to keep the lance cool. Water tempering can be used to reduce the blowing medium requirement. The relatively small amount of water injected forms a film along the lance tube wall and increases the inside heat transfer coefficient. A typical water flow would be 18–20 litres/min when blowing air. The effect on the boiler tube metal is similar to that described for water lancing.

Multi-nozzle semi-retractable lance sootblowers are employed on economisers. Half the lance remains permanently in the gas stream and when operated it moves into the boiler in the same manner as the long travel lance. These sootblowers can be built so that the lance oscillates through an arc of up to 120° instead of rotating, thus concentrating the blowing medium on the tubes when positioned on the top or bottom of the economiser tube bank.

Finned-tube economisers were sometimes fitted with rake sootblowers. It has been found in practice that finned-tube economisers can be effectively cleaned by long travel sootblowers and rake sootblowers are not now fitted to new CEGB boilers. A wallbox is provided in the boiler casing for the sootblower lance to pass through. This wallbox is designed to be fed with air to seal the boiler gases from the end of the lance.

Sonic sootblowers are on trial in a variety of positions on CEGB boilers. The device is a low frequency (20 Hz) sound generator designed to clean deposits from the boiler, economiser, airheater and precipitator. The device is driven by compressed air. Infrasound is generated at resonant frequency, producing waves. It is claimed that these waves will clean enclosures up to 4900 m^3. A few of the trials have been successful but others have not, and it is too early to predict the extent to which sonic sootblowing will be used, although it could prove a valuable addition to the available cleaning methods.

4.3 Sootblower design and installation

4.3.1 Lances

Lances operating in the high gas temperature zones of the boiler are usually made of 2.25% Cr 1% Mo or 1% Cr $\frac{1}{2}$% Mo steel. These materials have inadequate strength when subjected to the highest temperatures and 12% Cr steel is often used in these zones. Operation of 12% Cr above 800°C results in excessive material wastage and above 840°C a change of phase occurs making the material hard and brittle; 800°C can therefore be regarded as its working limit.

The total stress on a long travel lance is the static bending stress plus a dynamic component due to lance rotation. This cyclic loading is at the frequency of rotation of the lance and has an amplitude of plus or minus 15% to 20% of the static loading.

Loss of blowing medium when the lance is inserted results in rapid increase in temperature and unless the lance is retracted immediately, failure often results. If a blower comes to rest during its cleaning cycle, it must be retracted immediately; a static lance supplied with steam or air will erode any tubes in the path of the jets, mainly because the jets entrain gas and dust from the surrounding gas.

4.3.2 Nozzles

Although many lance nozzles have a convergent/divergent profile which would superficially appear to be satisfactory, their performance is often poor. This is because the nozzles have too sharp an inlet radius for the flow as it turns through 90° over the nozzle lip and the flow separates. The resultant turbulence and unsteadiness in the initial region of the jet leads to rapid dissipation of the jet and reduced cleaning effectiveness.

In work at Marchwood Engineering Laboratories, the nozzle inlet profile shown in Fig 3.18 was shown to prevent separation, provided the ratio of the approach area, Aa in Fig 3.19, to the total throat area (i.e., of both nozzles) was greater than 1.8 [10]. At the pressures commonly employed for sootblowing, full expansion of the flow demands an exit-to-throat-area ratio of between 1.8 and 2.7, but it is unusual to be able to accommodate such nozzles because of their length. Use of the recommended inlet profile and area ratio criterion usually enables at least a short conical discharge section with a divergence half angle of 10° to be installed.

Tests show that with an exit-to-throat-area ratio of 1.5 there is little loss in performance incurred

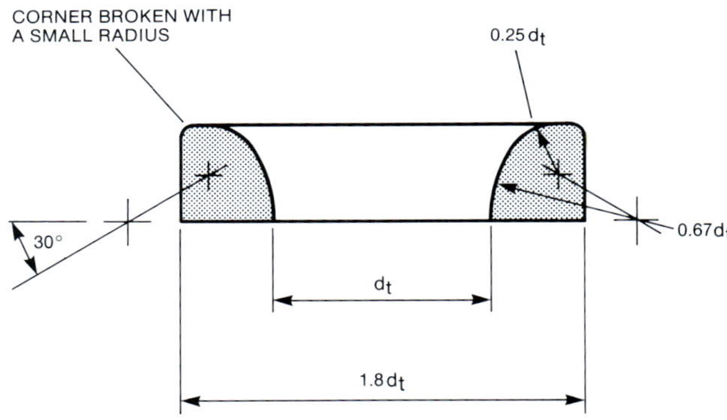

FIG. 3.18 Inlet nozzle geometry

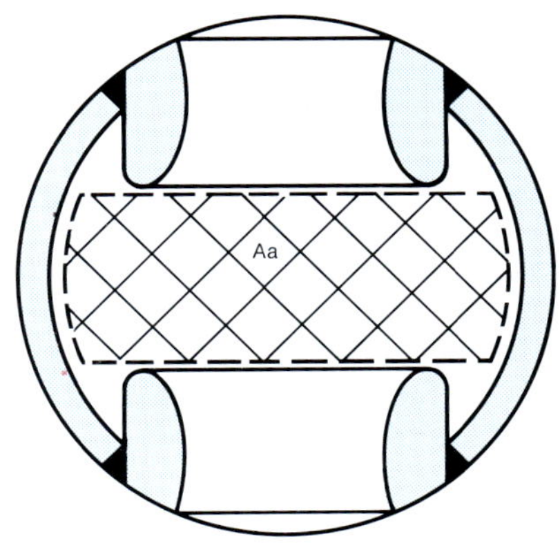

The nozzle inlet profile shown prevents separation provided the ratio of the approach area Aa to the total throat area is greater than 1.8.

FIG. 3.19 Section through typical sootblower tip

even when a ratio of 2.5 is required for full expansion. In situations where the lance tube is small for the size of nozzles, it is always better to accept little or no divergent section rather than compromise the nozzle inlet flow. If the inlet flow is compromised, the divergent section will not function properly anyway.

4.3.3 Sootblower installation on a 660 MW coal-fired boiler

The location and number of sootblowers installed in a 660 MW boiler burning coals with adverse slagging and fouling characteristics is given in Table 3.3 and shown diagrammatically in Fig 3.20. The table shows the duration of operation and air consumption of each sootblower group. Not all areas of the boiler require sootblowing every shift and the blowing cycle can be shortened to provide only essential cleaning with a saving in air or steam consumption. The frequency of operation of the sootblowers depends on the constituents of the fuel being fired and the efficiency of combustion.

TABLE 3.3
Typical sootblowing data for 660 MW coal-fired boiler

SOOTBLOWER LOCATIONS	TOTAL NUMBER PER BOILER	LENGTH OF LANCE metres	OPERATING PATTERN	DURATION OF OPERATION OF EACH BLOWER minutes	DURATION OF OPERATION OF GROUP minutes	AIR CONSUMPTION RATE PER BLOWER dm^3/s
Furnace	42	0.21	9 quads 2 pairs 2 singles	2.3	30	944
Superheater	24	12.8	2 quads 8 pairs	14	140	8 at 1652 6 at 1133 10 at 1038
Reheater	20	12.8	3 quads 4 pairs	14	98	6 at 1133 14 at 1038
Economiser	16	12.8	4 quads	14	56	1038
Air Heater	4	6.6	2 pairs	30	60	1355

Note: Total length of complete blowing cycle 384 min.

4.4 Choice of sootblowing medium

Until comparatively recently steam was used exclusively for sootblowing. Air was first used in the USA. In 1964 the first CEGB air sootblowing installation was ordered for the Rugeley *B* boilers and all boilers ordered since then have been equipped with air sootblowing.

The advantages and disadvantages of steam and air sootblowing are as follows.

4.4.1 Steam

Advantages

- The capital cost of steam pressure-reducing equipment and drains is less than the cost of compressors, motors, switchgear, and controls for an air system.
- The incremental capital cost of the boiler and water treatment plant is not significant.

Disadvantages

- Maintenance of the sootblowers, pipework, pressure-reducing equipment and drain pipework and valves is extensive due to thermal shock, condensate carry-over and corrosion.
- Effective sootblowing is not available at start-up and shutdown. This is particularly important with oil-fired boilers where airheater sootblowing is essential to avoid fires.
- Sootblowing is not available quickly due to the need to warm and drain the pipework.
- The use of auxiliary boilers for steam sootblowing is not attractive as the boilers must be of large capacity to withstand the size and instantaneous nature of the steam demand.

4.4.2 Air

Advantages

- Effective sootblowing is available at any boiler load and particularly on start-up and shutdown with oil-fired boilers.
- Condensate drains from individual sootblowers are not necessary, although the air pipework must be well drained to avoid corrosion from the moisture in the air.
- Maintenance of sootblowers is considerably reduced due to the absence of condensate, thermal shock and erosion on the sootblower mechanism, lance and nozzles. Gland repacking is reduced to a minimum.
- No warming through of the air mains is required.
- Insulation of the pipework is not required.
- The incidence of boiler tube erosion is significantly reduced, with consequent reduction of replacement generation costs.

Disadvantages

- The capital cost of the compressors, motors, switchgear, cables and compressor house exceeds that of the steam pressure-reducing system, etc.
- Maintenance of the compressors and controls. In general the compressors which are of screw or centrifugal design require a high standard of maintenance due to the fine clearances on which the efficiency

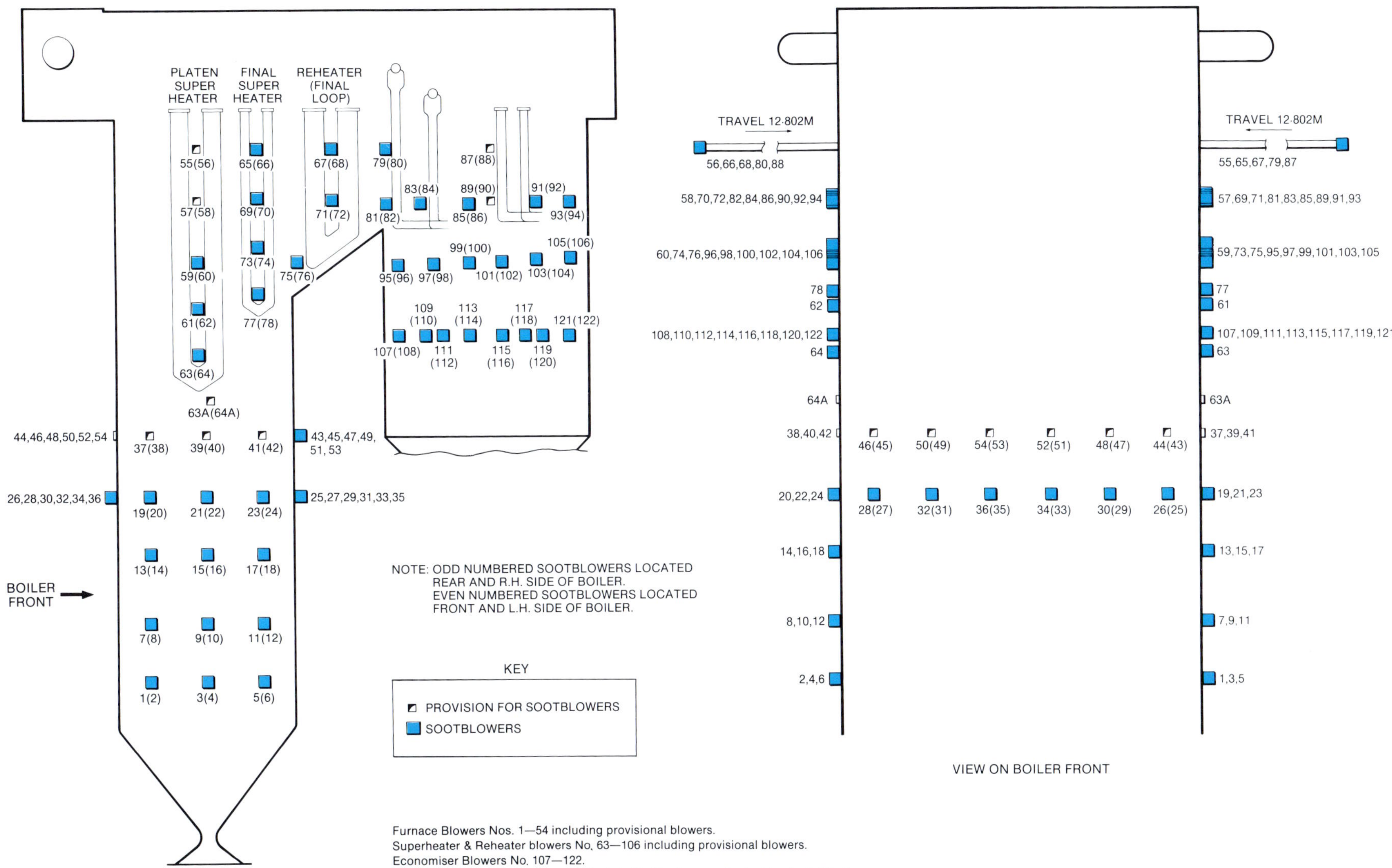

Furnace Blowers Nos. 1—54 including provisional blowers.
Superheater & Reheater blowers No. 63—106 including provisional blowers.
Economiser Blowers No. 107—122.
The airheaters and blowers are not included in the diagram. Each airheater is fitted with one sootblower on top of the matrix and one on the bottom.

FIG. 3.20 Sootblower locations for 660 MW coal-fired boiler

of the compressors depend. Maintenence on the compressors and sootblowers is less than that required on a comparable steam system.

4.4.3 Operating costs of steam and air sootblowing

Steam

Operational data from 500 MW boilers indicate that approximately 3.1 kg/s of steam is required for sootblowing. This sootblowing steam only makes a heat contribution to the boiler of about 0.06% of the total steam heat input, as most of the steam remains in the superheated state in the flue gas due to its low partial pressure. The net increase in boiler heat input during sootblowing is about 0.7%. With a blowing time of 4 hours per 8 hour shift, the average additional heat consumption is about 0.35%.

Compressed air

For one 500 MW boiler, a compressor of 3.3 m^3/s capacity at 17.25 bar is suitable for sootblowing; this will absorb about 1870 kW. Because this air is cooled, to cool the lances, none of this compressor power is recovered as useful heat.

Taking a 6 hour blowing cycle per 8 hour shift, the power consumption amounts to about 0.3% of the net unit output.

The similarity of the above ratios for steam and air as blowing media show that running costs are similar and the choice of medium has to be made on differences in maintenance costs attributable to the two media; the differential cost of compressors, pipework, etc., for the air blowers and the steam system for steam blowers, and the capitalised cost of loss of generation due to tube failures caused by sootblower erosion.

It is the CEGB's experience that air sootblowing has an overall economic advantage over steam and it is the policy to provide air sootblowing on new boilers. Investigations have shown, however, that there is no economic case for converting existing stations from steam to air blowing due to the stringent financial returns required for optional schemes of this nature.

4.5 Steam supply system

Steam is normally taken from the cold reheat pipes. While the pressure at this point is adequate at loads between 80% and 100% CMR, at lower loads the pressure is too low for effective sootblowing and another supply, usually from the outlet of the primary superheater, is provided.

A constant blowing pressure is maintained in the sootblower pipework system by pressure control valves. These valves are normally duplicated to allow on-load maintenance; the instantaneous opening and closing of the steam admission valves to each sootblower causes rapid wear on the control valves and periodic maintenance is required to avoid pressure surging in the system.

Adequate draining of the entire sootblower pipework is essential to prevent corrosion and damage to sootblower glands, feed tubes and nozzle assemblies. Thermostats operating drain valves must be correctly maintained and calibrated at all times.

4.6 Compressors for sootblowing

Air compressors for sootblowing are of screw and centrifugal design. Both are effective but the characteristics of these machines are completely different.

The screw compressor is a rotary positive-displacement machine. Basically it comprises two rotors, one having convex lobes formed helically along its length, whilst the other has corresponding concave flutes. Compression occurs as air is drawn into the inlet port and trapped between the rotor lobes and the casing. The air is progressively reduced in volume as the rotors rotate until it is expelled through the discharge port. The compression is very nearly adiabatic and the machine adjusts itself to the pressure existing in the system.

Figure 3.21 shows the power and volume of a screw compressor operating at fixed speed with a fixed suction pressure on a basis of discharge pressure. These characteristics give the screen machine flexibility and operational power saving.

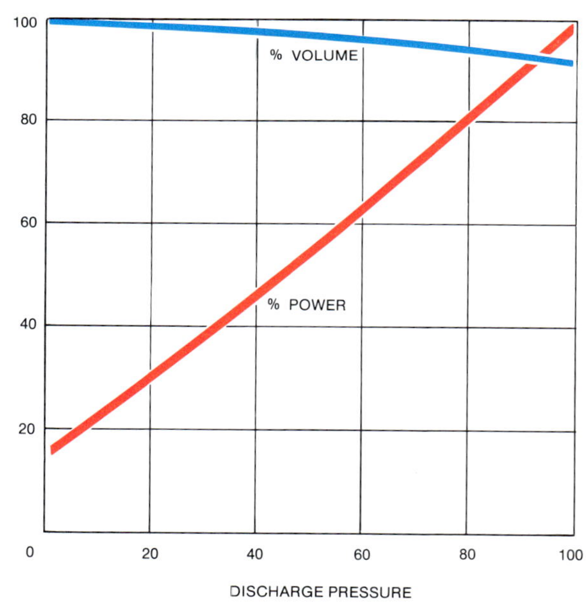

The graph shows the power and volume of a screw compressor operating at fixed speed with a fixed suction pressure on a basis of discharge pressure. These characteristics give the screw machine flexibility and operational power saving.

FIG. 3.21 Power and volume characteristic of a screw compressor

The rotors are fitted with precision-cut helical timing gears, see Fig 3.22. Due to the geometry of the rotor profiles and their meshing characteristics, the male rotor absorbs 85% of the compression power and the female rotor only 15%. As the drive is applied to the male rotor the timing gears absorb only 15% of the compression power. The timing gears ensure that the rotor lobes and flutes do not come into contact. As a result of this, no lubrication of the compression chamber is necessary and completely oil free compression is achieved.

Shaft seals are fitted at the inlet and outlet ends of the rotor bodies to ensure that the air being compressed is contained in the casing and that the rotor bearings are completely isolated from the compression chamber. Each seal comprises a number of carbon sealing rings contained in a sleeve and separated by lantern rings and spacers. The carbon rings each have a steel backing ring which is shrunk on. The interference between the steel and carbon is designed to ensure that the temperature characteristics of the bore of the seal match the expansion rate of the rotor shaft, thereby maintaining seal efficiency under all operating conditions.

The compressor casing incorporates a water jacket which surrounds the compression chamber. Cooling water, from a closed circuit cooling water system, enters the bottom of the casing and is discharged through a similar connection in the top of the casing.

Two-stage dry screw compressors are currently in use for sootblowing. However, should an increase in air pressure from the present 13 bar (g) be required then a three-stage machine will be required.

Control of the dry screw machine is by the opening of the automatic unloading valves situated in the interstage and final discharge pipework. The valves are arranged to ensure that the second stage compressor unloads before the first stage and that on reloading the first stage loads before second.

Figure 3.22 shows a section through a one-stage screw machine.

The centrifugal compressor (Fig 3.23) is an aerodynamic machine. It carries out the process of compression by generating a kinetic head in the air as it passes through the spinning impeller and converting the kinetic energy in the air to pressure energy in the diffuser. The air is then cooled and drained of moisture before passing to the second and subsequent stages. For sootblowing, five-stage centrifugal compressors are used. The discharge pressure capability of a centrifugal compressor is therefore directly related to the diameter of the impeller, its speed of rotation and the density of the gas it is handlng. As there is a clear air connection through the compressor from discharge to the suction via the impeller, changes in discharge conditions have a direct effect on the suction volume. There are no suction or discharge valves involved. Similarly there are none on a screw compressor.

Initial control of the compressor is by inlet throttle valve, this covers down to about 80% load. Below this load the unit automatically unloads by blowing off until the system pressure falls and permits a return to full load.

The performance curve in Fig 3.24 shows the capacity, pressure and power relationship. It will be noted that the dotted line represents the stability or surge limit at part load beyond which the compressor becomes unstable.

This instability arises when the pressure volume curve droops towards the zero capacity region after reaching peak pressure. When the capacity is reduced below the peak, the pressure in the discharge line exceeds that produced by the machine and the flow tends to reverse momentarily. As soon as the flow reverses, the system discharge pressure drops and the compressor resumes normal flow. This reversal of pressure and flow continues at high speed and sets up noise, vibration and high temperatures in the air which can damage the compressor if permitted to continue. A surge protection device is provided to give an alarm and unload the compressor. One side of the device measures the system pressure, the other side measures the same pressure but is restricted by a non-return valve and variable orifice. When surge occurs the pressure in the discharge line will fluctuate, but the variable orifice restricts the air flow giving the differential pressure needed to activate the compressor unloading control.

5 Feedwater arrangements and boiler fittings and mountings

5.1 Feed pipe arrangements

Figure 3.25 shows diagrammatically a typical boiler feed and circulation system for a modern boiler. The positions of the boiler circulation pumps are indicated for an assisted circulation design. For a natural circulation boiler, downcomer pipes would be installed without pumps or valves.

The economiser recirculating connections are for use during pressure raising or shutting down, to avoid temperature stresses or possible overheating due to the economiser not being full. This condition could arise if the feed flow is temporarily checked and some evaporation occurs in the economiser. The recirculating valves are closed when the boiler is in normal service.

The feed connections to the economiser are arranged to assist uniformity of water distribution to the tubes and the feed pipes from the economiser to the boiler drum are arranged to give a uniform distribution of feedwater in the drum.

5.2 Feed check valves

A typical feed check valve for a large modern boiler is shown in Fig 3.26. The inlet is under the valve

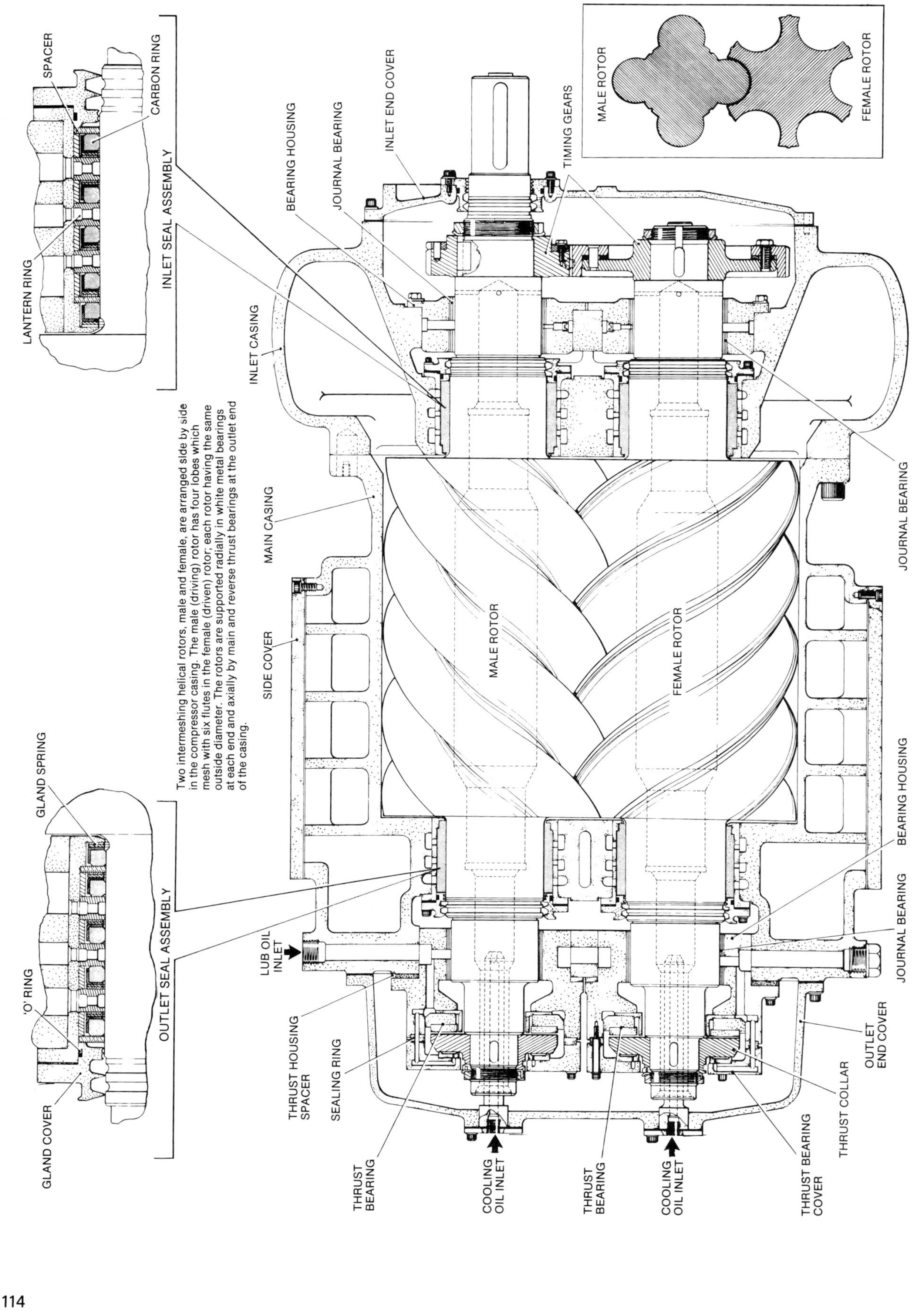

FIG. 3.22 Section through a one-stage screw compressor

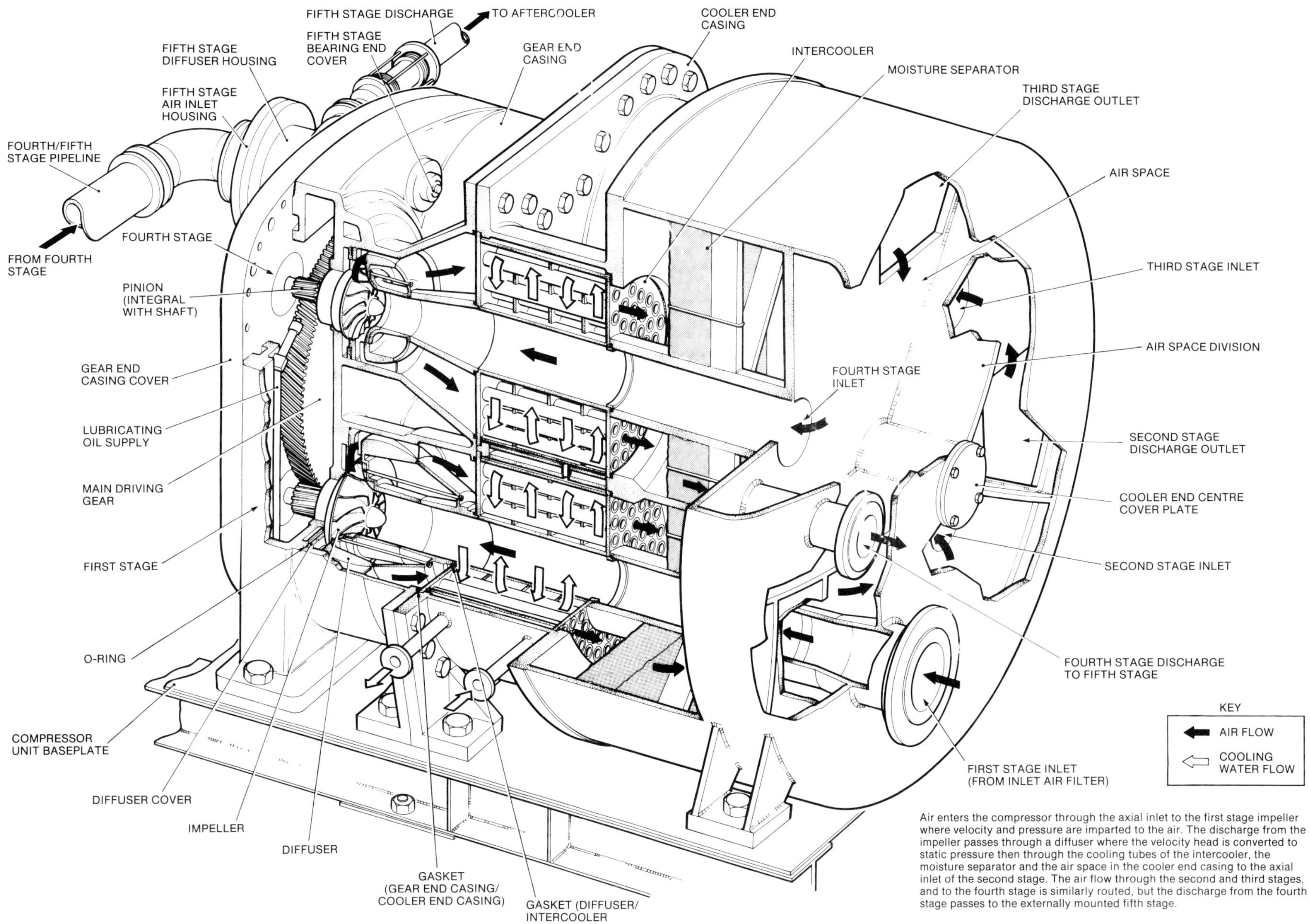

FIG. 3.23 Centrifugal compressor for sootblowing

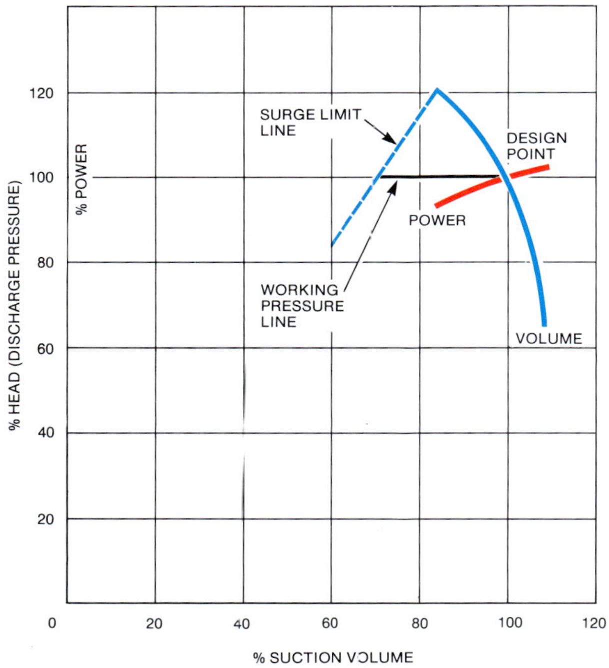

The effect of varying discharge pressure on suction volume is shown.

FIG. 3.24 Pressure, volume and power relationship at constant speed of centrifugal compressor

seat and the valve cannot be closed manually, but will close under a reversal of flow. The valves are sited at the economiser inlet. This type of valve is used where the feedwater regulating valves precede the high pressure feedheaters on the discharge side of the feedpumps and the avoidance of screwdown feed check valves allows the feedheater and subsequent pipework to be designed for boiler pressure instead of feed pump discharge pressure, with a consequent economy in cost. Combined stop and non-return valves are used in other installations where the feed regulating valves are on the discharge side of the heaters.

5.3 Fittings and mountings

To ensure the efficient operation of the boiler unit and to comply with the Factory Acts, the following fittings and mountings are specified for drum type boilers:

- An approved number of safety valves.
- An approved number of blowdown and drain valves for superheater, furnace walls, boiler and economiser.
- Air release valves, where necessary.
- Direct reading water level gauges and remote water level indicators.
- A high and low water alarm.
- Boiler water sampling equipment and the provision of connections on the steam drum for steam sampling.
- Main steam stop valves.
- Equipment for the injection of conditioning chemicals into the drum.

5.4 Safety valves

The number and size of safety valves fitted is such that, with the valves fully open, the maximum evaporation of the boiler can be discharged without a rise in drum pressure of more than 10% of the highest set pressure. The valves are fitted on the drum and superheater outlet, with those on the latter being sized to pass at least 20% of the maximum evaporation of the boiler to provide coolant steam and prevent overheating of the superheater tubes.

In selecting the set pressures for the drum and superheater safety valves the following factors have to be considered:

- The margin between normal operating pressure and set pressure to minimise lifting of the valve due to normal operational pressure fluctuations.
- Relative set pressures of drum and superheater valves to ensure that superheater valves lift before drum valves, and drum valves reseat before superheater valves.
- Reseat pressure should be higher than normal operating pressure.

Safety valves on the reheater must be sized to pass maximum reheater flow without a rise in reheater inlet pressure of more than 10% of the highest set pressure. Protection of the reheater tubes is afforded by having the valves with the lowest set pressure on the outlet side and with a capacity of not less than 20% of the reheat maximum flow.

The types of safety valve used are:

- Direct loaded safety valve — a safety valve in which the loading due to the fluid pressure underneath the valve disc is opposed by direct mechanical loading. The two types used are the spring type (see Fig 3.27) and the torsion bar (see Fig 3.28).
- Supplementary loaded safety valve — a safety valve which has, until the pressure at the inlet of the safety valve reaches a set pressure, an additional force (supplementary load) to increase the sealing force which may be provided by means of an extraneous power source which is reliably released when the pressure at the inlet of the safety valve reaches the set pressure; also it attains its certified

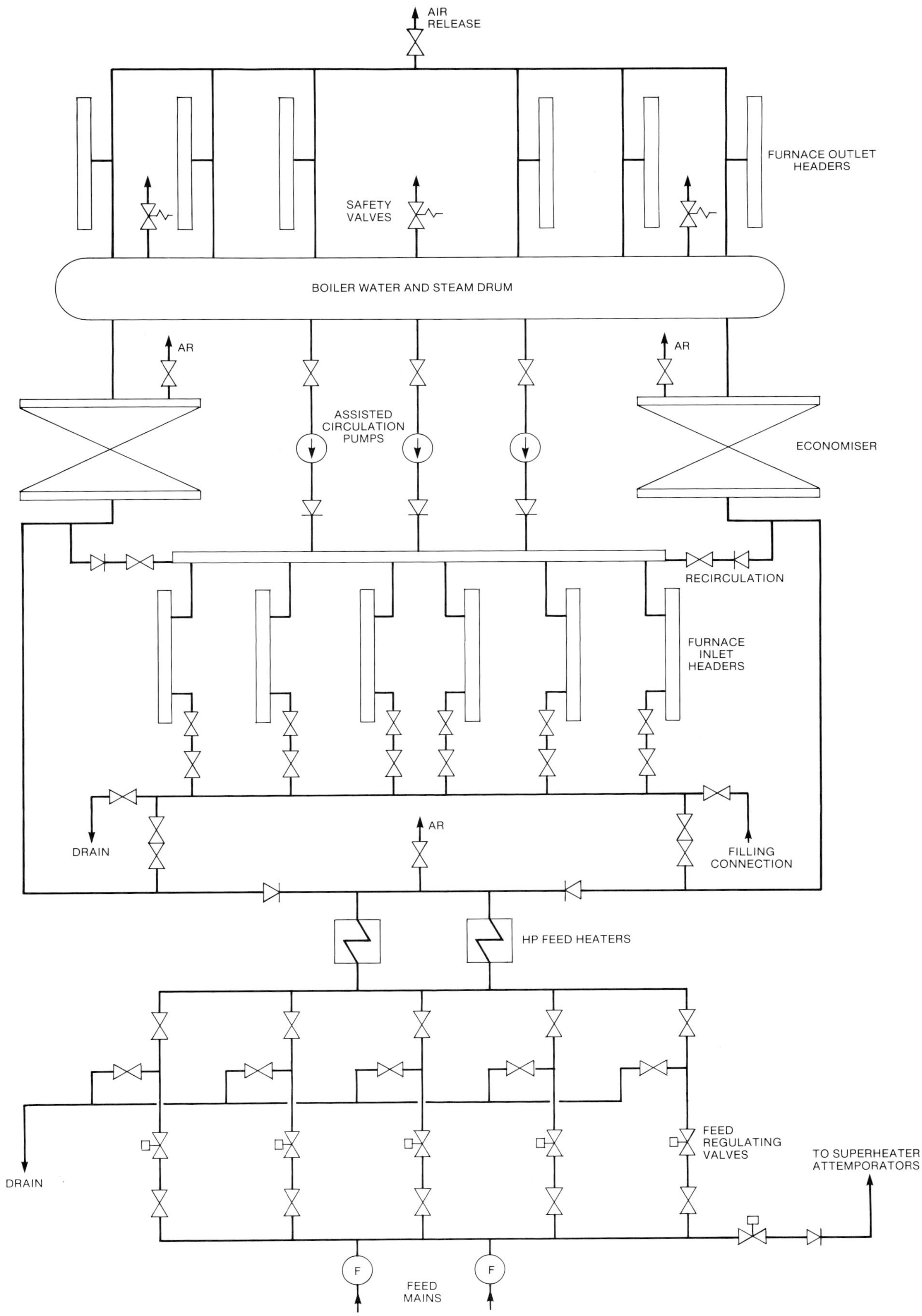

FIG. 3.25 Typical boiler feed and circulation system

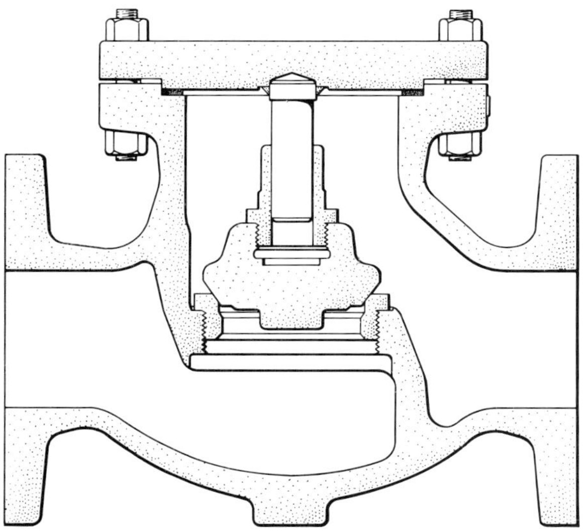

FIG. 3.26 Typical feed check valve construction

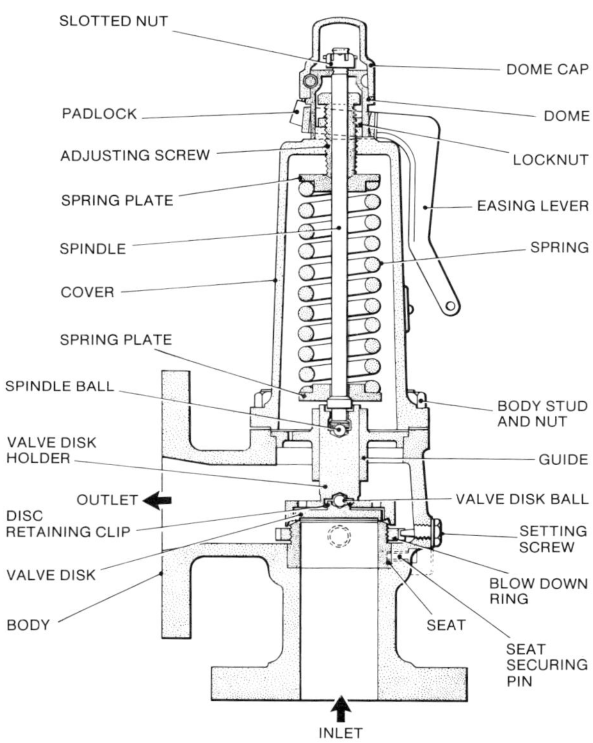

FIG. 3.27 Typical spring type safety valve construction

discharge capacity in the event of the supplementary loading not being released at an overpressure not exceeding 115% of the set pressure.

- A pilot-operated safety valve (indirect loaded safety valve) the operation of which is initiated and controlled by the fluid discharged from a pilot valve which is itself a direct loaded safety valve.

It is now CEGB practice on high pressure boilers to use supplementary loaded valves since these prevent simmering which can occur with direct loaded valves when the fluid pressure approaches the valve set pressure. Simmering results in steam flowing through a small area of opening and this often results in seat damage. The use of supplementary loaded valves permits the set pressure to be approximately 5% above normal operating pressure without incurring problems of inadvertent lift or simmering during operation. A typical supplementary loaded valve is shown in Fig 3.29.

The safety valve escape pipes are led to the nearest convenient point on the boiler house roof and where necessary incorporate silencers to minimise noise nuisance when the valve lifts. Provision needs to be made in the escape pipework to accommodate the relative expansion between the valve and the escape pipe. Drains are required to handle the condensed steam in the valves and piping.

5.5 Steam stop valves

The boiler main steam stop valves are located as near as practicable to the superheater outlet, allowing for accessibility.

A typical design of the parallel slide type of valve used is shown in Fig 3.30 and has a venturi throat to reduce the diameter of the valve discs and seats. This results in a small force acting on the valve seats and makes for easier operation together with reductions in the body diameter and thickness, spindle diameter, control gear, overall height, weight and cost of the valve. Such advantages serve to compensate for any increase in steam pressure drop over a valve with a constant bore.

The small spring between the discs keeps the discs on the seat when there is no pressure in the line and prevents small particles of debris coming between them. It is of the utmost importance that no scoring of the seats or discs should occur, or leakage will result. To prevent this, the discs and the seats have a facing of a very hard alloy deposited on to them. These faces are ground smooth and lapped so that at least 75% of the width of the faces are in contact all round the circumference of the face. The small spring has no effect in sealing the valve as this is maintained by the steam pressure within the body of the valve.

A small bypass valve, with interconnecting pipework between the inlet and outlet branches, is used to reduce the pressure differential across the main steam stop valve when opening. This reduces the starting torque and enables a smaller motor to be used. The valve internal fitments, however, are suitable for the full differential pressure. The main valve and the bypass valve are motor operated, with the opening or closing of the bypass valve being arranged in sequence with the main stop valve.

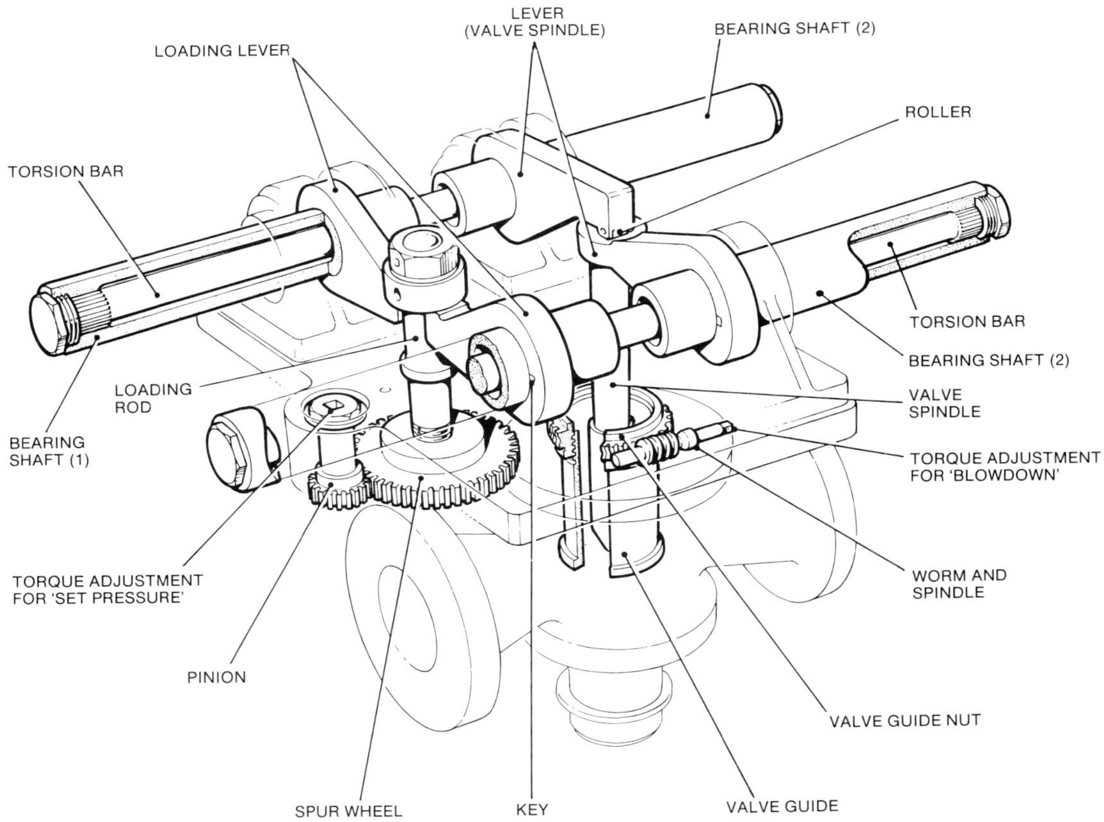

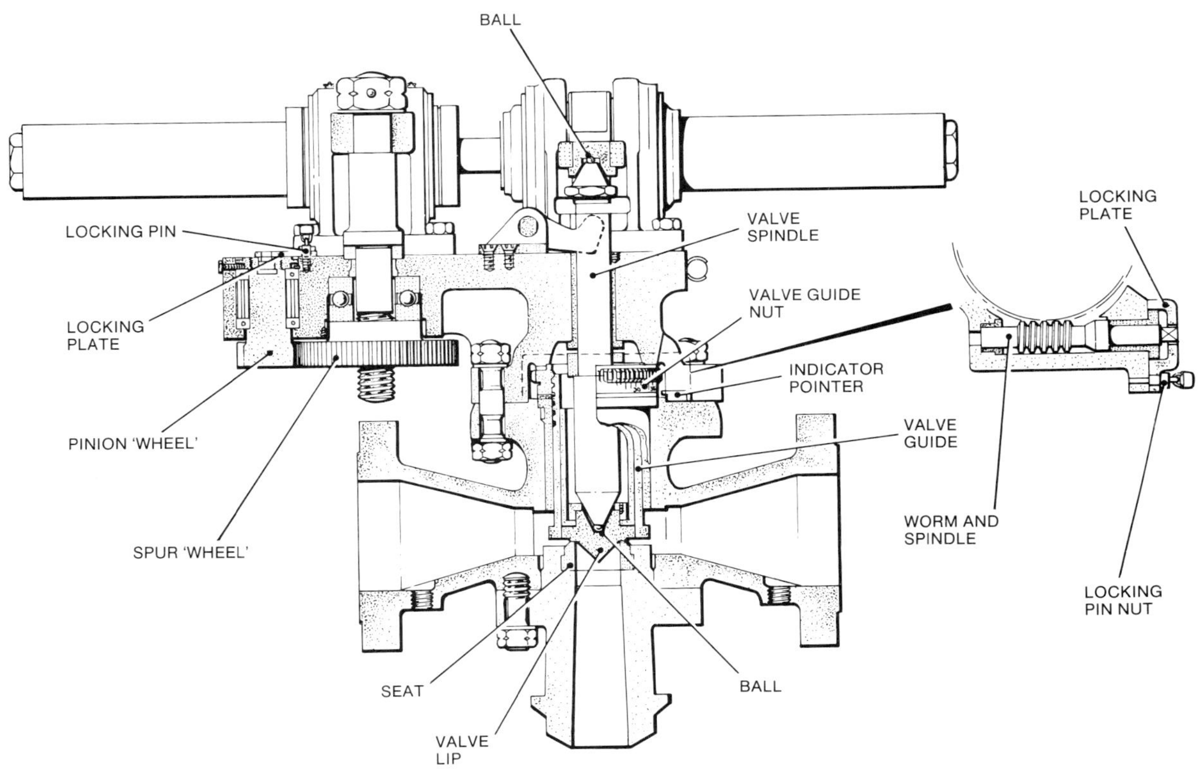

FIG. 3.28 Typical torsion bar safety valve construction

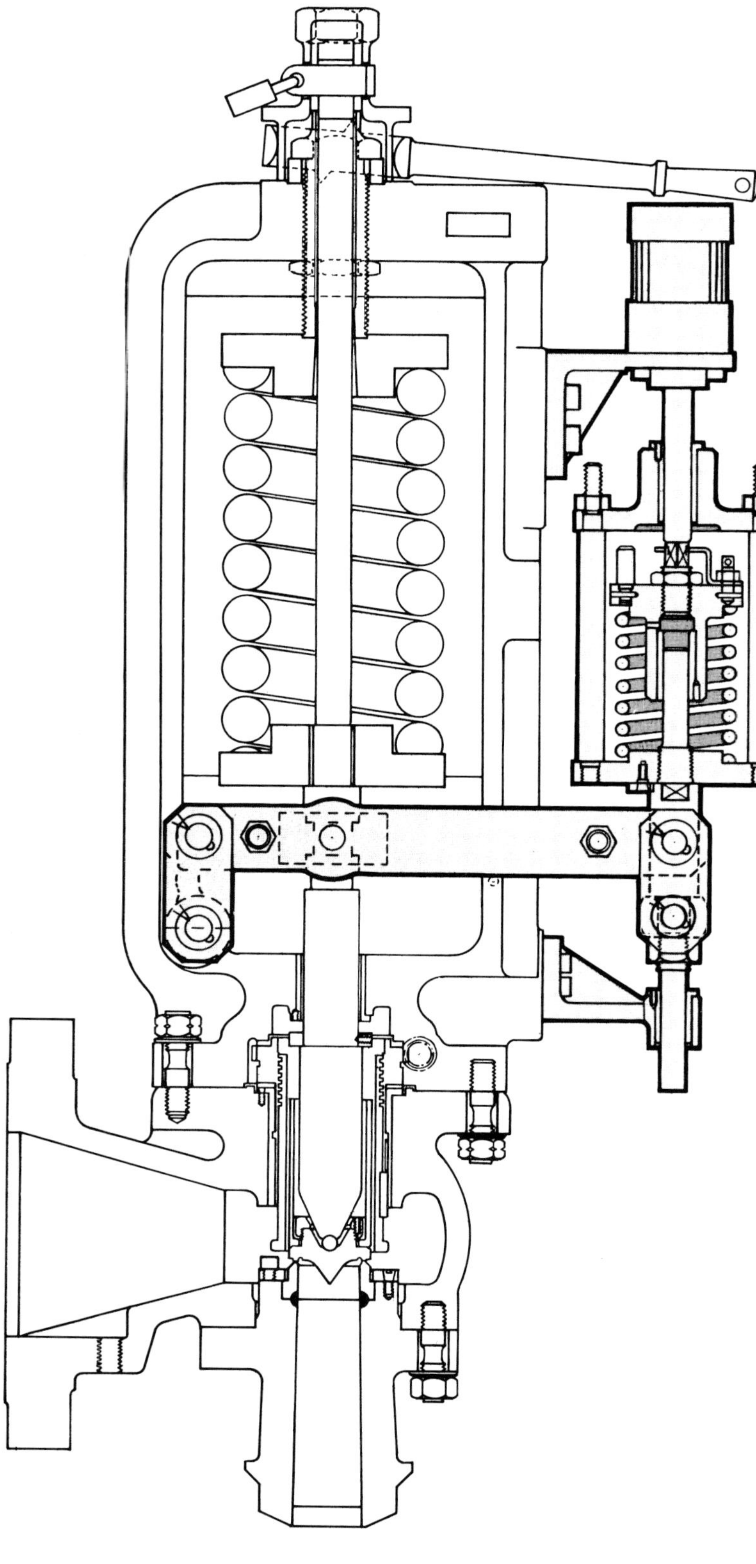

FIG. 3.29 Typical supplementary loaded safety valve construction

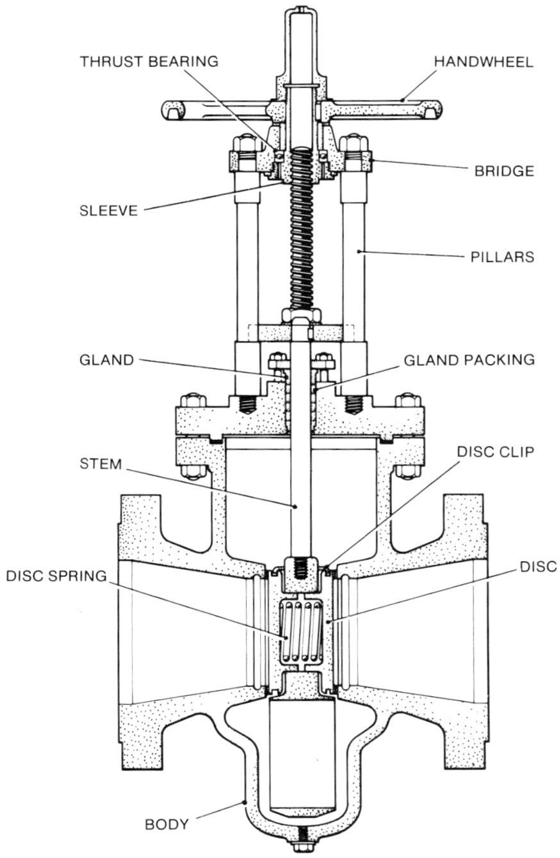

FIG. 3.30 Typical venturi parallel slide valve construction

5.6 Blowdown and drain valves

Blowdown valves are fitted at strategic points on boiler drums to discharge, as necessary, boiler water containing undesirable concentration of chemicals, so assisting in the control of the boiler water within the desired chemical limits. Drain valves are fitted to all points on the boiler from which the water is to be drained for emptying purposes. Those on the superheater are used to drain off water which accumulates on cooling down the boiler and are opened during steam raising to ensure a flow of steam through the superheater.

Both blowdown and drain valve connections are placed at a convenient point for operation and are arranged with two valves in series to minimise wear on the valves and to facilitate maintenance. When operating either blowdown or drains, the valve on the pressure side (known as the master valve) is opened first, followed by the outer valve (known as the martyr valve) and the outer valve is closed first, followed by the valve on the pressure side. In this way the master valve is to some extent protected and the heaviest wear is taken by the martyr valve, which is more readily replaced if faulty.

The master valves used for blowdown or draining are of the parallel slide type similar to that shown in Fig 3.31. The martyr valve should be of a type such as a globe valve, suitable for throttling.

Both blowdown and drain valves may be motor operated and remotely controlled from the unit operating desk.

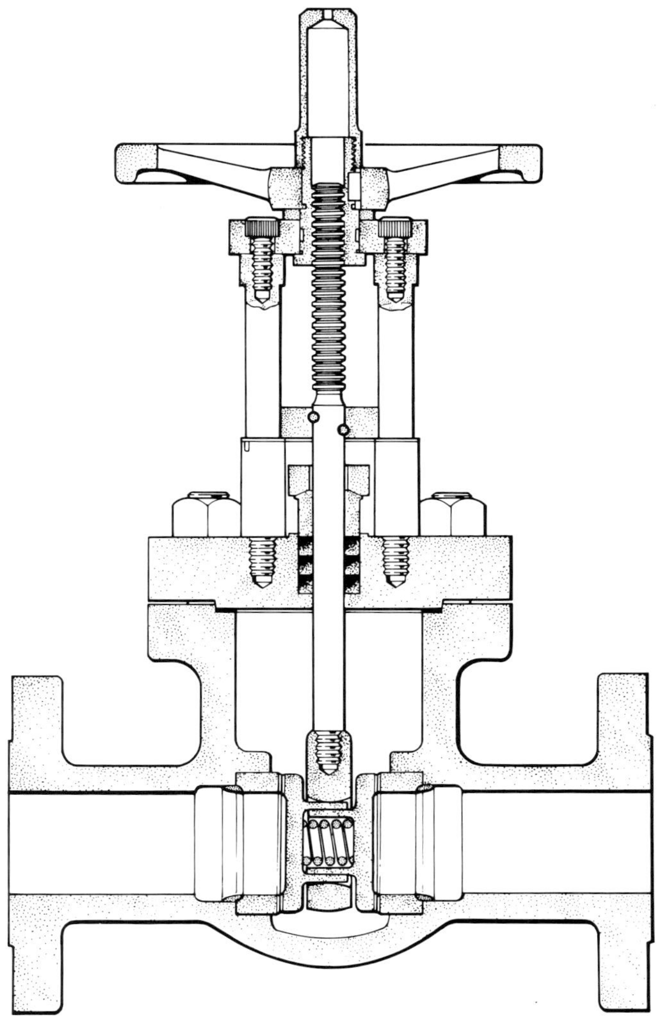

FIG. 3.31 Typical parallel slide blowdown valve construction

5.7 Air release valves

When a boiler has been opened up for maintenance, or even when standing cold, air will accumulate internally and as pressure is raised will be forced to the top of the steam drum and into any pipework before the boiler stop valves. It is important to secure the removal of all air from the steam space and small bore air release valves are fitted to the highest points for this purpose.

The valves, usually of the parallel slide pattern, are opened when steam raising begins and are not closed until the boiler pressure reaches about 4 bar.

The discharge from the valves is usually piped to waste but in some cases may be allowed to blow to atmosphere.

Air release valves may be motor operated and remotely controlled from the unit operating desk.

It is important to use the air release valves to draw air in when emptying a boiler and also when filling the boiler for a hydraulic pressure test to ensure that no air is trapped. Trapped air constitutes an explosion danger during the hydraulic test should a failure of any of the pressure parts occur. Where multiple connections such as from drum-to-superheater are fitted it is not always practicable to vent all of them and any remaining air may need to be removed prior to a hydraulic test by vacuum exhausting.

6 References

[1] The Open Channel Air Preheater: James Howden and Co. Ltd, Glasgow

[2] ESDU 80013: Heat Pipes — General Information on their Use, Operation and Design: Engineering Sciences Data Unit: August 1980

[3] Humphries, P.: High Pressure Jetting of Power Stations Regenerative Airheaters: CEGB Report. TPRD/M/TL1023/M85: February 1985

[4] Kays, W. and London, A. L.: Compact Heat Exchangers: McGraw-Hill: 2nd Edition 1964

[5] Chew, P. E.: Matrix Element Performance Rig — Experimental Technique and Method of Calculating Results: CEGB Report TPRD/M/TL385/M82: November 1982

[6] Hausen, H.: Improved Calculations for Heat Transfer in Regenerators: VDIZ Beiheft Verfahrenstechnik 1942 No 2: The Iron and Steel Institute Translation No 138: 1942

[7] Chew, P. E. and Chojnowski, J.: Method of Assessment of Power Station Regenerative Airheater Performance: CEGB Report RD/M/R119: 1971

[8] Marsh, A. R. W. and Chew, P. E.: Computer Program for the Prediction of End Temperatures and Pressure Drops in a Power Station Regenerative Airheater (REGMOD 3L): CEGB Confidential Report MM/COMB/TL317: October 1981

[9] Ryan, L. C.: A Program to Analyse the Performance of Regenerative Airheaters (REDUB 24): CEGB Report CS/C/P395: December 1974

[10] Sarjeant, M.: Sootblowers, jet aerodynamics and nozzle design: CEGB Report RD/M/R188: September 1973

7 Nomenclature

A — fine flow plan area of rotor on air or gas side, m^2

C — constant in Stanton number relationship

D — constant in Fanning friction factor relationship

n — index in Stanton number relationship

p — index in Fanning friction factor relationship

Re — Reynolds number

Pr — Prandtl number

St — Stanton number

f — Fanning friction factor

h — convective heat transfer coefficient, w/m^2 °K

C_p — specific heat of fluid, J/kg °K

μ — dynamic viscosity of fluid, kg/m s

dh — hydraulic diameter, m

G — mass flow of fluid per unit free flow area per unit time, kg/m^2 s

CHAPTER 4

Dust extraction, draught systems and flue gas desulphurisation

Contents

1 Introduction

1.1 General considerations

It is the policy of the CEGB not only to comply with regulations but to minimise the effects of emissions at all times in order to play its full part in the national effort to reduce atmospheric pollution.

There is a further incentive for the CEGB to keep dust emission to a minimum and that is its importance when new sites for power stations are discussed. These are difficult to obtain and negotiations would be hindered if instance of excessive pollution from existing stations could be cited.

1.2 The need for efficient gas cleaning

Some 80% of the CEGB's generating capacity is from power stations burning coal. The environmental problem of cleaning waste gases from these stations can be understood by appreciating that since they burn pulverised coal some 60% to 80% of the ash in the coal would be emitted from the chimney if gas cleaning equipment were not installed. On a typical 2000 MW station, the dust that would be emitted when burning coal having an ash content of 20%, would be in the order of 110 t/h. This emission has to be reduced to 0.55 t/h to comply with the required emission standard of not more than 0.115 g of dust in each cubic metre of gas at standard temperature and pressure (STP), 16°C and 1 bar.

In considering the possibility of nuisance being caused by dust emission, both quantity of dust and size of particles are taken into account. Under correct operating conditions the quantity of dust discharged and the particle size would be sufficiently small to be adequately dispersed. However, faulty operation or maintenance of dust collectors will allow greatly increased quantities to be emitted, either in the form of smoke or fine dust which will produce a visible plume, or larger particles and agglomerates which will be deposited locally, any of which may give cause for complaint.

The monitors described in Section 2.4 of this chapter are used to check performance. On coal-fired stations a standard for the observation monitor is set for each station and actions are suggested should this reading be exceeded. The greater the excursion, the more drastic the action required.

1.3 Legislation

The control of most types of atmospheric pollution is enforced by statute. The main legislation is as follows:

- Alkali and Works Regulation Act 1906, as extended by the Alkali Act
- The Health and Safety (Emissions into Atmosphere) Regulation 1983 No.943
- The Public Health Act 1936
- Clean Air Acts 1956 and 1968
- Regulations made under the Clean Air Acts

The provision of the Alkali Act, concerned with the regulation of processes carried on in certain chemical and industrial works specified in the Act or in the Order extending its application, have the greatest impact on the generation of electricity. In 1958, as a sequel to the Clean Air Act, the scope of the Alkali Act was greatly extended to embrace many additional operations, including the generation of electricity, in which the control of smoke, grit and dust presented problems of special technical difficulty.

The enforcement of the Alkali Act and Orders is the responsibility of Her Majesty's Industrial Air Pollution Inspectorate (HMIAPI) through the devices of Improvement and Prohibition Notices under the Health and Safety at Work Act 1974.

The Clean Air Acts are an integral part of the public health code, and their enforcement, in respect of industrial and domestic premises, is the responsibility of local authorities. This work is normally undertaken by the Chief Public Health Inspector of each local authority.

There is no statutorily defined limit for emission from generating stations but the Health and Safety at Work Act 1974 states that *'it shall be the duty of the person having control of any premises of a class prescribed to use the best practicable means for the reduction of noxious or offensive substances and for rendering harmless and inoffensive such substances as may be so emitted'*. The requirement is a pre-requisite of registration for a station.

Registered works are subject to routine inspection and the owner is required to provide all necessary facilities for entering, inspecting, examining and testing and must also disclose in confidence such details of the processes as are required by the Inspector.

In addition to the statutory requirements, the following condition is attached to consents for new coal and oil-fired power stations obtained under Section 2 of the Electric Lighting Act 1909:

'The layout of the Station shall be designed so as to permit the installation of such plant as may be reasonably practicable for the prevention of a discharge of sulphur and its compounds into the atmosphere and the Board shall if so required at any time by the Minister install such plant'.

The CEGB therefore has an obligation to use the best practicable means to prevent or minimise the emission of dark smoke, grit, dust and noxious or offensive gases from all its generating stations. It is also liable under common law, and proceedings may be taken against it for nuisance.

The words 'best practicable means' are interpreted to refer not only to the provision of efficient methods and appliances, but also to the manner in which such appliances are used and to the proper supervision of any operation in which the gases are evolved. The obligation to adopt the best practicable means is continuous and this could entail alterations to plant and methods as new techniques become available.

2 CEGB practice

2.1 Overall system of gas flow

The gas from the furnace passes through the superheater and the reheater and then on to the economiser and the air heater. At each stage heat is given up until at the exit from the airheater the gas temperature has been reduced to 120°C for coal firing (Fig 4.1).

From the airheater the gas is ducted to the inlet manifolds of the electrostatic precipitators where the velocity is reduced before entering the precipitator casing. The cleaned gas is then conveyed to the induced draught fans and from there to the chimney.

On the majority of large boiler units there are three precipitator casings with crossover ducts at the inlet and outlet to enable an equal amount of gas to flow through each. The latest practice is to have only two precipitators, one for each airheater, an arrangement that eliminates the need for the crossover ducts.

Figure 4.2 shows a schematic arrangement of the gas and air system for a 500 MW coal-fired boiler.

2.2 Standard of gas cleaning

The standard of particulate removal from the boiler gases has been raised by the CEGB over the years. Prior to 1965 the collecting efficiency required was in the order of 98% but for stations built after this the standard was raised to an efficiency of 99.3%. After 1978 the requirements were changed from collecting efficiency to a limit of the dust burden of the gases leaving the chimney and is not to exceed 0.115 g/m^3 when the measured burden is corrected to a gas condition of a temperature of 16°C, a pressure of 1 bar and CO_2 content of 12%. These corrections are necessary to give a standard of dust burden since all three have an affect on the gas volume.

The standard had to be raised because of increasing public awareness in environmental matters but in any case the emission from a typically sized power station could not be allowed to increase. For example, a station built prior to 1960 could have had 4 × 120 MW units equipped with gas cleaning equipment with a collecting efficiency of 98% giving a total dust emission of 0.85 t/h.

A modern installation of 3 × 660 MW units with the same collecting efficiency of 98% would result in an emission of 2.2 t/h which clearly would be unacceptable. The present standard of 0.115 g/m^3 at STP requires a precipitator efficiency of 99.5% which reduces the total emission to 0.55 t/h.

Figure 4.3 shows the Peabody Sturtevant flat-plate wire and weight electrostatic precipitator used on the 660 MW boilers at Drax power station. Each of the three casings per boiler has six zones and two levels of collector plate rapping.

Close attention to the design of modern precipitators has to be given to ensure that the required collecting efficiency is maintained during the life of the station. Major overhauls are normally carried out only every three years and since a clean precipitator works more efficiently than one where dust has accumulated on the electrode system, it is important that the rapping gear is effective and reliable. It is for this reason that the efficiency test to check compliance with the specification is carried out after the first 12 months of operation without maintenance.

The record of the CEGB in dealing with the emission problems of power stations is of a very high order, both in money spent and results achieved. Extra to the many millions of pounds spent on dust collecting plant, which costs approximately £10 million for a 3 × 660 MW station, both money and effort are spent on research to improve dust collector design and performance as well as investigations on matters concerning air pollution generally. This work is carried out either in laboratories or on suitable sites.

Some power stations have been emitting unacceptable quantities of dust and these difficulties have been overcome, either by decommissioning or by reduced usage of such plant as stations of higher thermal efficiency are commissioned. Where it is necessary to keep the station operational, unsatisfactory dust collectors have to be modified or replaced.

2.3 Agreements with HM Industrial Air Pollution Inspectorate

Most of the requirements of HMIAPI are covered by the foregoing Sections 1.3 and 2.2 which respectively give the requirement and the CEGB's method of complying, but there are other matters that have been agreed.

Although most of the CEGB's stations have to comply with the required 99.3% collecting efficiency or the more recent requirement of an emission of 0.115 g/m^3 at STP, it was agreed that some older, small-to-medium size stations, should comply with a maximum emission of 0.460 g/m^3 at STP. This limit is not regarded as permissible emission from plant capable of attaining a lower figure, nor can it ensure freedom from complaint.

On the CEGB's major power stations *approved particulate monitors* are installed either in the chim-

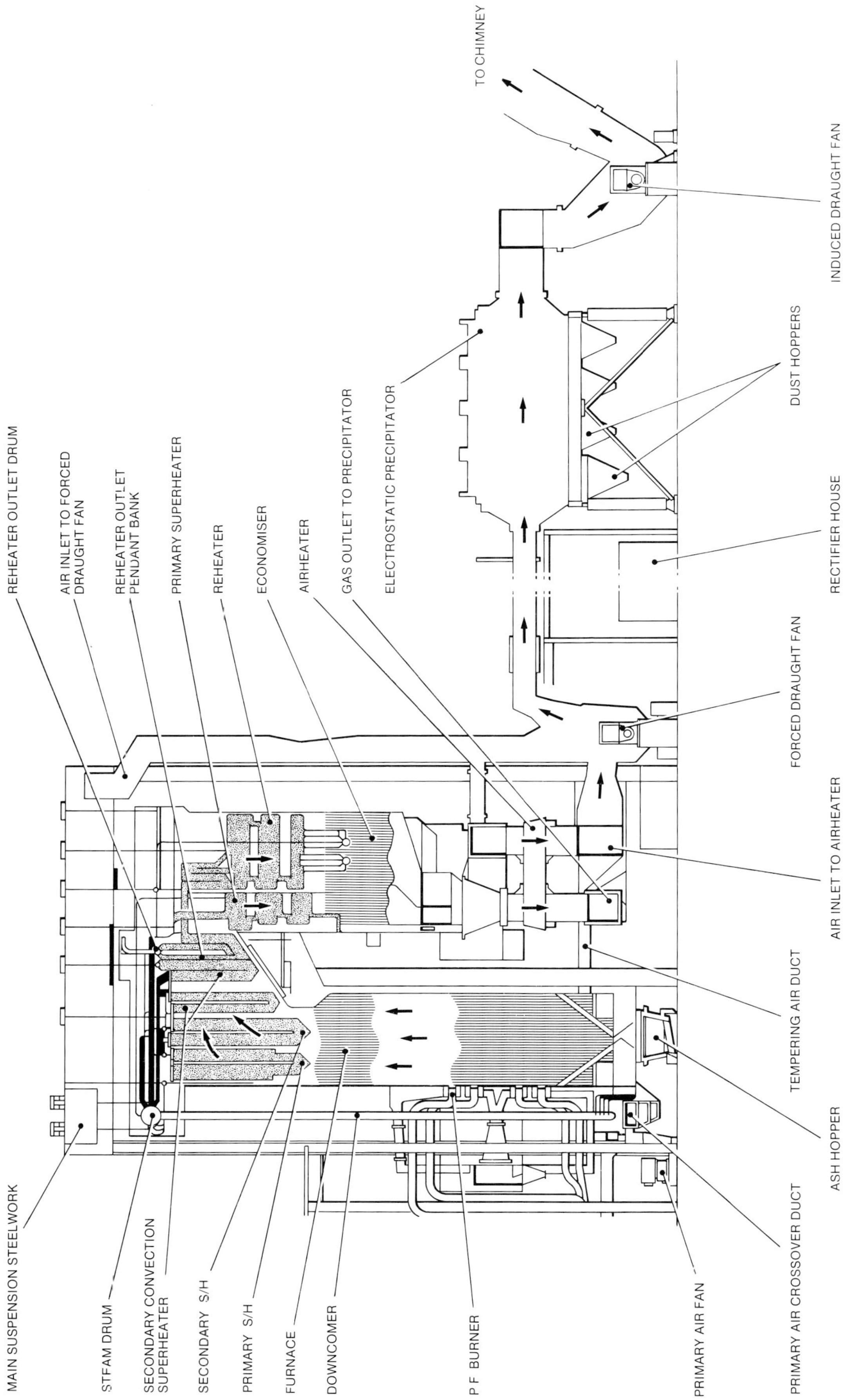

FIG. 4.1 Boiler gas flow system

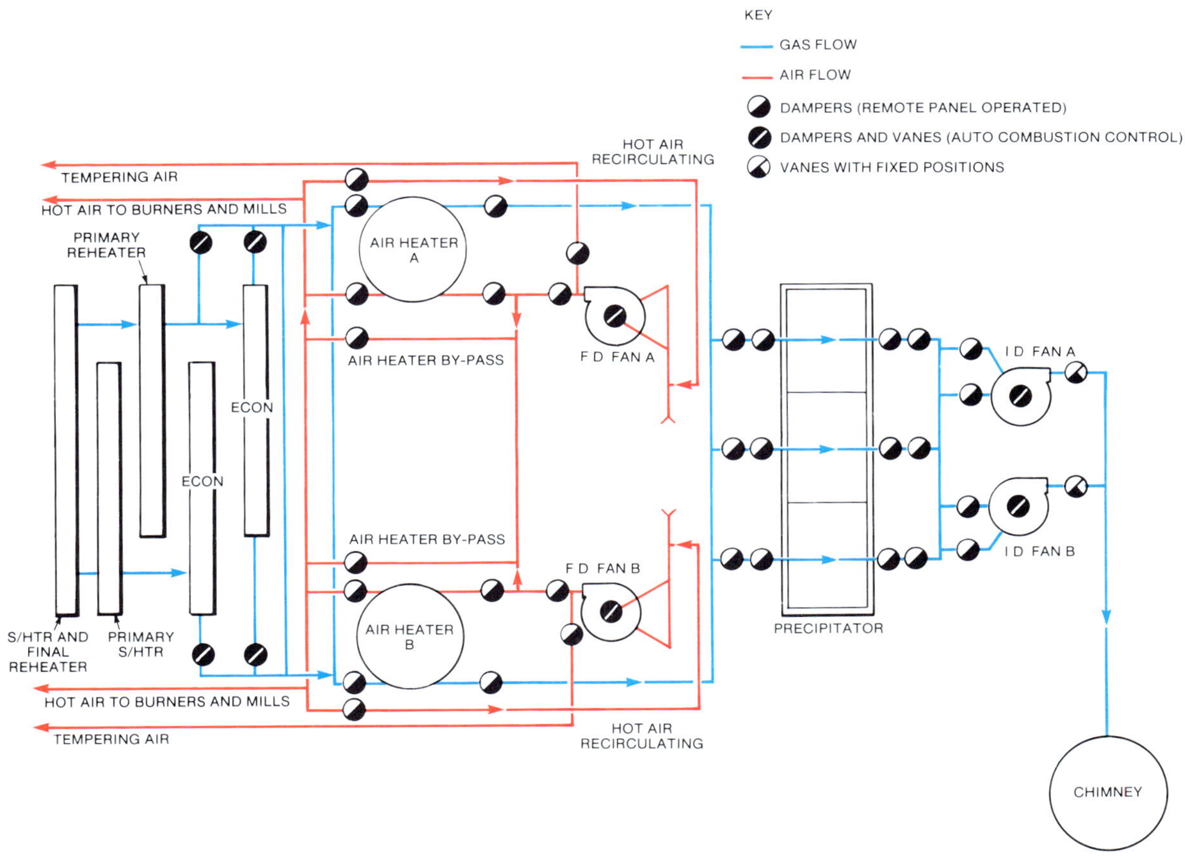

FIG. 4.2 Gas and air system schematic

ney which is the ideal position, but if this is not possible, in a representative location in the ducts.

Trials have been carried out by the CEGB and, as a result, the Erwin Sick and the Serop Series III have been approved by HMIAPI, as well as by the CEGB.

Installation of the approved particulate monitors in most cases meets the need previously met by coarse dust monitors (CERL Mk.II and III) and thus they can be removed after consultation with HMIAPI. The approved particulate monitors also meet the requirement of smoke monitors which were previously fitted and therefore these can also be taken out of service.

The use of continuous particulate monitoring, especially by the approved monitors, can render it unnecessary to carry out regular annual emission tests, the reduced frequency of testing being subject to local agreement with HMIAPI. In these circumstances the roll charts from the monitors would be retained for at least one year, and secondary records and monthly averages would be retained indefinitely.

Any complaints from members of the public or on behalf of outside bodies have to be dealt with as a matter of urgency and a record kept of the particulars of such complaints.

Wherever particulate monitoring is not available, operating staff have to make the necessary arrangements for the observation of chimney emission. In the event of abnormal emissions of dark smoke the particulars are recorded in the station log.

Except during permitted periods such as during start-up or during an emergency, no boiler plant should continue to be operated with emissions likely to give cause for serious complaint. In the event of such an emission, immediate investigation is undertaken and, if rectification of the problem cannot be effected within a reasonable period, the reduction of boiler loading would be considered as a first step to restoring normal chimney emissions. Should such action prove ineffective, arrangements would be made for the plant to be shut down as soon as possible after consultation with System Operation. Priority is to be given to identifying and correcting the cause of the trouble and the plant is not to be returned to commercial operation until it can be assumed that the emission will be reduced to a satisfactory level.

2.4 Checking performance and monitoring instruments

The efficient operation of all dust arrestor plant is essential if the objectives are to be achieved. Day-to-day performance can be judged from the records

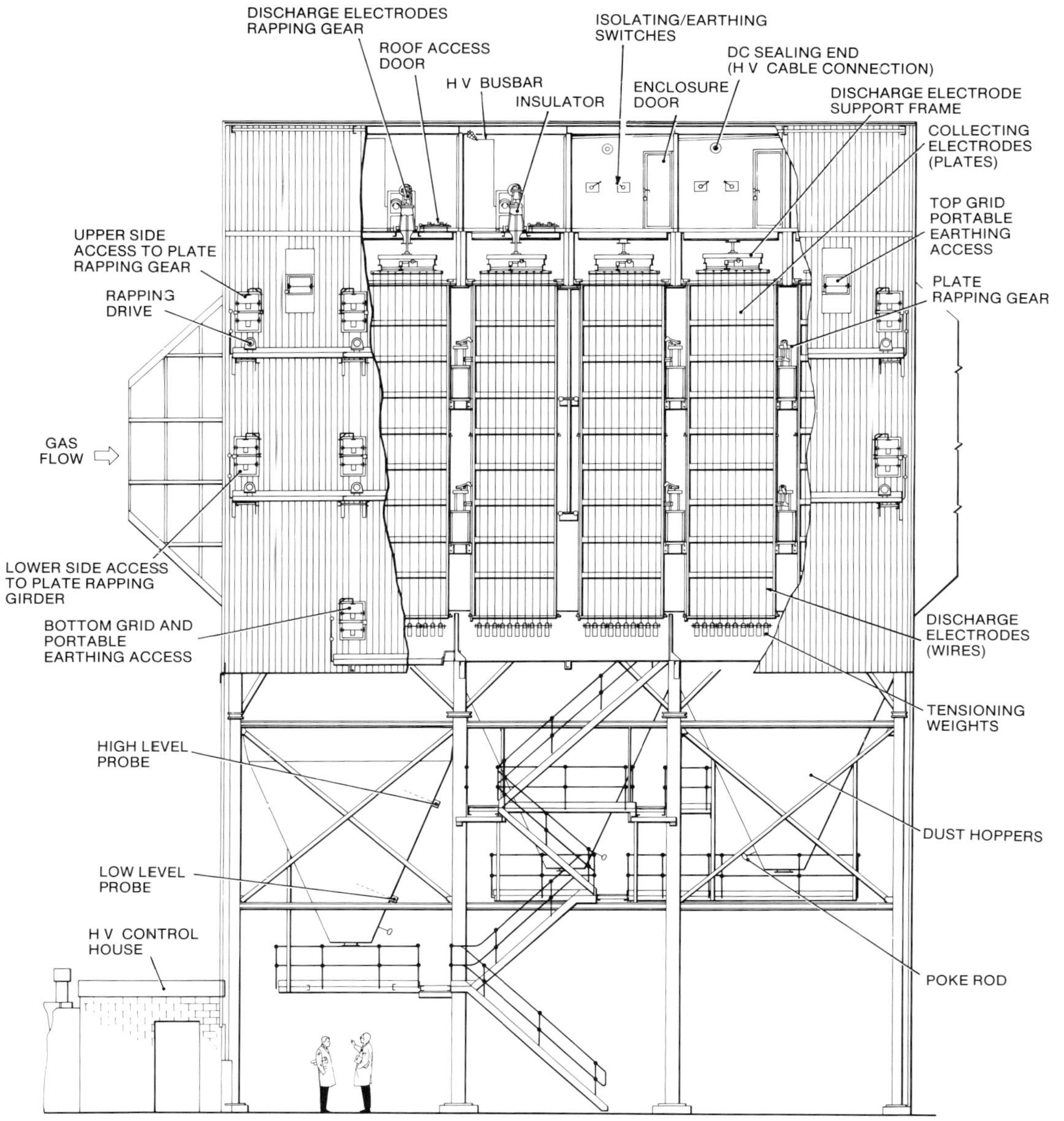

FIG. 4.3 Electrostatic precipitator general arrangement

of the approved particulate monitors and the electrical conditions of the precipitators. The monitors should be used as indicators of trends in the emission levels and station operating instructions should define the action to be taken when emission levels exceed the normal operating value.

The required action should include investigation, diagnosis and reporting of the incident in the station log. Where such an incident occurs or could be anticipated, telephone calls to the Local Authority and the HMIAPI District Inspector are made to improve understanding of the CEGB's problems.

2.4.1. Gravimetric testing

At the present time there is only one way to assess the quantity of dust that is being carried in a gas stream within a duct and this is by gravimetric testing. The test position selected should be at the greatest distance possible downstream of any bend or change of section in the ductwork. The position selected for new plant is checked for satisfactory flow patterns on a model of the precipitators, see Section 5.8 of this chapter.

The duct is considered to be split into a grid of between 18 and 36 sampling positions depending upon the size of the duct. For instance, eight sampling holes in one side of the duct would enable the sampling probe to be entered and testing carried out in four positions in the depth giving a required 32 positions. In a test, the first activity is to establish, using a pitot tube, the gas velocity at each of the grid positions.

At the start of the test, usually of four hours duration, a dust sampling probe is placed in each position for equal periods and a sample of the gas with the entrained dust extracted at the velocity recorded from the previous survey. The dust is filtered out in fabric bags and the total quantity of dust is related to the total gas flow to give a dust burden, usually measured in grammes/cubic metre at STP.

Figure 4.4 shows the CEGRIT automatic dust sampler designed by CEGB for the measurement of emission in PF boilers.

To check the collecting efficiency of a precipitator, all inlet ducts and all outlet ducts are simultaneously sampled and the burdens at inlet and outlet compared. All testing is carried out to BS893 [1] when the boiler conditions are steady.

2.4.2 Obscuration monitors

As a result of the work of a joint working party, including extensive site trails, the Erwin Sick RM 41 and the Serop Series III monitors were approved by HMIAPI and the CEGB and are termed *approved particulate monitors*. Both of these monitors record the obscuration of a light beam passing across the gas stream. The Erwin Sick is the more sophisticated instrument and was preferred since it is more reliable and requires less maintenance than the Serop.

The Erwin Sick monitor (Fig 4.5) has a light source and a photocell housed on one side of the duct and a light reflector on the other, allowing the loss of light due to the dust concentration to be registered and recorded on a chart. Special features include a reference beam to counteract changes in lamp brightness and a periodic zero check which is recorded. Both housings have air purging to prevent the ingress of dust.

The Serop monitor (Fig 4.6) also employs a light beam and provides a continuous record, but in this case the light is housed at one end of a 2.3 m long tube with a photocell at the other end. The tube, which has a 1.37 m slot on both sides, through which the gas passes, is installed in the duct with the photocell situated outside the duct. Air purging keeps the light source receiver clean. It has the advantage of only requiring access to one side of the duct.

2.4.3 Coarse dust monitors

On many stations the CERL Mk III (Fig 4.7) coarse dust monitors are fitted. These have a vertical tube in the duct with an opening facing the gas stream and dust entering the tube drops down and is deposited on a glass plate. Periodically the density of the deposited dust is recorded by passing light through the layer to a photocell. After an interval of time the dust is blown clear and a wiper mechanism cleans the glass.

On coal-fired stations these monitors can, with the agreement of the local inspector, be removed if obscuration monitors have been fitted. On oil-fired stations these monitors are recommended in addition to the obscuration monitors, because they give a good indication of poor atomisation in the burner which gives rise to the emission of large agglomerates.

2.4.4 The Ringelmann chart

The visual assessment of smoke emission, for the purposes of the Clean Air Act, is by means of the Ringelmann chart, the darkness of the smoke being compared with standard shades of grey on the chart. The shades of grey are obtained by cross-hatching in black on a white background so that a known percentage of the white is obscured. The chart is placed at such a distance from the observer that the black lines merge into the white background and reproduce for each shade a uniform grey. The number of shades (Ringelmann numbers) range from 0 (all white) to 5 (all black) in stages of 20% obscuration of background. The British Standards Institution has now produced a standard Ringelmann Chart, BS2742C, and also instructions for the use of this chart [2].

Although the Ringelmann Chart is the legal criterion it is only practicable for spot checks. The more usual method is by continuously recording smoke density instruments.

2.4.5 High voltage instrumentation

Since each of the zones on a precipitator normally has independent high voltage sets for energisation, the instruments can be used as a check on performance. The current and voltage readings will vary depending upon the zone being observed, but if the fuel supply to the boiler does not change, any variations in the instrument readings can be taken as a guide to performance. For instance, if the obscuration monitors show an increase in emission it could be that there is a fault in a particular zone and this could be identified by a change in the instrument's readings.

2.4.6 Hopper level indicators

Automatic dust level indicators are fitted in the precipitator hoppers. The high level indicator is essential since it ensures that the dust in the hoppers does not build up to a height where contact would be made with the electrode system causing a short-circuit to earth. It is useful to have a low level indicator to show that after the emptying process, the hopper has been emptied.

There are various types of indicator currently in use including:

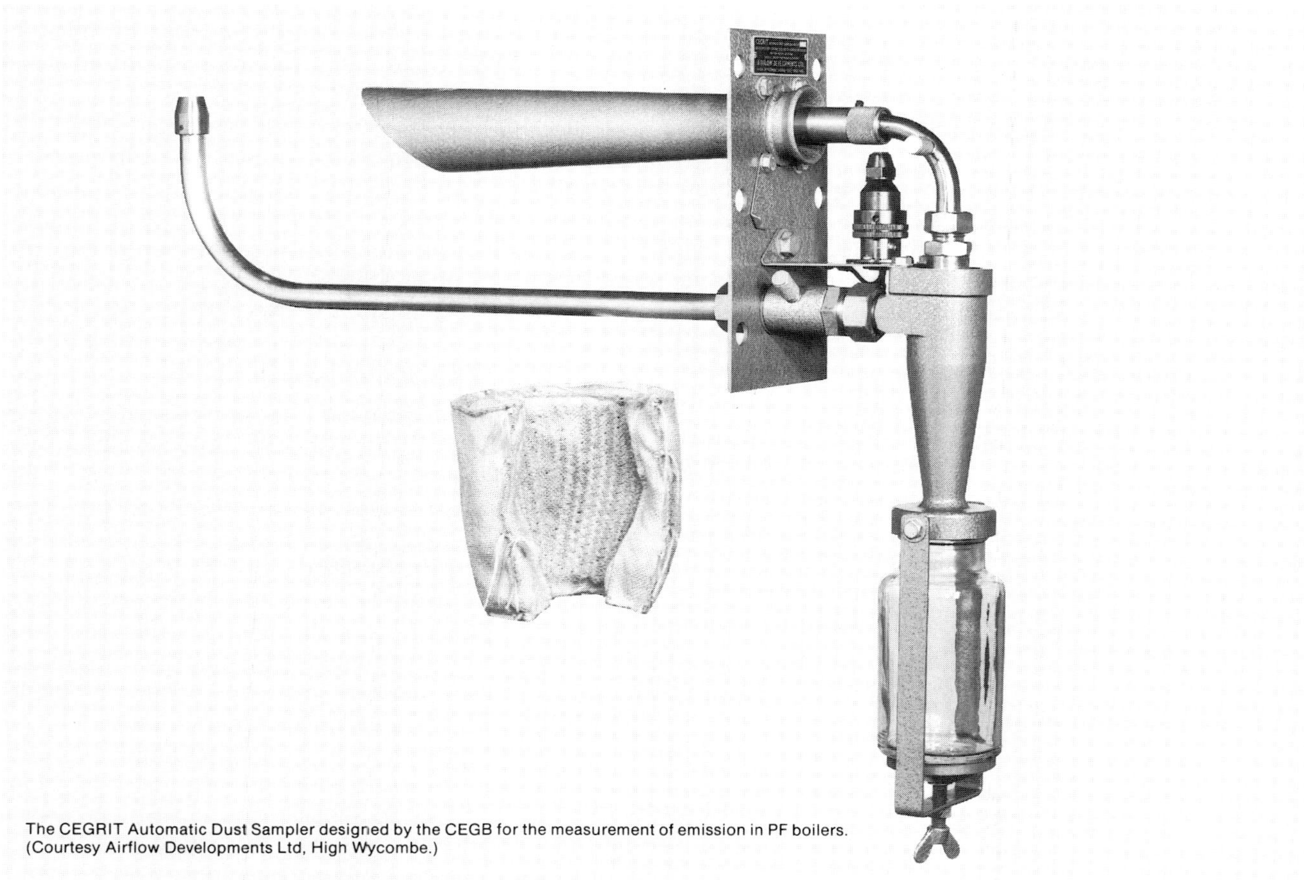
The CEGRIT Automatic Dust Sampler designed by the CEGB for the measurement of emission in PF boilers. (Courtesy Airflow Developments Ltd, High Wycombe.)

FIG. 4.4 Automatic dust sampler

- The capacitance type which is activated by the dust completing an electrical circuit in a probe.
- A tuning fork kept activated until the vibration is damped by surrounding dust.
- A suction tube where the negative pressure in the precipitator sucks air into a tube until closed off by the dust, the lack of flow causing an indication.

2.4.7 Maintenance

Continuous satisfactory performance can only be achieved by regular inspection and maintenance of the dust collecting plant. Major overhauls are normally only possible every three years and it is therefore necessary to adjudge the life of the items subject to wear or deterioration from other causes, in order that replacement can be planned and thus prevent unnecessary outages.

The rapping gear of a precipitator has to work in a hostile environment and in particular is subjected to the abrasive nature of the dust. The design therefore should readily allow for replacement of wearing surfaces such as in the bearing and swivel points.

During major overhauls the state of the discharge and collecting electrodes as well as the rapping gear needs to be assessed to ensure that no major replacement is required prior to the next outage. The alignment of the electrodes is extremely important in order to maintain the correct electrical clearances, and at the same time the support insulators and the general state of the plant needs assessing to ensure it is mechanically and electrically satisfactory.

2.5 Other sources of dust nuisance

Serious nuisance can also be caused by windblown coal or ash carried from coal stores or ash disposal plant. This type of dust is particularly objectionable, being larger in size than normal chimney dust. Particles up to 500 microns can be carried into neighbouring premises especially where there is no intervening open ground.

If there is a local complaint about dust emission, it is important that the station manager is able to confirm that the fault is at the station and, if so, the source of the dust. The dust can be identified by a combination of microscopic examination, chemi-

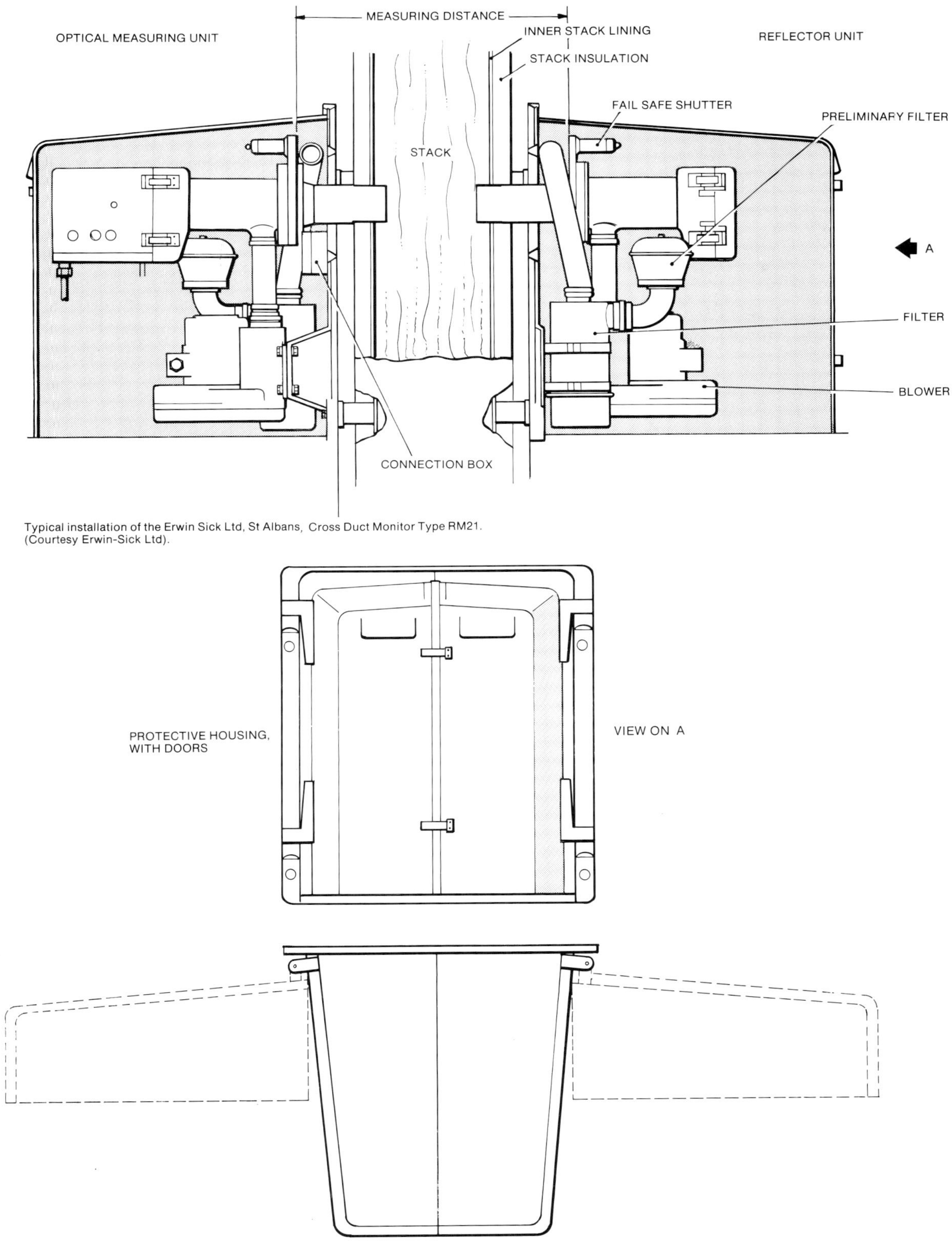

FIG. 4.5 Erwin Sick cross-duct monitor

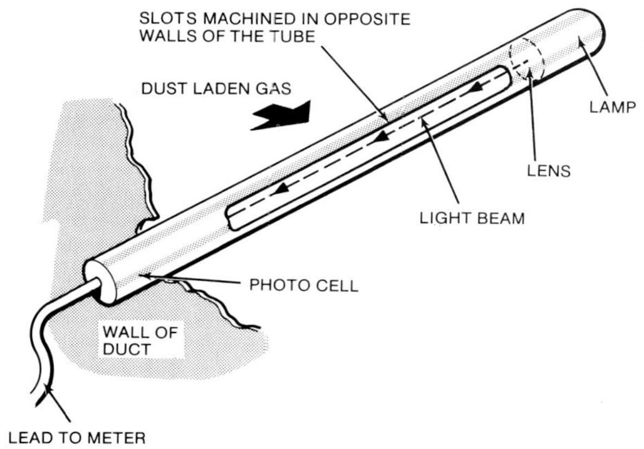

FIG. 4.6 Serop dust monitor

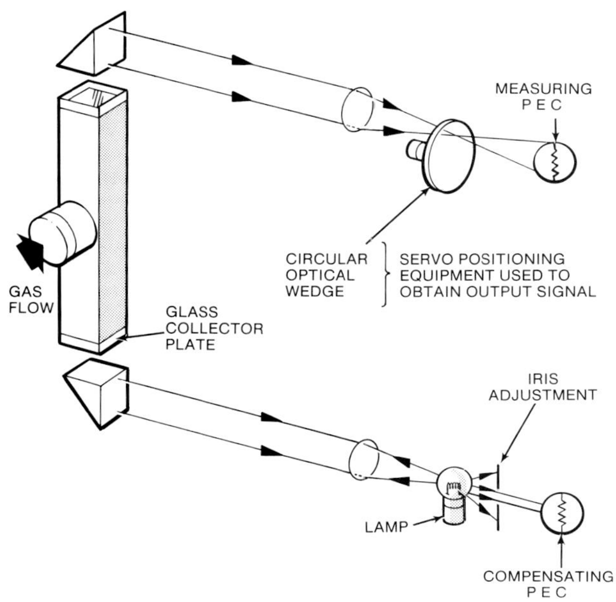

FIG. 4.7 CERL Mk III coarse dust monitor

cal analysis and dust gradings at various distances from the station.

3 Gas cleaning requirements for various types of boiler

3.1 Pulverised fuel firing

All the CEGB's modern coal-fired stations use pulverised coal which is burnt in suspension within the boiler and to obtain high boiler efficiencies the mills have to grind the coal to a high degree of fineness. The requirement is that 70% of the pulverised coal shall be less than 200 mesh and that only 1% shall be in excess of 52 mesh. Burning this fuel produces a very fine dust consisting mostly of the ash content of the coal since it is unusual for the carbon content, that is the unburnt coal included in the dust, to be in excess of 5%.

The sizes of particles emitted from the combustion process are conventionally regarded as falling within the following limits:

Smoke	—	less than 1 micron
Dust	—	1 to 76 micron
Grit	—	greater than 76 micron

The term *dust* is commonly used when referring to the total emission of particulate matter.

That part of the dust which is termed *grit* will fall out of the gas stream and be collected in the furnace bottom and in the economiser hoppers where fitted, whilst the remainder, termed dust, is carried forward to the electrostatic precipitators. For the purposes of precipitator design the *carryover*, as it is termed, is taken as 80%, but dependent on many factors can vary between 60% and 85%.

As stated in Section 2.2 of this chapter, the dust burden is measured at STP and 12% CO_2 in the gas. Since most modern boilers work with a CO_2 in gas of 14%, it follows that correcting the volume to 12% increases the gas volume and thus decreases the dust burden. It can be seen that this correction is important when discussing chimney emissions. The emission can be expressed in grams or milligrams.

With a low ash fuel of 14%, the dust burden entering the electrostatic precipitator would typically be 11 g/m^3 at STP and for a high ash fuel of 25% would be 23 g/m^3 at STP, both figures corrected to 12% CO_2 and assuming 80% carryover. If high carbon in the dust is anticipated, it would be necessary to take account of this in calculating a dust burden.

Although the dust emitted from pulverised fuel boilers is basically very fine, agglomeration of fine particles may occur which, if emitted from the chimney, can cause local deposition and possible nuisance. If the agglomerates are formed in the ducts prior to the precipitator they are collected, but if they are formed after the precipitator or the walls of the ducts and the chimney, they will be periodically emitted, particularly during start-up.

A modern power station precipitator can be designed to give the required dust emission standard, taking into account the dust burden, the fineness of the dust and its chemical composition.

3.2 Oil firing

The particulate emission from oil-fired boilers is of a low order and on some stations can be as low as 0.070 g/m^3 at STP although this naturally depends upon the design of the boiler and, in particular, on the design of the burners.

With the use of heavy oils which have a high sulphur content, the combustion process produces sulphur dioxide, some of which is converted to sulphur trioxide which then combines with water vapour from the air and the oil to form sulphuric acid.

The dust emission consists for the most part of very fine soot particles. Because the ash content of oil is very low, typically around 0.1%, the dust burden consists of 80% carbon, the remainder being attributed to the ash, various sulphates and other chemical combinations. However, some of the emission consists of large coke particles formed from partially-burnt oil in the order of 100 microns in size. Dust deposited on the elements of the airheater or on the walls of the flues and the chimney, also tends to agglomerate. The deposited dust absorbs the sulphuric acid and when dislodged gives rise to the emission of large acid smuts which can cause a local nuisance and damage to surrounding property.

The major problem is therefore the emission of acid smuts. The first defence is to reduce the amount of acid by reducing the available excess oxygen in the furnace which combines with the sulphur. Modern practice is to control excess oxygen to 0.5%, but care has to be taken in the achievement of this because if there is insufficient to burn the oil there would be an increase in carbon, which constitutes an increase in the dust burden.

Another feature of design is the importance of keeping the gas and the swept surfaces at a temperature above the acid dewpoint since any cooled surface causes the dust to accumulate. The gas temperature is maintained above 150°C and all swept surfaces are lagged to prevent cooling.

Because these means are not completely successful in eliminating acid smuts, it is current practice to inject magnesium hydroxide (trade name Lycal) into the gas stream prior to the airheater. This neutralises the acid, thus reducing the risk of emitting acid smuts and also reducing corrosion of the airheater elements and the flues.

On some earlier stations, mechanical collectors (also called *grit arrestors*) were installed, but this is no longer done because their collecting efficiency was very low and they did little to alleviate the problem of acid smuts.

4 Types of dust extraction equipment

4.1 Electrostatic precipitators

On all modern CEGB power stations where pulverised coal is being fired, the removal of the dust from the gas stream is by means of electrostatic precipitators. The dirty gas is channelled between collecting plates where the dust particles are electrically charged and collected on the plates. Electrostatic precipitators are favoured because they can be designed to remove a high percentage of the dust in the gas stream. The required efficiency is dependent upon the ash content of the coal and the standard of emission required and this usually means a collecting efficiency of between 99.3% and 99.8%.

The gas passes through a precipitator at a very low velocity and consequently the resistance to flow is less than 12 mm of water gauge, thus keeping the power absorbed by the induced draught fans to a minimum.

4.2 Mechanical collectors

Mechanical collectors consist of a bank of tubes aligned to the gas flow with a vane at the entrance of each to impart a swirl to the gas. The centrifugal force so created, throws dust to the wall of the tubes where it is ejected through a small slot, whilst cleaned gas continues towards the chimney. These collectors have a relatively low efficiency, but it was thought that if they preceded an electrostatic precipitator the removal of some of the dust would reduce the load on the precipitator and allow its size to be reduced. Unfortunately, the results obtained from these combined plants were disappointing, and later information showed that the performance of the precipitators was considerably reduced in dealing with the finer dust resulting from the use of a mechanical precollector. It was therefore decided in 1964, that on all future power stations electrostatic precipitators only would be installed.

4.3 Bag filters

Bag filters, although often fitted to the vents of dust handling systems, are as yet unused as main collectors in CEGB stations. In this form of dust collector the dust-laden gases are passed through a woven cloth or fabricated fibrous medium which filters out the dust. The filter fabric is usually woven in the form of long bags or stockings and a considerable number of bags are connected in parallel in relation to the gas stream (some 10 000 bags would be needed for a 500 MW unit). The dust collected on the filter fabric is removed periodically by shaking, rapping, reverse air flow or jets, and falls into the dust collecting hoppers below. Very high efficiencies, in excess of 99.5%, can be achieved.

The difficulties of obtaining a fabric having the necessary mechanical, chemical and thermal characteristics suitable for power stations has largely been overcome, but the problem with the bags becoming blinded with dust causing high pressure drops, has not. This problem is being experienced both in the United States and Australia where they are used on some stations. The design pressure drop of 152 mm water gauge, which is very high, can be exceeded by a multiple of two, and the only remedy is to replace the bags which is a costly procedure.

4.4 Wet dust collectors

Wet methods for the collection of dust are not favoured by the CEGB. Unless they have a very high resistance to gas flow or a very high water pressure, both of which greatly increase the operating costs, they cannot give the required collecting efficiency. They need large quantities of water which has to be treated before being recirculated or discharged. It is necessary to cool the gas before the collector and, unless it is reheated afterwards, the plume from the chimney will not rise and thus will cause a local nuisance.

Since none of these disadvantages apply to the use of electrostatic precipitators, it can be seen why they are favoured for use on power stations.

5 Electrostatic precipitators

5.1 Principles of design

The principles upon which an electrostatic precipitator operates are that the dust-laden gases pass into a chamber where the individual particles of dust are given a negative electric charge by absorption of free ions from a high voltage (HV) DC ionising field. Electric forces cause a stream of ions to pass from the discharge electrodes to the collecting electrodes and the particles of dust entrained in the gas are deflected out of the gas stream onto the collecting surfaces where they are retained. The particles have a negative charge and are thus held on the collecting plates which are earth positive. The particles give up their charge and are held by molecular attraction until they are removed by intermittent blows applied by rapping gear. This causes the dust particles to drop into dust hoppers situated below the collecting electrodes.

Immediately before an electrostatic precipitator, the ductwork has a tapered section so that the gas velocity is reduced from around 15 m/s in preparation for its passing through the precipitator casing at between 1.0 m/s and 1.8 m/s. Figure 4.8 shows the plan arrangement of ductwork and electrodes in a typical three-stage precipitator.

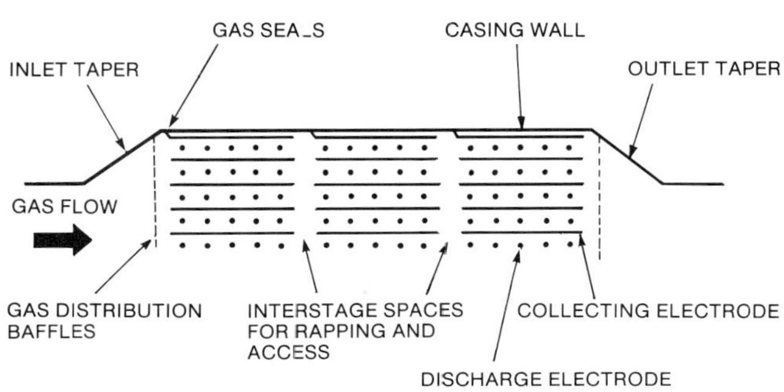

FIG. 4.8 Three-stage precipitator ductwork and electrodes plan arrangement

Within the casing are suspended collecting electrodes which are in the form of flat plates, suitably stiffened to prevent their becoming unstable in the gas stream. They are between 6 m and 15 m in height and are spaced generally between 200 mm and 300 mm apart although, not as yet used by the CEGB, this spacing can be up to 400 mm. Thus, in the casing there are formed a number of gas lanes from one end to the other.

Between the collecting electrodes there are, suspended from suitable insulators, discharge electrodes on which there is impressed the high tension voltage.

The design of a precipitator depends on the velocity a particle of dust effectively travels towards the collecting electrode. By *effective* it is meant that a particle may be collected and then entrained in the gas stream before it is finally trapped. This velocity is known as the *effective migration velocity* usually represented by w and is expressed in cm/s. It is derived from the following equation:

$$\text{Efficiency} = 1 - e^{-wk}$$

where: Efficiency is expressed as decimal

e = Naperian log base

k = factor

w = effective migration velocity in cm/s

This expression is commonly known as the Deutsch formula where the k value took account of the many variables that influence the performance of a precipitator, but today k is generally known as the *specific collecting area* (SCA)

$$k = \frac{A}{Q}$$

where A = projected area of the collecting electrodes in m^2

Q = quantity of gas to be treated measured at the operating conditions in m^3/s

This expression can thus be transposed as follows:

$$w = \frac{1}{k} \log_e \left(\frac{1}{1 - \text{efficiency}} \right)$$

It can thus be seen that w is purely a convenient way to bring together these three parameters, efficiency, gas volume and plate area. The value of w derived from the data obtained during tests on installed plant can then be used as a design basis for new plant.

In some cases it is desirable to investigate the performance for a particular dust, on a pilot plant, to establish the value of w. If the maximum current density is used then extremely high migration velocities can be attained, anything up to 20 cm/s whilst on a full scale plant the figure could be less than 8 cm/s. The pilot plant has therefore to operate on the current density that is thought to be attainable on the full scale plant.

The inlet dust burden is dependent upon the ash content of the coal and the carryover from the boiler. The outlet dust burden will be dependent upon the legislation governing environmental matters and might also take into account impending legislation over the life of the plant. From these two figures, obviously the required collecting efficiency can be established.

Careful consideration needs to be given to the decision on the collecting efficiency required; too low necessitating costly additions later and too high adds considerably to the first cost. A precipitator designed to have an efficiency of 99.3% would have to be increased in size some 10% to attain 99.5% and 30% for 99.8%. This is less surprising when it is considered the slip has been decreased from 0.7% to 0.5% and 0.2%. The effect is worsened by the fact that the migration velocity decreases with the larger SCA values that are required for more efficient plant.

Thus having established the required collecting efficiency, the gas volume to be treated and the characteristics of the coal and ash, it is a simple matter to calculate the required collecting plate area, bearing in mind the dust is collected on both sides of the plates.

5.2 Design and arrangement

Unit sizes in operation on major stations are 500 MW and 660 MW but future stations will probably have 900 MW units. Currently, there are three precipitators in parallel for a unit of 500 MW or 660 MW with interconnecting ducting at the inlet and outlet ends to enable a balanced flow of gas through each of the precipitators.

On future stations, there will only be two precipitator casings per unit. This arrangement eliminates the need for interconnecting ducting because the gas would come from the two airheaters through two precipitators and on to the two induced draught fans. This more economic arrangement has been made possible because structural engineers can design the much larger casing required. It was also thought desirable to use three flows so that in the event of a fault only one third of the plant would be lost. With the greater reliability of precipitators this is no longer so important, and in any case the precipitators are split into many electrical zones so that in the event of a fault only a small part need be isolated.

The precipitators for a 660 MW unit would handle a gas volume of around 900 m^3/s of gas at 120°C and would be designed for a collecting efficiency of 99.5% to reduce the dust burden from 23 g/m^3 at STP to 0.115 g/m^3 at STP. If in the future the required emission standard is reduced to 0.070 g/m^3 or 0.050 g/m^3, then the efficiency would have to be increased to 99.7% or 99.8%.

To attain the present standard of 99.5% the precipitators would have a total width of approximately 63 m and, with four to six zones in series, have a total length of approximately 17 m excluding the taper from the ductwork at the inlet and outlet. The overall height including upper chambers and the hoppers would be around 27 m.

5.3 Discharge and collecting electrodes

The discharge electrodes are of various types, depending on the particular manufacturer, but generally fall into two categories, wires or barbed electrodes. Wires have a square or square with hollow-sided cross-section of dimensions between 3.2 mm and 4.8 mm with sharp corners to promote corona discharge. Barbed electrodes have some form of pointed metal protrusion, either attached to a vertical member or formed from that member.

The assembly of the discharge electrodes also varies. They can be strung from the top support grids straight down for the full height of the plates to the bottom grids. A 32 mm tube can be used which is supported from the top grid and steadies at the bottom and has, at intervals, cross-tubes through which the wire is threaded and secured to give electrodes either side and parallel to the main tube (or mast as it is generally known). In other designs, tubular frames are suspended between the plates and the electrodes fixed to the frames. These are split into sections by cross-bars up the height, so that each electrode is 3–4 m long instead of the full height of up to 15 m.

The collecting electrodes, or plates as they are commonly called, are virtually flat plates, suitably stiffened to give the required stability, or a series of vertical channels between which the flat plates are assembled.

Another type is called the *strip electrode*; this is a series of strips each of which is rolled to a suitable section to give the rigidity required for the full height. These formed strips are about 0.5 m in width and there are about three to ten strips per zone.

It is necessary to ensure that the aspect ratio of the whole treatment zone does not exceed unity, that is, the height of the plates must not exceed the total length of the treatment zones. If this ratio is not adhered to, there is a risk that dust released from the top of the plate will not have time to reach the hoppers before reaching the outlet.

For the efficiencies normally required for a modern power station, the number of zones in series would not be less than four and can be as high as eight. The number of zones in series with the number of casings in parallel give the total self-contained zones per boiler, each having its own HV set and independent rapping. If this does not give sufficient performance with a zone (or zones) out of service because of a fault, it is necessary to subdivide the zones electrically by means of isolating switchgear.

5.4 Rapping gear

Although the majority of the dust collects on the plates, a certain amount adheres to the discharge electrodes and to promote the maximum current flow, both must be kept clean. Rapping gear is therefore installed for both. The majority of precipitators have some form of hammer which strikes the plate and electrode assemblies, the point of impact being within the precipitator since if the hammer strikes a shaft which is designed to transmit the energy to the required position, losses are incurred which makes the rapping gear ineffective, see Fig 4.9(a).

For the discharge electrode rapping, the hammers can be on a shaft across the width of the plant which are raised and released at the required intervals, all hammers striking virtually simultaneously the electrode support frames or grids, see Fig 4.9(b).

In another system, rods are lifted by cams and dropped. Although with this system the cams can be in the gas stream, they are usually housed above the gas chamber and the rods raised through a suitable gas seal and then released to drop and strike the electrode grids.

The rapping of the collecting electrode plates has to be arranged so that only a small area is rapped at the same time to avoid an unduly high carryover of dust. In this case hammer levers are spaced along a rotating shaft, each successive one being radially displaced so that each hammer in turn falls to give the required impact. With flat plates the hammer will strike on the leading or trailing edge and there can be more than one shaft within the height of the plate to give extra rapping. On strip plates, the hammers strike a bar connected to each electrode strip.

In another system, flat plates are struck vertically downward by rods which have been lifted by cams and released. The cams arranged along the shaft are displaced radially so that only one rod drops at a time.

The plate rapping gear has to be set to rap more frequently on the inlet zone, where there would be a heavier dust burden, than on the latter zones where most of the dust has previously been removed. There is so little dust collected on the discharge electrodes that these are usually rapped with the same interval at each stage. On most installations, the interval and the period of rapping is controlled by individual time clocks. On recent designs a microprocessor is used which facilitates adjustment and ensures that the rapping is sequential as it is detrimental to performance if more than one zone in one casing is rapped simultaneously.

5.5 Casings

For the most part, the casings for electrostatic precipitators are constructed in 6 mm steel plate suitably stiffened, but on older installations reinforced concrete was used. Unless there were particular local reasons, reinforced concrete would not now be used because it results in a very heavy casing requiring extra foundations; the time for erection on site is longer and, should corrosion occur later, it is difficult to repair.

The casings and support structures are designed to withstand the suction of the induced draught fans and take the loads imposed by the mechanical equipment as well as the weight of dust adhering to the plates and collected in the hoppers. The height of the casings makes any allowance for wind loading extremely important. The gas in the casing is usually 120°C for design purposes, but can rise well above this, and the construction has to allow for considerable expansion in all directions. The support structure is usually rigid and the casings are supported on this by rollers or sliding pads.

The casings are lagged to ensure that the gas does not come into contact with cold surfaces which would then cause condensation and acid attack from the sulphur in the coal. All the lagged vertical surfaces are protected by cladding and care is taken to ensure there is no *heat bridge* where the cladding support structure passes through the cladding. The roof has more dense lagging material with chequer plate on top for walkway and working areas, the whole, in most cases, being roofed over to prevent ingress of water and to give a protected area for maintenance.

5.6 Dust hoppers

The hoppers situated under the collecting plates are in most cases designed to hold a 24 hour accumulation of dust and because some 80% of the total dust is collected in the first hoppers, these have to be considerably larger than for the subsequent zones. All hoppers are designed to give an angle of slope in the four corners, known as the *valley angle*, of 55° from the horizontal to ensure that dust does not accumulate. The capacity of the hoppers is calculated with a low bulk density to allow for the aerated state of the collected dust, but the strength of the hopper and its support is based on a heavier density of dust that has compacted prior to its removal. Dust emitted from pulverised fuel boilers is free flowing while it is still hot, but when it cools in a hopper it can become damp and sticky. For this reason the bottom 1 m of the hoppers are electrically trace-heated and the whole of each hopper is lagged. The area round the hoppers for each casing is enclosed to prevent local nuisance from escaping dust and also to prevent wind-cooling of the hoppers.

The dust handling system for the collected dust, has to include a valve to prevent air being drawn into the casing via the hoppers.

In some cases the dust is removed continuously, a system that allows the first zone hopper to be made

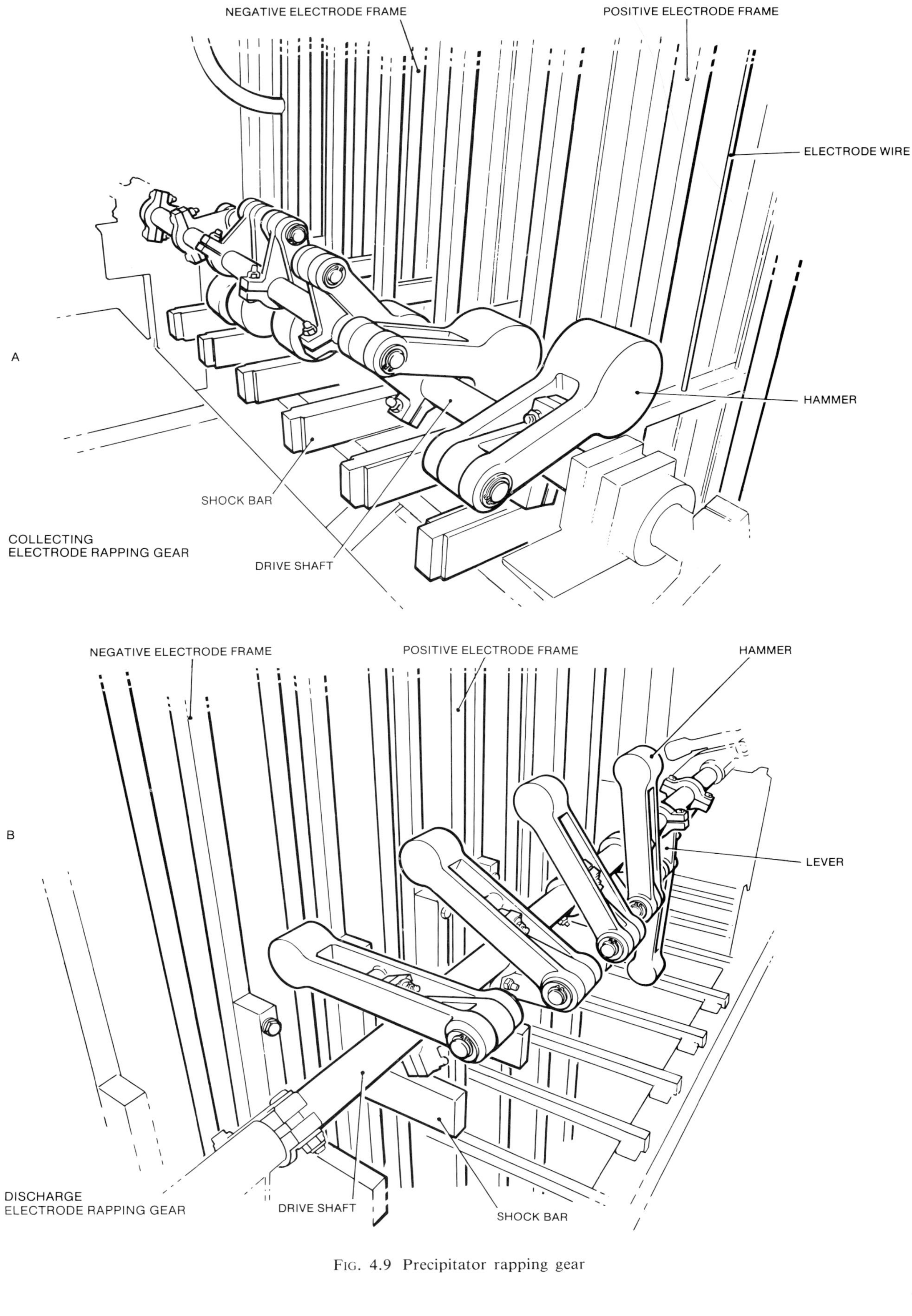

FIG. 4.9 Precipitator rapping gear

smaller but still having capacity to allow for the repair of dust handling plant before the dust level rises too high. A further advantage is that there is little likelihood of the dust cooling and being difficult to remove.

The hoppers are fitted with high level indicators which are some distance from the bottom grids of the discharge electrodes and give warning before the dust rises to a level where the electrodes are shorted to earth. In many installations, low level indicators are fitted to show that hoppers have been satisfactorily emptied during the ashing period.

In most cases a poke rod is fitted near the hopper mouth to break up any bridging of the dust that might occur. Chains from each top corner of the hopper can be attached to the end of the rod, which, when agitated, help to remove accumulated dust. Aeration pads to assist the flow of dust can also be used.

5.7 High voltage sets and control

On a particular station burning a consistent fuel, the collecting efficiency of precipitators would normally vary directly as the amount of power supplied to the electrodes, provided excessive flashover does not occur or other faults which would cause leakage paths to earth. Obviously, a certain minimum voltage must be applied in order to ensure a healthy corona discharge. The optimum operating voltage will be dependent upon the design of the precipitator, the gas condition and the characteristics of the dust to be collected.

The electrodes are energised from a high voltage transformer/rectifier system which converts the normal AC power to the required high voltage DC. They are usually designed on a basis of 50 kV, or 40 kV if barbed discharge electrodes are used with a space between the plates of 250 mm. The voltage would be decreased if narrower plate spacing is used, or increased with wider spacing. The rectifiers which are housed in the transformer tank have silicon diodes.

A slight reduction from the optimum power fed to the discharge electrodes, results in a significant reduction in the collecting efficiency and it is therefore important that sophisticated control systems are used.

The voltage to the precipitator is regulated by the firing of thyristors which increase the voltage from zero at a steady rate until a spark occurs. At this point the voltage is slightly reduced and then again ramped-up until there is a further spark. This process is continuous and thus the optimum is always sought. The rate of increase is adjustable. If, on the commencement of ramp-up after a spark, there is a further spark within a period of 180 ms, it would be considered as an arc and the voltage would be reduced to zero for an adjustable period up to 100 ms, this period being necessary to allow ionised gas within the precipitator to disperse. If the voltage remains at a very low value for a pre-set period in spite of attempts by the control to raise it, then the equipment would be tripped out and an alarm initiated.

An alternative system is by means of a controller which increases the primary input energy at pre-set intervals. It then compares the potential existing on the discharge electrodes before the raising step with the value obtained after. If the high tension voltage has, in fact, increased, the potential remains at this higher level until the next step. If the potential should, however, be below the original value, the controller reduces the primary input energy. In this way, the maximum high tension voltage which can be applied to the discharge electrodes is continually searched for and is maintained.

It is possible for all the functions of control to be by means of a microprocessor which can be programmed to carry out all the above activities. This has the advantage of lending itself to further sophistication if it is required in the future.

Because of the size of modern power station precipitators, it is important to reduce the power consumption whilst retaining the required collecting efficiency. A recently established means of doing this is by using only part of the waveform, which reduces the power consumption and causes a rather more peaky electrical input to the precipitator. Considerable savings can be made under certain operating conditions and there is a possibility that the modified impulse to the precipitator can assist in the collection of difficult dust.

A method of imposing high voltage pulses, superimposed on the normal DC output to the precipitator, has been developed, but before considering the use of this technique, it must be established that the extra cost can be justified by the increase in the collecting efficiency.

5.8 Gas distribution

The total gas quantity from the boiler must be split to give equal portions in each of the precipitators and, having done that, gas control devices have to be installed to give uniform gas flow in all parts of each zone. It is also essential to ensure that the gas cannot by-pass the treatment zone via the hoppers.

Because of the size of modern installations it is not possible to carry out all the work required by the installation and adjustment of splitters, guide vanes and baffles to obtain good gas distribution. Each change would result in an unacceptable amount of work and also add considerably to the time to commission a new plant. The work is therefore carried out on a scale model of between one eighth and one twelfth full size where adjustments can speedily be made. The model includes all the gas flues from the airheaters to the induced draught fans.

Measurements in the ducts are made with a pitot tube whilst in the precipitator, where the gas velocity is less, small diameter anemometers are used. Pieces of wool or smoke are used to visually check the true direction of the gas flow.

To obtain a satisfactory distribution of gas through the treatment zones of a precipitator, it is first necessary to obtain an even distribution of the gas in the ducts upstream of the inlet taper, this being done by installing splitters and vanes, particularly where there are bends in the flue. Having obtained this, it is then possible to install one or two perforated baffles in the inlet taper of the precipitator to spread the gas satisfactorily. The free area of various parts of the baffle are adjusted to give the desired result. Less elaborate baffles are required at the outlet end of the precipitator to prevent the flow contracting towards the duct.

Control vanes are added to the baffles as well as baffles between zones to reduce any gas flow into the hoppers.

A high standard of distribution is required, this being that 90% of the test points in a plane should be within ±25% of the average velocity in that plane. This is necessary because the loss in collecting efficiency caused by high velocity is not compensated by the gain in the low velocity areas. It is possible to obtain the required standard without incurring a high pressure loss which would put unacceptable load on the induced draught fans. The pressure drop across the precipitator is usually less than 12 mm water gauge.

Before precipitators are fully commissioned, the induced draught fans are run and gas velocities within the precipitators are checked. If necessary, adjustments are then made to bring the velocity distribution to the design limits.

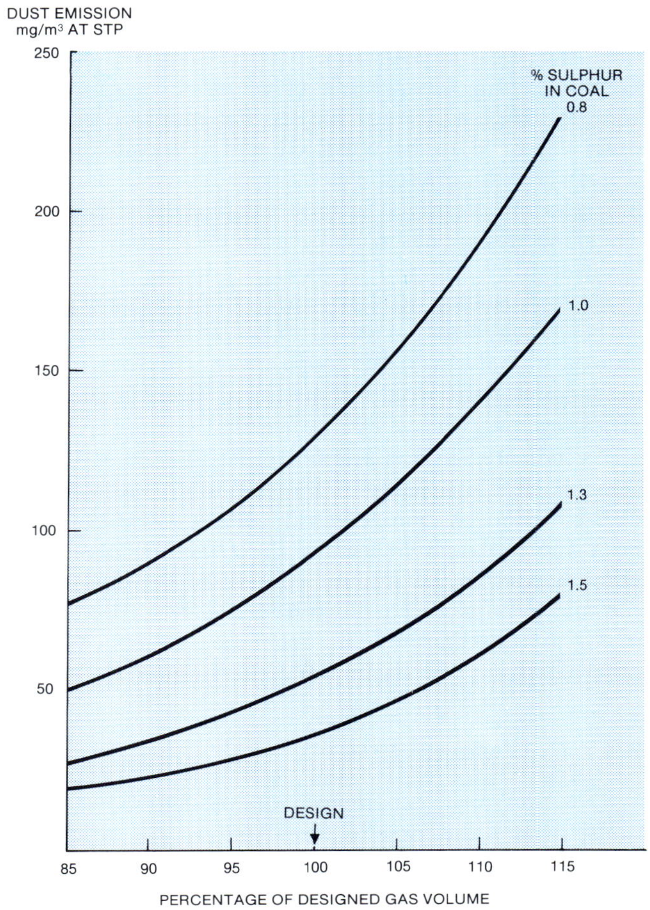

FIG. 4.10 Effect of sulphur and gas volume on dust emission

6 Factors affecting the design and performance of electrostatic precipitators

6.1 Gas volume

The gas volume entering a precipitator is expressed in terms of the quantity of gas in m^3/s at the operating temperature. For design purposes it is vital that careful consideration is given to this figure. In terms of the total cost of a station, the gas cleaning equipment is a small part, but with the strict environmental standards now required, its shortfall can result in the costly necessity of reducing load. If the precipitator has been designed on too small a gas volume, there is no simple solution to the problem. In some cases it might be possible to make improvements to the existing precipitators, however, because of site constraints, such improvements are always extremely expensive.

Figure 4.10 shows the typical effect of changed gas volume on dust emission plotted for various values of sulphur in coal.

The gas volume has to be calculated from the boiler heat output required and the range of coal that is to be fired. To this figure must be added adequate margins to allow for leakage, particularly across the airheater. Account should be taken of the possible deterioration in the performance of the boiler and turbine over the life of the plant, and any foreseen changes in the coal supply, since modern plant is designed for a life of 30 or 40 years and even more.

6.2 Gas velocity

For power stations the gas velocity is preferably between 1 m/s and 1.8 m/s in the treatment area, higher velocities tending to scour the collected dust off the plates. The required velocity is obtained by the correct sizing of the cross-sectional area of the precipitator whilst the required total plate area is obtained by altering the length of the plant.

Research work has shown that up to a gas velocity of about 1.5 m/s, the effective migration velocity tends to increase, but beyond this velocity the migration velocity falls. Since many precipitators are designed on velocities in excess of 1.5 m/s, it follows that excess gas volume will cause a drop in efficiency.

6.3 Specific collecting area

If high efficiency is required for a given installation, then the specific collecting area (SCA) will also be high to attain this with a given effective migration velocity. High efficiency means that more of the fine dust will have to be collected and, because of this, w (effective migration velocity) is reduced since the precipitator decreases in performance on fine dust. Thus the design has to take account of the SCA in its effect on w as well as its effect on the efficiency.

If, for instance, a plant with 99.6% efficiency was to be designed and data on effective migration velocities were only available from lower efficiency plants, use of this data would result in a shortfall in plant efficiency.

As indicated in this section, many factors influence the collecting efficiency of a precipitator and thus the dust emission. A change in gas volume alters both the SCA and the gas velocity and these factors, combined with changes in the sulphur content of the coal, have the major impact. To illustrate the effect, the curves in Fig 4.10 assume that no other factors are changed other than the gas volume and the sulphur.

6.4 Gas temperature

Within the range experienced on modern power stations, where the precipitators follow the airheaters and the gas temperature is between 120°C and 150°C, the gas temperature has little affect on performance. However, if the temperature is higher than envisaged in the design, the gas volume would have increased.

6.5 Coal analysis

The coal analysis obviously affects the gas volume, an important factor as already mentioned, but other characteristics of the coal are of particular importance to precipitator performance. The most important is the sulphur content of the coal, which can vary from 0.2% to 3.0%, because it affects the electrical characteristics of the dust as indicated in Section 6.6.2 of this chapter.

The ash content of the coal dictates the quantity of dust carried in the gas stream and thus the required collecting efficiency of the precipitator. The average ash content is about 20%, but it can be as high as 30% and down to 5%. High dust burdens from high ash fuels reduce the collecting efficiency for the reasons given earlier in Section 6.6.2.

For reasons that are not fully understood, high volatile coals give a dust which is precipitated more readily.

6.6 Characteristics of dust

6.6.1 Fineness

The fineness of dust entering a precipitator has a considerable affect on the collecting efficiency because larger particles are more readily collected than the finer ones.

A typical grading of PF dust entering the precipitor would be as follows:

Micron size (less than)	%
100	86
80	82
60	74
40	64
20	52
10	34
5	15

For the purpose of precipitator design this is not altogether a convenient definition of the dust size and therefore it is an advantage to express it by a single figure representing the specific surface area (SSA). A simple method for obtaining a reasonably accurate value for the SSA from a dust size grading, expressed as cm^2/g of dust, is given in Table 4.1.

It will be seen that the weighting factors are larger for the finer dust fractions and thus the SSA reflects the degree of difficulty in collection. A coarse dust grading could have an SSA of 2000 cm^2/g whilst a fine dust could be 3500 cm^2/g and, if nothing else changed, a precipitator handling the typical dust given above emitting 0.115 g/m^3 would emit 0.80 g/m^3 with the coarse dust and 0.160 g/m^3 with the fine dust. These figures are given to illustrate the effect of dust grading which has to be taken into consideration in the design stage.

The overall design of a precipitator has also to allow for a lower migration velocity in each subsequent stage as the dust becomes finer due to the preferential removal of the larger fractions, until the last stages are handling dust for the most part of less than 5 micron.

The measurement of the size distribution of this fine dust presents difficulties. Three methods are available and each is thought to give a grading that would be applicable to the conditions when the dust is in the gas stream. Unfortunately, all the methods need a certain amount of expertise to get consistent results from a particular instrument, and in any case the three methods give slight variations when the results are compared.

The cascade impactor has the advantage of collecting and sizing the dust in the gas stream. The particle-laden gas is drawn through a series of orifices of decreasing diameter at each stage, depositing particles within a given size range upon a target plate. Smaller particles remain entrained in the gas stream to be deposited in succeeding stages. After a period of ex-

TABLE 4.1
Calculation of specific surface area of dust based on typical dust

(1)	(2)	(3)	(4)
Micron Size	% in Range	Weighting Factor	(2 x 3)
+ 20 20 to 10 10 to 5 — 5	48 18 19 15	950 2200 4000 7500	45600 39600 76000 112500
		(5) Total	273700
Specific Surface Area $= \frac{(5)}{100} = 2737$ cm^2/g			

posure during which the gas duct velocity must be matched by the velocity through the impactor, it is dismantled and the weight of each size fraction in the various stages is weighed.

In experienced hands the instrument can yield reliable results in the range of dust below 20 micron, but care has to be taken to operate above the dewpoint of the gas in order that condensed acid and water does not affect the weight determination. Its main disadvantage lies in the expertise required and the time-consuming nature of each determination.

Malvern laser diffraction sizing is used on recovered samples which can be obtained in self-aspirating probes for the small quantities of dust required, provided the dust is less than 10 microns in diameter. Larger particles have to be obtained by isokinetic sampling in the gas stream. The recovered samples must be resuspended in a liquid phase. Water containing a non-ionic dispersant is a suitable medium for fly ash. Ultrasonic agitation for a period of some minutes is found to effect satisfactory redispersion of most fly ash materials. After agitation, the dispersion is transferred to the measuring instrument and measured without undue delay. The instrument depends on the diffraction of a low power laser beam as it penetrates the suspension of particles and once the sample is prepared results can be obtained within a minute.

The Coulter counter is also used for recovered dust and the same procedures as for the Malvern apply to the collection and dispersion of the samples. This instrument depends upon the electrical effect created by the passage of the small resistive particles dispersed in liquid between electrodes in a narrow bore capillary. Measurements take rather longer than with the Malvern laser diffraction device as each particle is examined separately during its passage through the capillary.

Both the Malvern and the Coulter devices are available with sophisticated microcomputer software which expedites the display of the results.

6.6.2 Resistivity

Since a precipitator is an electrical circuit, it follows that the resistivity of the dust to an extent regulates the flow of current. A layer of dust on the plates having a high resistivity will reduce the current flow and consequently there will be a reduction in the collecting efficiency. The prediction of the resistivity of the dust from a particular coal is at the moment not possible so the figure cannot feature in the design process, but its measurement can be useful in diagnosing the cause of poor performance on operating plant.

The measurement is to record the resistance of the collected dust layer under operating conditions and is more to do with the surface conditions of each particle rather than the chemical composition of the individual particles. Dust, such as silica which forms a large part of PF dust, has a very high resistance, but if its surface is coated with a conductive layer such as sulphuric acid, the resistivity of the dust layer could be low. It is now apparent that checking the resistivity of dust previously removed from the gas stream has little or no value because the surface condition changes very quickly once it is exposed to the atmosphere.

From the foregoing it will be seen that the sulphur content of the coal, the SSA of the dust and the quantity of dust, to a large extent regulate the resistivity. The sulphur in the coal combines with oxygen to form sulphur dioxide and some becomes sulphur trioxide which can combine with moisture to form sulphuric acid. The effectiveness with which this quantity of acid coats the particles is regulated by the quantity of dust, determined by the ash content of the coal and the SSA of the dust, the combination of which gives the total surface that has to be coated with the available sulphuric acid. Difficulty anticipated with a low sulphur coal might be offset if the ash content and the SSA were low.

If there is a high carbon content in the dust caused by incomplete combustion, this would reduce the resistivity, but it is not significant on modern boilers where the figure is normally less than 5%.

Normal resistivity figures are between 10^9 and 5×10^{11} Ωcm. High resistivity is generally considered to be in excess of 5×10^{11} Ωcm. High values not only reduce the current flow but can cause what is known as *back ionisation*. Normally the negatively charged particles are attracted to the earth positive plate where they give up their charge, thus the surface of the collected dust is positive and continues to attract negatively charged particles. With high resistivity dust, the dust layer prevents the newly deposited dust giving up its charge; thus the surface becomes negatively charged and repels the negatively charged dust particles.

6.6.3 Chemical analysis

In recent years investigators have been considering the chemical analysis of the ash presented to a precipitator since the characteristics of the coal did not always explain the performance of the precipitator. The work is still in progress, but it is generally considered that the more important constituents are as listed in the following, however, as more knowledge is gained, other factors may be found to have an influence:

Iron oxide (Fe_2O_3) is important since it acts as a catalyst which assists in the conversion from SO_2 to SO_3 in the gas and thus has an influence on the surface condition of the dust particles.

Calcium oxide (CaO), particularly if the percentage is high, because it tends to absorb the available surface acid and thus causes a reduction in the collecting efficiency of the precipitator.

Sodium oxide (Na_2O) is considered to be important, particularly where the sulphur content of the coal is low, say less than 1%.

In the majority of cases, the ash analysis becomes less important if the sulphur in the coal is high, say above 2% and becomes of primary importance below 1%.

6.7 Gas conditioning

High resistivity dust causes a fall in the effective migration velocity of the precipitator and, for a particular collecting efficiency, a larger precipitator has to be installed. High resistivity dust often results from firing low sulphur coal, which, as mentioned in Section 6.6.2 of this chapter, results in the dust not being coated with the conductive sulphuric acid. On large plants the extra cost can be considerable, particularly if the sulphur content of the coal is less than 1%.

In most cases this problem can be overcome by injecting sulphur trioxide (SO_3) into the gas stream, which has the effect of reducing the dust resistivity. The injection point in the duct has to be upstream of the precipitator and at a position to give sufficient time for the SO_3 to mix with the gas before entering the precipitator. The SO_3 is injected in very small quantities, between 12 and 18 parts per million is sufficient to raise the collecting efficiency very considerably. These small quantities do not increase the quantity of sulphur emitted from the chimney because it is converted to sulphuric acid and coated on the dust particles which are then collected. On a recent full scale test it was shown that as the injection rate increased, so did the acidity of the dust, but the SO_3 in the gas remained minute until the dust was fully coated. At this point, the acidity of the dust remained constant whilst the SO_3 in the gas increased. This illustrated the importance of having the right injection rate, but even if too much is injected it would not significantly increase the total emitted sulphur when it is considered that most of it is in the form of sulphur dioxide (SO_2) and, typically, is at concentrations of 1000 PPM in the gas.

7 Common causes of unsatisfactory performance

7.1 Excessive gas volume

One of the major reasons for a precipitator not operating with the required collecting efficiency, is an excess of gas volume. This can be due to a shortfall in the efficiency of the boiler, turbine or generator necessitating the overloading of the boiler to give the required electrical output from the generator; the firing of coals needing more excess air than allowed for in the design; air leakage into the system, particularly via the seals of the rotary airheater or high gas temperatures which expand the gas.

7.2 Effectiveness of rapping gear

It is most important that the discharge wires and the collecting electrode surfaces are kept clean, since one of the most frequent causes of the fall in performance can be traced to ineffective rapping. Ideally both of these surfaces should be completely clean but this is rarely possible, however, satisfactory performance can be attained with no more than a light dusting on the discharge electrodes with perhaps a 1.5 mm layer of dust on the collecting electrodes. If the dust is highly resistive even this can cause a fall in performance. On the other hand if the dust is very conductive, plants can operate satisfactorily with very much larger deposits of dust. It is therefore essential that satisfactory rapping gear is installed and that it is adequately maintained, particularly if difficult dust is to be collected.

7.3 Failure of discharge electrodes

The discharge electrodes are subject to failure and in the past have been a frequent cause of precipitator outage. The most common cause of failure has been due to fatigue of the wires adjacent to the fixing points and has been for the most part eliminated with improved designs. The other cause has been due to spark erosion and, again, this has been greatly reduced because of improved design and by the fact that, with the advent of thyristors in the control systems, the quick response time eliminates excessive sparking. With modern power station precipitators there are a large number of zones, so that when a zone has a faulty electrode it can be isolated without an undue increase in dust emission, until the fault can be rectified.

7.4 Change of coal

A precipitator is designed to collect the dust from a specified range of coals, but if a coal outside this range having characteristics which make for poor precipitation is fired, then the emission will increase. Even if the coal results in the same collecting efficiency, an increase in the ash in the coal will increase the inlet dust burden and thus the emission.

7.5 Automatic control

The various adjustments to the automatic HV control are set by the manufacturer or other competent person. If the equipment is incorrectly set, then the control will not respond to the conditions within the precipitator and the efficiency will fall.

7.6 Air ingress

Small quantities of air being sucked into the dust hoppers can cause re-entrainment of dust, therefore leakage via the dust handling system has to be eliminated. Air leakage anywhere on the precipitator can cause a local drop in temperature which can result in rapid corrosion of adjacent metal parts.

7.7 Overfilled dust hoppers

Overfilling of the dust hoppers will allow the dust level to rise until it reaches the bottom of the discharge electrodes and cause an electrical trip of that particular zone, thus the emission will increase until the fault is rectified. The high dust level can cause damage to the discharge electrodes, and if the level continues to increase so that it builds up within the gas paths between the collecting plates, then the all-important alignment of the electrode system will be affected.

7.8 Maintenance

The maintenance of a precipitator is obviously important, particularly when it is considered that many of the working parts are in a hostile environment. Modern plants are designed to operate without undue maintenance during the three years between major outages. During such outages the alignment and state of the discharge electrodes and the plates needs to be checked; the rapping gear needs inspection and in most cases any bearing surfaces subject to wear, need to be replaced.

Insulators need to be cleaned and inspected for cracks and replaced if necessary. To clean the precipitator, large quantities of water at low pressure are preferable to high pressure jets which can damage the plant and not wash away all the accumulated dust. Following the cleaning an inspection of the casing for any signs of corrosion is essential.

8 Draught plant

8.1 Introduction

The CEGB uses aerofoil-bladed centrifugal fans for induced draught, forced draught, primary air and gas recycling service on boilers. Flat-bladed fans, with heavily reinforced wearing surfaces and casings known as exhausters, are used to draw coal/air mixtures from suction type pulverised fuel mills and propel the mixture to the burners.

Variable-pitch axial flow fans have a significant share of the international boiler fan market. This type of fan is not in CEGB service although suitable fans are now available from UK manufacturers. The CEGB acting as Consultants for the Castle Peak *B* (Hong Kong) 660 MW boilers and Rihand (India) 500 MW boilers has approved variable-pitch axial fans for forced draught service.

8.2 Outline design

8.2.1 Centrifugal fans

Induced draught (ID) fans and forced draught (FD) fans have wide impellers and run comparatively slowly at 490, 590 and 740 r/min because of the large volumetric capacity and moderate pressure requirement. The primary air (PA) fans have narrow impellers and run at 1490 r/min to generate the high pressures required by the system resistance of the mills, pulverised fuel pipework and burners.

The gas recycling (GR) fans are nearly as large as the FD fans due to the high gas temperature and hence high specific volume of the gas at the economiser outlet. The fans run at speeds up to 1490 r/min to generate even the moderate pressures required for the service.

Figure 4.11 shows a typical large centrifugal fan for ID and FD service, and Fig 4.12 shows a typical primary air fan.

8.2.2 Variable-pitch axial fans

The components of a variable pitch axial fan suitable for forced draught duty are shown in Fig 4.13. For a given speed the hub diameter indicates the pressure rise capacity, and the blade height controls the volumetric capacity. For each hub size, therefore, there is a maximum and minimum blade height, the actual sizes being limited by aerodynamic and mechanical design considerations. Both the casing and the rotating assembly are considerably lighter than a centrifugal fan of comparable duty. A typical axial fan for FD boiler duty runs at 980 r/min.

8.3 Selection of fans

8.3.1 Duty and establishment of performance

Typical fan duties for a 660 MW coal-fired boiler are shown in Table 4.2. The performance of large fans is predicted by testing models in the manufacturer's laboratory. The models have impellers up to 1 m diameter and are accurately made and tested for performance and power consumption. Those tests which are relevant to the CEGB are witnessed and the results calculated independently.

The tests are used by the CEGB to check performance claims submitted in draught plant tenders. The manufacturers use the test data to predict the design and performance of full size fans capable of meeting the specified duties at a competitive price. Further information is derived from the full scale tests carried out on fans after manufacture. These provide essential information to the designer as predictions from model tests are not infallible, particularly for higher pressure fans.

For CEGB orders, a comprehensive performance test is carried out on the first fan of each type using the manufacturer's 2000 kW factory test rig with standardised airway to BS848 [3]. This ensures that the fan is capable of meeting the specified duties and that the power consumption at these duties is within the agreed tolerance. When assessing the relative merits of fans, it is essential to examine the absorbed power at comparable duties rather than loosely-worded efficiency claims which may be based on total rather than static pressure. Total pressure can conceal diffuser losses which are not recoverable in practice.

Site testing is used to determine the performance of the fan when connected to the boiler. However, it should be understood that a site test is inevitably inferior to the works test for the following reasons:

- The boiler output must be held steady so that pitot traversing can be carried out. One pitot traverse for a large duct would have at least thirty measuring points.
- Leakage in the boiler casing and ducts will affect the result.
- Bends in the boiler ducts both upstream and downstream of the fan will adversely affect the fan performance when compared to the straight airway to BS848 used on the works test.
- Determination of motor power is less accurate than is possible with the calibrated motor used for works testing.

Should a site test have contractual implications, the tolerances on power and performance and also the positions of the test points must be agreed in advance of the test.

8.3.2 Control of centrifugal fans

There must be a margin in pressure and volume between boiler continuous maximum rating (BCMR) duty and the design duty of the fan to cover the following factors:

- The varying constituents of the coal including ash, chlorine, sulphur and moisture. These can lead to slag build-up on high and low temperature surfaces, resulting in higher draught loss and increased volumetric flows.
- Leakage of air from the FD system across the airheater seals and into the ID gas stream.
- Fan casing seal leakage.
- Operation of the boiler with more excess air than the design intent.
- Changes in flue gas final temperature affecting pressure drops in the gas system, and increasing the volumetric flow of the ID fan.
- The draught produced by the chimney varies with ambient temperature.

Under ideal conditions, therefore, BCMR operation of the fan does not occur at the highest fan efficiency. The margin from BCMR to fan design duty is normally 20% volume with the appropriate 44% increase in head. Under the most adverse conditions this can be virtually all absorbed.

Two methods of fan control are employed; inlet vane control using radial vanes in the casing inlet (see Fig 4.14), and variable-speed control.

Inlet vane control

In some applications the driving motors are designed for two-speed operation so that the fan capacity at

FIG. 4.11 Typical centrifugal fan for FD and ID duty

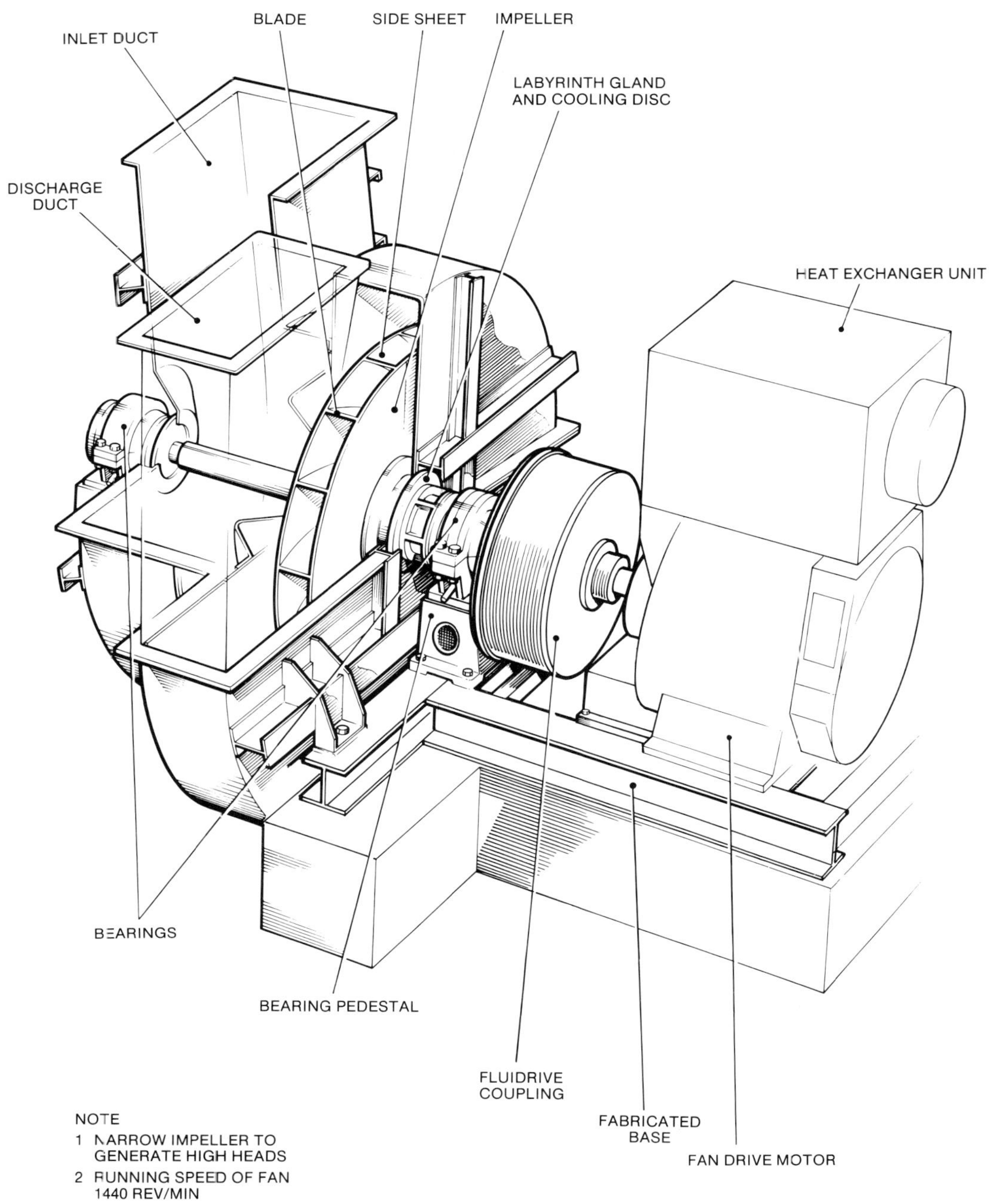

FIG. 4.12 Typical centrifugal fan for PA duty

low speed is sufficient to carry the BCMR of the boiler. The higher speed provides the margin mentioned. Almost all the two-speed motors are of 12 pole and 10 pole design, thus the speeds are about 490 and 590 r/min which result in this same margin. In some cases operation at BCMR is not achieved at low speed and the fans are run at top speed, thus negating the advantage of the two-speed drive. Figure 4.15 shows a set of fan performance curves with the vanes at various settings, the system resistance line shows the fall-off in fan efficiency as the flow and pressure decrease.

Variable-speed control

The efficiency of a fan controlled by varying the speed remains within close limits over a range of duties, but the speed changing device, whether by hydraulic coupling or motor, will have a varying efficiency characteristic. When selecting a variable-speed drive, care must be taken to ensure that its dynamic response characteristics meet the requirements of the boiler control system.

Speed control by hydraulic coupling is now used mainly for gas recycling fans and sometimes for primary air or exhauster drive. The coupling can be

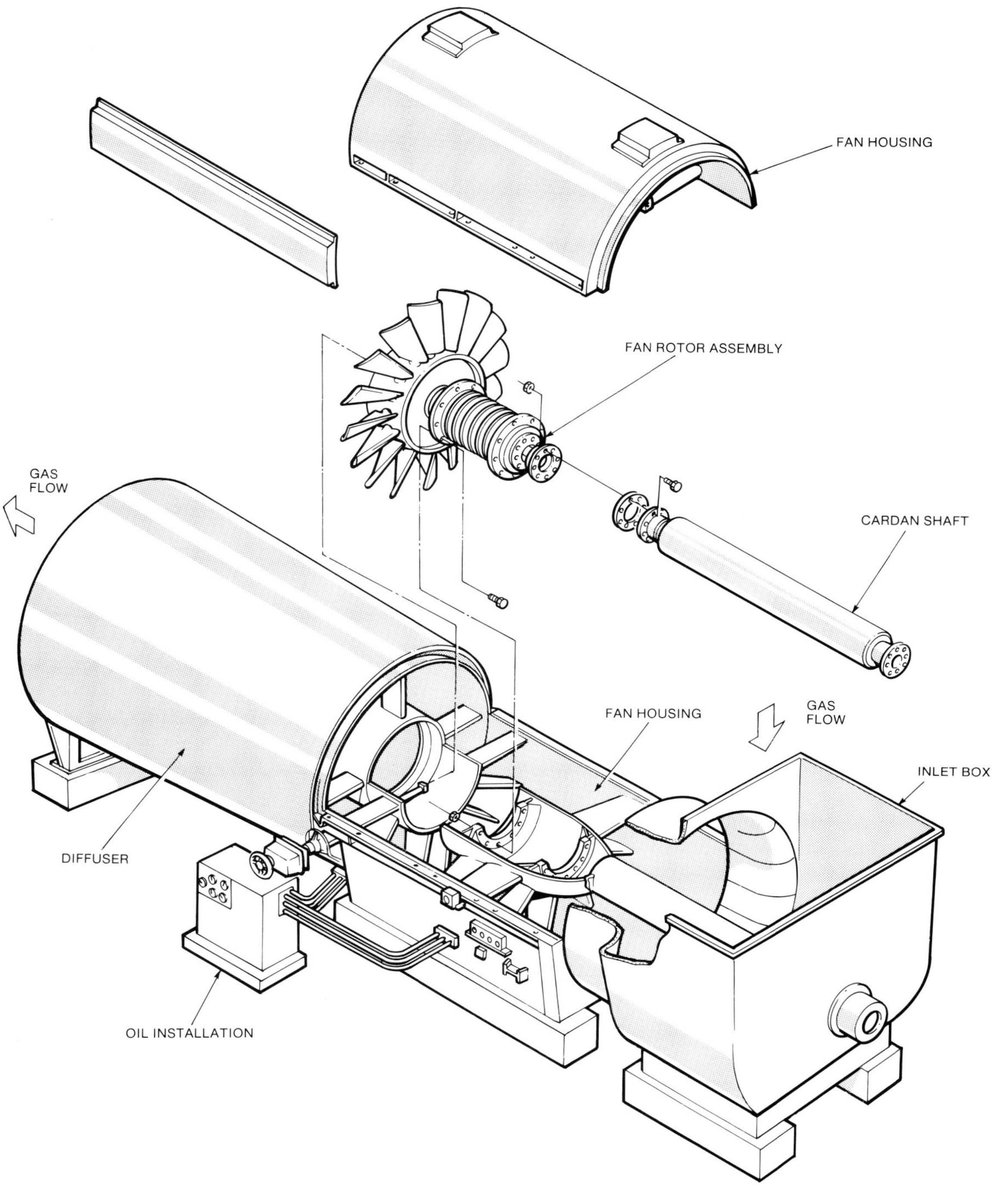

FIG. 4.13 Variable-pitch axial fan for FD duty

considered as two hollow bowls with radial vanes inside. One bowl is connected to the driving motor and the other to the fan, see Figs 4.16 and 4.17. If the driving bowl is filled with liquid and rotated by the motor, then under centrifugal forces, the liquid is flung outwards, constrained by the vanes and the bowl periphery. It will pass across the gap between the two elements and impinge on the vanes of the driven bowl. The forces created by this action appear as a torque at the output shaft. After circulating through the vanes of the driven bowl, the liquid returns to the driving element and the process is repeated. It should be noted that:

- There is no mechanical connection between input and output members.
- The power is transmitted by a ring of oil continuously circulating between these input and output members.

TABLE 4.2
Typical fan duties for a 660 MW coal-fired boiler

FAN PARAMETER	UNITS	BCMR DUTY	FAN DESIGN DUTY
Induced Draught Fan — two per boiler			
Volume	m^3/s	410.2	492.3
Pressure — static	mbar	35.97	51.3
Operating temperature	°C	120	120
Speed	r/min	590	590
Power absorbed at coupling	kW	1950	2900
Impeller diameter	m	3.92	3.92
Forced Draught Fan — two per boiler			
Volume	m^3/s	292.8	355.7
Pressure static	mbar	36.3	47.5
Operating temperature	°C	32	32
Speed	r/min	590	590
Power absorbed at coupling	kW	1390	2000
Impeller diameter	m	3.44	3.44

- If sufficient load is applied to the output shaft to stall the coupling, the driving member continues to impose the torque on the output shaft. When the overload is removed, the driven member will accelerate smoothly.

This difference between input and output speeds is normally expressed as a percentage of the input speed and referred to as *slip*. Couplings are selected so that the slip under normal full-speed working conditions is usually between 2% and 4%. In general terms, this factor can also be used for an estimate of the efficiency of the coupling. One of the advantages of the hydraulic coupling is the turn down ratio. The speed can be varied over a range of up to 5:1. Speed control of the hydraulic coupling is by varying the quantity of oil circulating between the impeller and runner.

Variable-speed motors which maintain a high efficiency over a wide range of speeds by means of inverter-converter control are now available in powers up to 10 MW. Motors and their control equipment can have capital costs up to three times that of single-speed motors. The speed response of the motors must be adequate for boiler control. An inlet damper or inlet vane control is sometimes fitted in addition to give a rapid initial response until the speed control meets the required conditions.

8.3.3 Control of variable-pitch axial fans

Control of variable-pitch (VP) axial fans is effected by a hydraulic cylinder located in the hub which connects to the pitch-change mechanism in the rotor assembly. Each blade is fixed to its blade shaft which forms part of the hub assembly, see Fig 4.18.

A typical performance curve for a VP axial FD fan is shown in Fig 4.19. It will be noted that the lines of constant efficiency have their longer axes approximately in line with the system resistance line, giving a greater range of high efficiency capacity than the centrifugal fan. This characteristic gives a power advantage over the centrifugal fan at part loads. However, with the CEGB station connected to the national grid, boilers spend only comparatively short periods at loads below 85% CMR.

8.4 Mechanical design and operation

8.4.1 Centrifugal fans

Finite element analysis

The impeller of a centrifugal fan is a highly stressed assembly, most of the stress arising from its rotation, and the rest from the aerodynamic and acoustic forces. Today with advanced computer techniques, finite element analysis enables the stresses in a particular impeller design to be calculated and plotted. This ensures that stress levels can be kept within acceptable limits, and assists in the production of an optimum rotor design.

Critical speed

A fan with an operating speed too close to its critical speed will be over sensitive to out-of-balance forces. Computer programs are now available to facilitate the calculations which determine the critical speed not only of the shaft and impeller but also the other items which must be taken into account in the assessment. These include the bearing oil film,

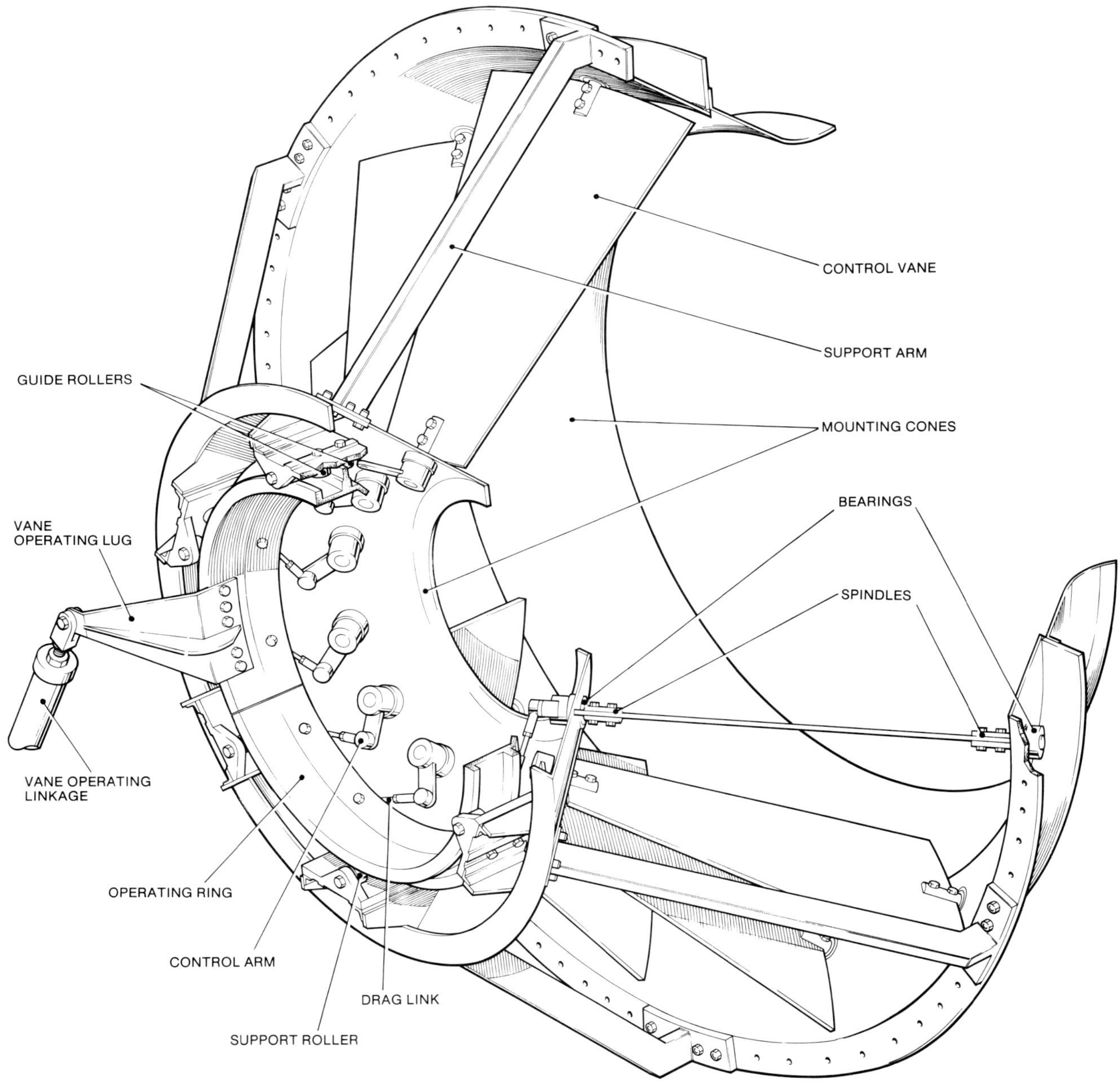

FIG. 4.14 Radial vane control gear for output control of a large centrifugal fan

bearing housing, pedestal, grillage, concrete plinth and piling. In general, the critical speed should be a minimum of 20% above the operating speed, and for fans in dirty and arduous service, such as mill exhausters, the critical speed margin must be considerably increased.

Bearings and bearing support stiffness

It is essential for the smooth running of a fan that the items which support the bearings from the foundation piling up are sufficiently stiff. Accelerometers used in conjunction with real time analysers (RTA) can be mounted on each part of the support system. From the vibration signature so obtained, a plot of the amplitude of vibration against frequency can be made, and any lack of stiffness in the components corrected. Measurements from existing installations enable adequately stiff components to be designed for new installations. Finite element analysis of foundations and plinths is available. This is a valuable technique for checking and optimising foundation designs.

Bearings for centrifugal fans are of split-sleeve construction, the white metal shells fit into spherical seats in the main bearing housing (see Fig 4.20). Lubricating oil is pressure fed from the oil pump through filters and air blast coolers to each bearing. For fan starting and rundown, high pressure jacking oil pumps are fitted to lift the shaft and impeller and so establish an oil film. The jacking oil is supplied at approxi-

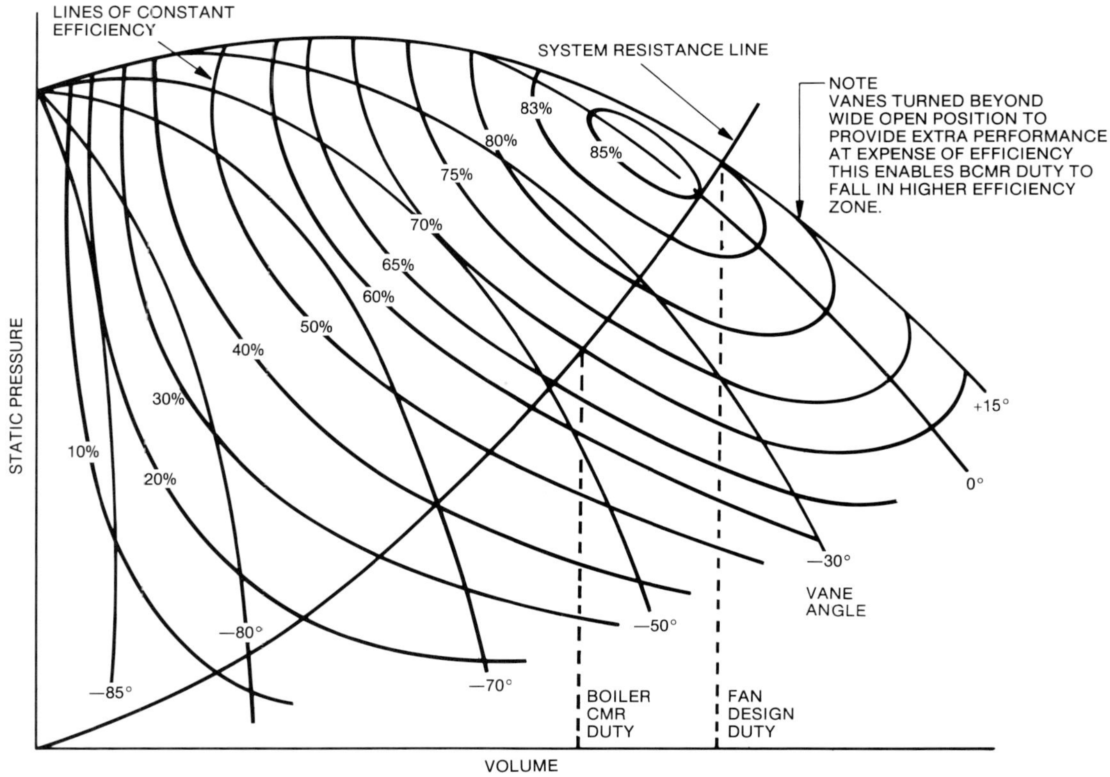

FIG. 4.15 Typical performance curve of centrifugal fan with inlet vane control

mately 50 bar. In addition to prolonging bearing life, the starting torque required from the motor is considerably reduced.

An electromechanical brake is fitted to the ID fan to ensure that when the driving motor is switched off the fan does not continue to turn at low speed under the influence of chimney draught, with consequent damage to the bearings.

Vortex bursting

Large fans controlled by inlet guide vanes can produce severe pressure pulsations when the inlet guide vanes are 40–60% open. The partly open vanes increase the strength of the pre-swirl vortex into the eye of the impeller. The vortex becomes unstable and produces pulsations in the gas flow. These pulsations can be overcome by fitting 'dorsal fin' plates to the casing, the plates are arranged axially along the shaft and direct the flow into the impeller eye.

Rotating stall

Another flow phenomenon takes place when the fan is operating at low load and is known as *rotating stall*. A local flow disturbance stalls one of the blade passages, in consequence the gas flow is diverted and increases the angle of attack on the blade nearest to the suction side. This in turn stalls, leading to flow recovery in the original stalled passage. In this way the stalled passage moves around the impeller giving poor flow conditions and vibration. Rotating stall is recognised by its characteristic of two thirds running speed fundamental frequency. The vibrations set up can lead to serious duct vibration downstream of the fan which can shake off the insulation and eventually crack the duct welding and plates. As in the case of vortex bursting, dorsal fins generally provide a solution to this problem.

Erosion and vibration

Many problems associated with boiler fans are due to erosion and vibration. Erosion arises from the ash carried in the gas stream and mainly affects hot primary air fans and to a smaller extent ID fans. Hot primary air fans are affected by dust picked up from the air preheater plates, and the impellers and casings must have adequate scantlings to ensure an economic life whilst maintaining the structural integrity between overhauls. If the output of these fans is controlled by varying the speed, some reduction of erosion is obtained.

ID fans for coal-fired boilers have hollow aerofoil blades with solid noses and protective liners where the blade is attached to the centre plate. This liner protects the blade attachment welding.

In the case of hot PA fans the impeller erodes in irregular patterns and the balance is upset, leading

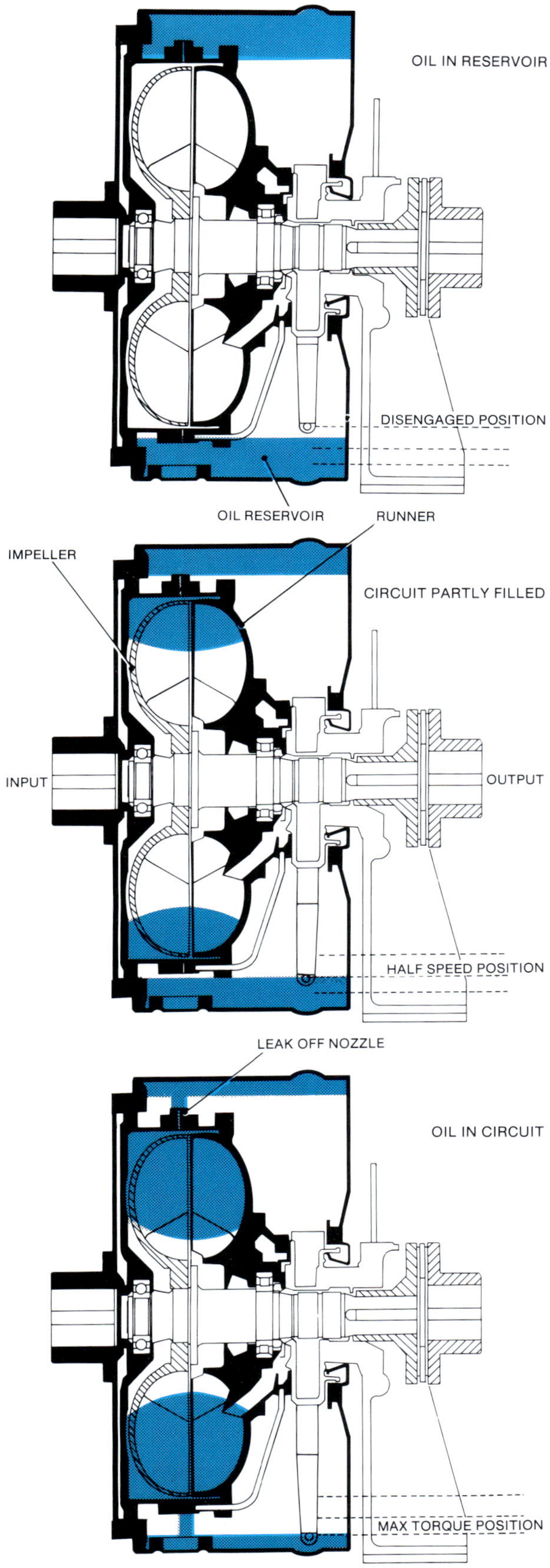

FIG. 4.16 Hydraulic coupling — principles of operation

to vibration. The high heads of up to 115 mbar and air temperatures of 250°C require fan maximum speeds of about 1480 r/min. The fan casings are fitted with renewable liners in areas liable to erosion.

Cold primary air fans and FD fans are not subject to erosion as the air is drawn from the area around the top of the boiler, thus contributing to the ventilation of the boiler house and giving preheat to the combustion air. For design purposes this air temperature is taken as 32°C; although in summer this can be exceeded.

FD fan circuits are sometimes fitted with a recirculation duct which allows hot air from the air preheater outlet to be drawn into the FD fan suction duct. Dust pick-up from the airheater can sometimes foul the FD fan under particularly adverse conditions.

Other problem sources

Other sources of trouble include incorrectly set shaft seals which cause the fan shaft to rub on the fan casing as the fan temperature rises to its normal level. Badly adjusted seals allow the emission of noise from the casing as well as leakage. Coupling misalignment can cause premature bearing failure.

Materials

Materials for centrifugal impellers must include in their properties, cold forming ability, weldability and high strength at both elevated and low temperatures. The low temperature properties, i.e., the Charpy V notch tests, are significant because a cold fan starting at zero or sub zero temperatures will reduce the air temperature even further due to the negative pressure. Steels now in use for impellers are low alloy, rolled, quenched and tempered and are designated RQT500 and RQT700 series.

It is essential that the quality of the plate for manufacturing impellers is maintained to the specification. Rigid quality control at all stages of manufacture of the impeller, the welding and the final annealing is also essential if the impeller is to be of the requisite quality.

8.4.2 Variable-pitch axial fans

Materials for axial fans depend on the service and operating temperatures. For forced draught service aluminium blades are used, this minimises the centrifugal load on the bearing which carries the blade spindle and on the hub.

Axial fans are not as robust as centrifugal fans and require maintenance on a scheduled basis to ensure their availability. The roller bearings which carry the main shaft, the thrust bearings, and the bearings which carry the blades all require replacement at not more than three-yearly intervals. Erosion of the blades will inevitably result in a loss of performance due to the increased clearance between the blade tip and the

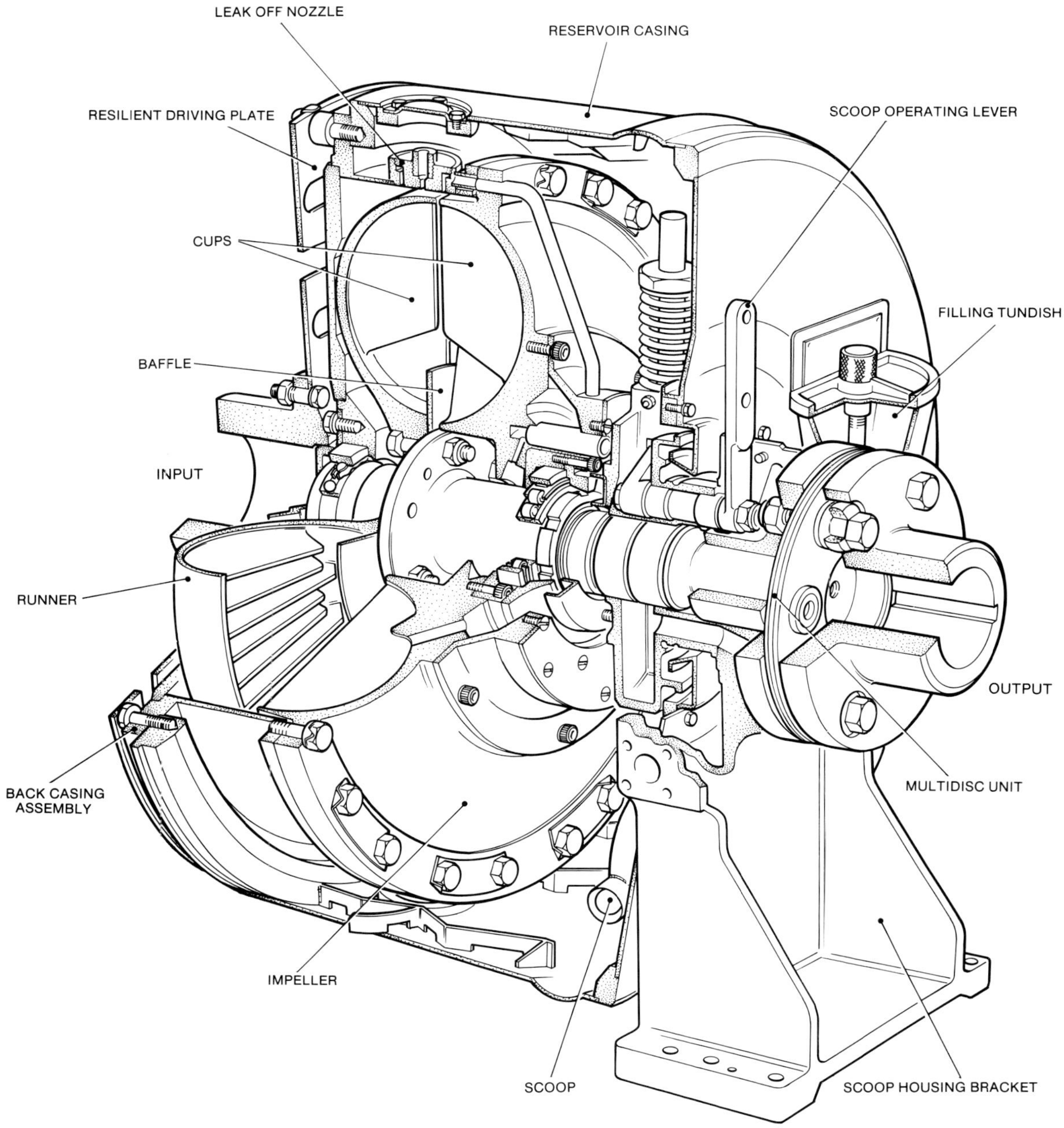

FIG. 4.17 Hydraulic coupling construction

casing and the change in blade profile. The hydraulic unit which actuates the blade pitch-change mechanism requires overhaul on the same basis as the bearings. Because of the considerably lighter weight of the axial fan rotating assembly it can be removed from the fan casing and taken to the workshop for maintenance in a comparatively short time. The hydraulic units of axial fans operating under conditions of constant load are known to stick. To avoid this it is usual to move the hydraulic blade-pitch mechanism through its full travel once a day.

8.5 Future trends

There are no particularly significant trends in draught plant for CEGB boilers; axial fans for forced draught service are now available and will be assessed on their merits when new plant is required.

New materials and improved design will enable a greater performance to be obtained from a smaller and lighter fan, but only small improvements in efficiency can be foreseen and these will be at premium cost.

The use of a single fan for FD or ID service in place of the usual two is feasible and will show a

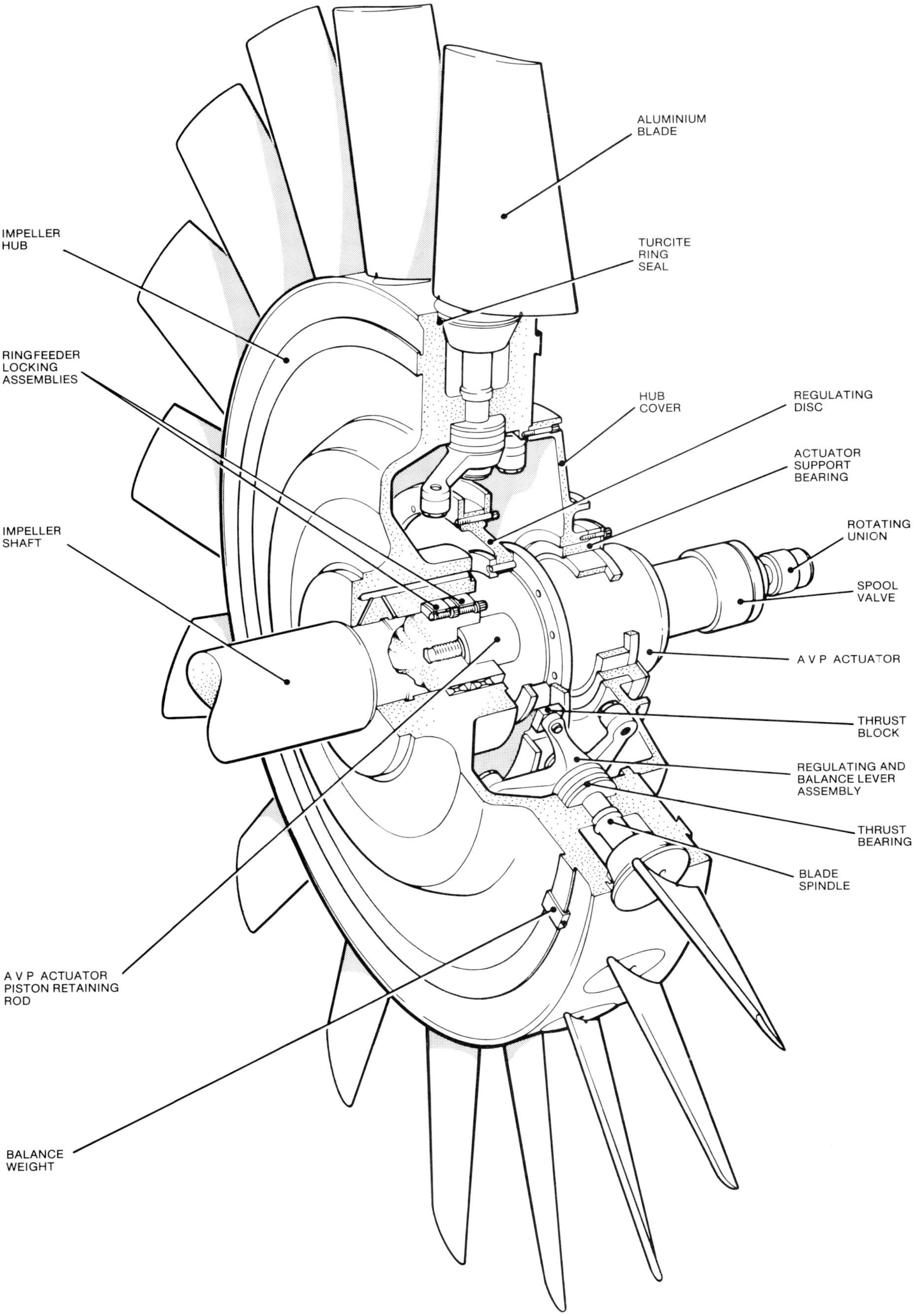

FIG. 4.18 Axial fan rotating assembly

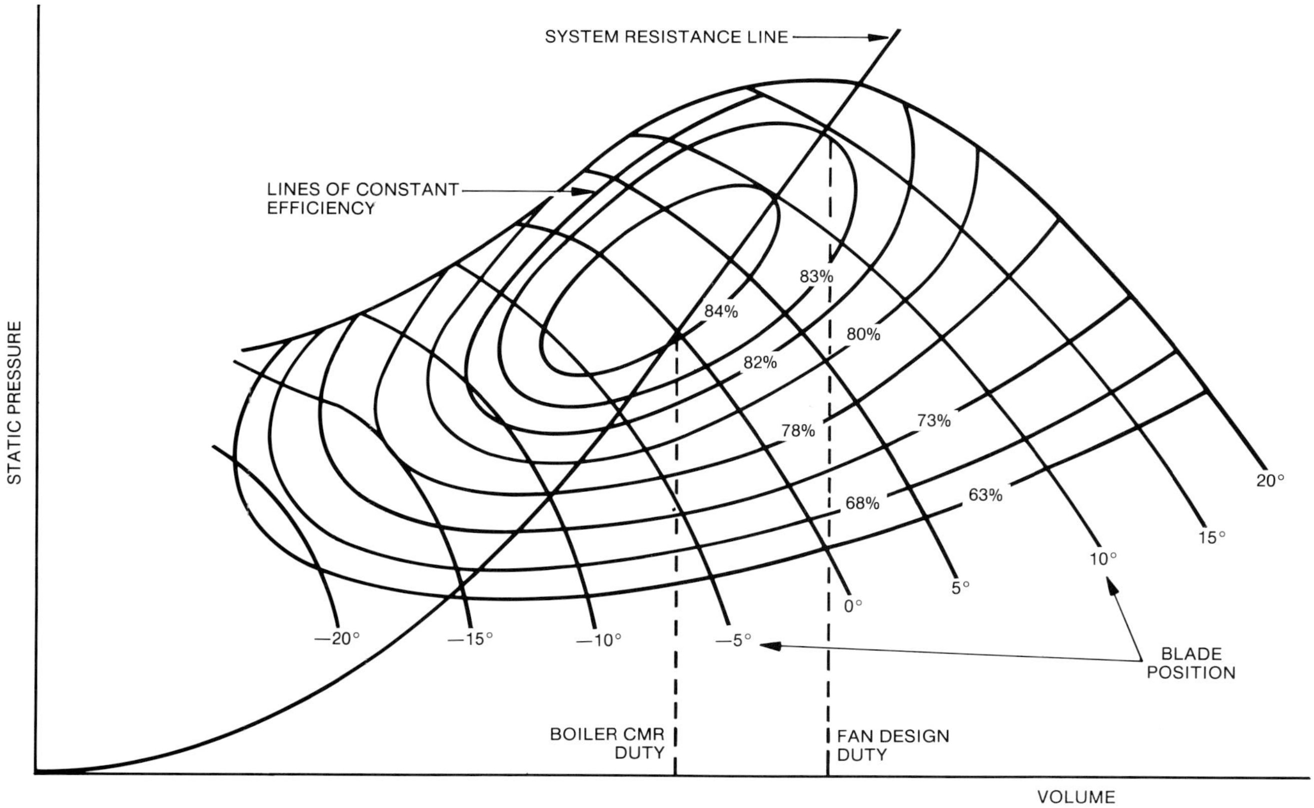

FIG. 4.19 Typical performance curve for variable-pitch axial flow fan

direct capital saving, but the cost of complete boiler outages while repairs are carried out to a fan or motor will soon negate the initial saving if the repairs are unscheduled. Very few large boilers have single fans.

9 Chimneys

9.1 Environmental considerations

Pollutants emitted from a chimney must first clear the area of turbulent air created by the wind around the chimney top. Nearby buildings will also set up turbulent areas and if the chimney exit is located within one of these then the pollutants may be drawn downward, even if the exit velocity is high. The first stage in chimney design is therefore to ensure that the exit is sufficiently high above the boiler house or nearby buildings and that the gases are emitted with sufficient upward velocity.

After leaving the top of the chimney, the gases are carried higher by the heat they contain and the thermal buoyancy that this heat gives them. This buoyancy can be maximised if all the gases from the power station boilers are emitted from a single chimney. For ease of maintenance a multi-flue chimney can be used. As they rise, the gases disperse and are moved along by the wind, eventually reaching the ground or local buildings. The air concentration of pollutant experienced at the ground will depend on the amount of pollutant emitted, the rise achieved after leaving the chimney and the height of the chimney. The calculation of this concentration is technically complex and requires sophisticated mathematical models. These models must take into account the topography surrounding the power station, local meteorology, the presence of tall buildings and other power stations or sources of pollutants. In severe cases, wind tunnel experiments may be needed.

The concentration of pollutants close to the chimney will, on a long term basis, be very small. They will rise quickly to a peak within several chimney heights and then decrease slowly. With a properly designed chimney the annual average concentration, even at the point of maximum concentration, will be much smaller than the total pollution experienced at the same distance from the power station caused by the many other sources surrounding that point.

Pollutants may travel long distances. As they travel they can be converted by chemical reactions into different chemical compounds. Eventually they will be returned to the ground, to buildings or to other surfaces such as trees, either directly by dry deposition as a gas or by wet deposition via rain, snow or fog. This happens regardless of the height at which the pollutant

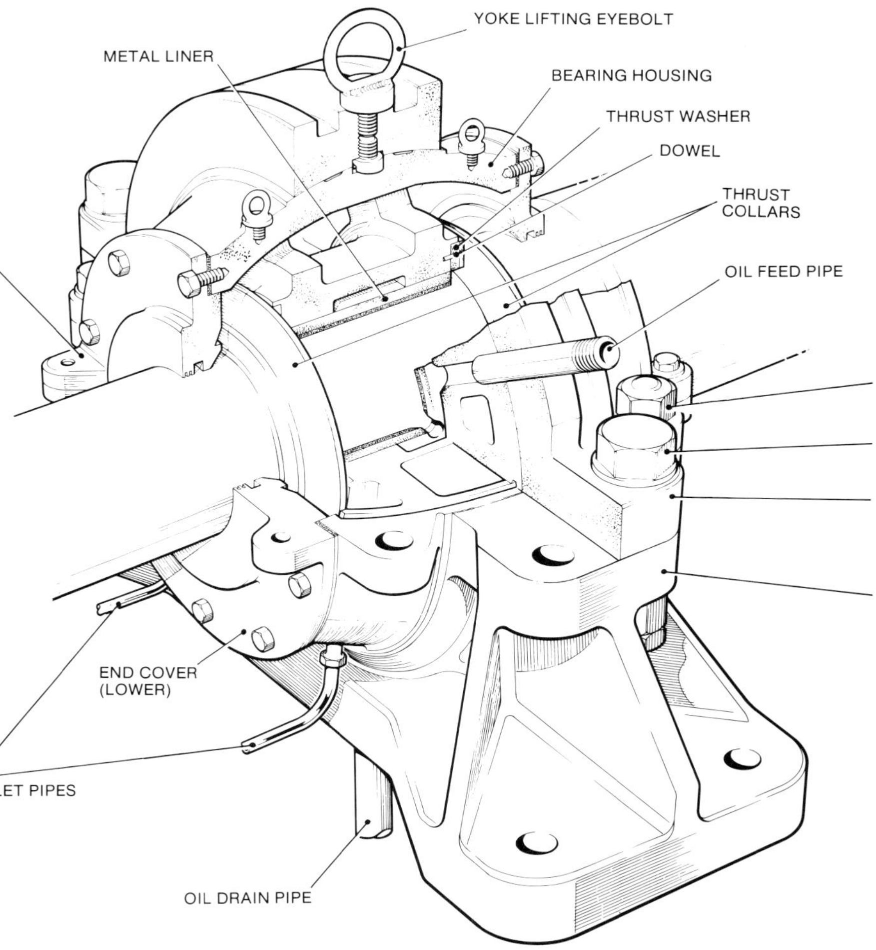

Note: The bearing shows the thrust collars which are only fitted at the driving motor end of the fan shaft. The shaft is free to expand towards the outboard bearing.

FIG. 4.20 Typical bearing for FD and ID fans

was emitted, and the height of the power station chimney is largely irrelevant in the consideration of international long range transport.

Emissions to the air from coal and oil-fired power stations in England and Wales are controlled by the HMIAPI within the Department of the Environment. Framework legislation for the control of air pollution from large industrial plant has recently been agreed for the European Community. UK legislation is being adapted to take this into account.

Power stations must obtain authorisation from the HMIAPI before they can operate and it is an offence to operate without this authorisation. The Inspectorate requires that the power stations employ the 'best practicable means' (BPM) to control air pollution and it sets out what this means in practical terms in 'Notes on Best Practical Means for Electricity Works'. The latest version of these Notes was published in 1975, although amendments have been agreed since that date between HMIAPI and the power station operators. A new version is currently under discussion.

The Notes cover a wide variety of design and operational aspects including criteria to be applied in the design of chimneys. They set emission limits for solid particulate matter and dense smoke. Future versions, when agreed, will include emission limits for sulphur dioxide and nitrogen oxides. HMIAPI also requires that the design of the chimney should be adequate to ensure acceptable ground level concentrations of gaseous pollutants. To assist in this it specifies that the design efflux velocity at the chimney exit should be not less than 15 m/s at maximum continuous rating, that flues should have adequate insulation to minimise the condensation of acidic gases and that, where practicable, emissions from separate boilers be combined into one single or multi-flue chimney.

One section of the Notes deals with oil firing and the need to reduce the emission of acid smuts. These are flakes of carbonaceous material which adhere to the interior surface of the chimney or ductwork. If the gases cool below a critical temperature, acid is deposited onto this material, which can then build up and be emitted from the chimney under conditions of changing load. HMIAPI requires that all surfaces in contact with the flue gases should be maintained above the critical temperature.

Although all new power stations will be fitted with systems for removing sulphur dioxide from the flue gases and burners which produce relatively low levels of nitrogen oxides, other air pollutants will remain in the flue gas and at certain times, such as start-up, the desulphurisation system may not be operational. There will still therefore be a need to ensure adequate dispersal of flue gases and the determination of an appropriate chimney height. In most cases it is unlikely that the chimney at a power station would be less than two and one half times the height of the associated boiler house. In difficult cases, or where more than one chimney is used, greater height may be needed.

The flue gas desulphurisation system cools the gases and a heat exchanger or some other device is needed to reheat the gases in order to provide sufficient buoyancy. For coal-fired units a temperature of not less than 80°C is adequate.

9.2 Design and construction

The function of a chimney is to discharge flue gases to the atmosphere at such a height and velocity that the concentration of pollutants, such as sulphur dioxide, is kept within acceptable limits at ground level. Brickwork makes a suitable structure for free-standing chimneys up to about 60 m high, but for taller chimneys the overturning moment due to increased wind load can be more economically resisted by a reinforced concrete shaft.

Flue gases from an oil-fired boiler have a sulphur content of 3–4% and a temperature of about 150°C; at or near to dewpoint, this results in a dilute solution of sulphuric acid being produced. It is therefore necessary to provide a lining to protect the concrete shaft internally from heat and acid attack. The lining is normally in the form of free-standing acid-resisting brickwork about 100 mm thick, which is self-supporting for a maximum height of 10 m. Consequently, the lining is built as a series of truncated cones carried on corbels inside the concrete shaft at 10 m intervals (see the inner chimney detail in Fig 4.21). There is a cavity 50 mm wide between the concrete shaft and the brickwork lining which may be filled with an insulating material or left as an air gap; the junction between successive sections of lining is sealed with glass fibre and lead to exclude flue gases from the cavity.

The lining is usually specified as dense acid-resisting brickwork to BS3921 [4], laid in potassium silicate mortar. Where alkaline or wet conditions may be experienced (as at the top of the chimney), a synthetic resin should be used instead of mortar to avoid softening of the joints which are normally kept as thin as possible, say 3 mm to 5 mm. Linings are usually about 100 mm thick for flues up to 6 m in diameter, but in the lower levels of the chimney and around gas entry points, a lining of about 200 mm thick is provided.

Coal-fired boilers have lower gas temperatures in the range 115–120°C and the sulphur content of the coal is lower but the construction is similar.

9.3 Multi-flue chimneys

The column of hot gases rising inside a chimney continues to rise as a plume without appreciable dispersion for some height after leaving the top of the chimney and so increases the effective height of emission; for example, with a chimney 200 m high the effective height of emission could be 500 m.

From the expression:

$$C \propto \frac{Q}{H^2}$$

where C = the concentration of pollutant at ground level

Q = the rate of emission of flue gas

H = the effective height of emission

It can be seen that an increase in the effective height of emission has considerable effect in reducing the concentration of pollution at ground level. Research has shown that the plume rise (which governs the effective height of emission) is largely dependent upon the heat content of the plume; therefore, for a power station with several boilers (each with its own flue to avoid reduced velocities when all boilers are not in service), the plume rise can be maximised by closely grouping the flues to concentrate the heat into one plume. This has led to the development of the multi-flue chimney in which several flues are enclosed within a circular reinforced concrete windshield, and since 1961 all power stations with a capacity of 1500 MW or greater (i.e., with three or more units of 500 MW or larger) have been provided with a multi-flue chimney having one flue for each boiler.

Multi-flue chimneys are of two main types. Figure 4.21 shows a reinforced concrete windshield enclosing

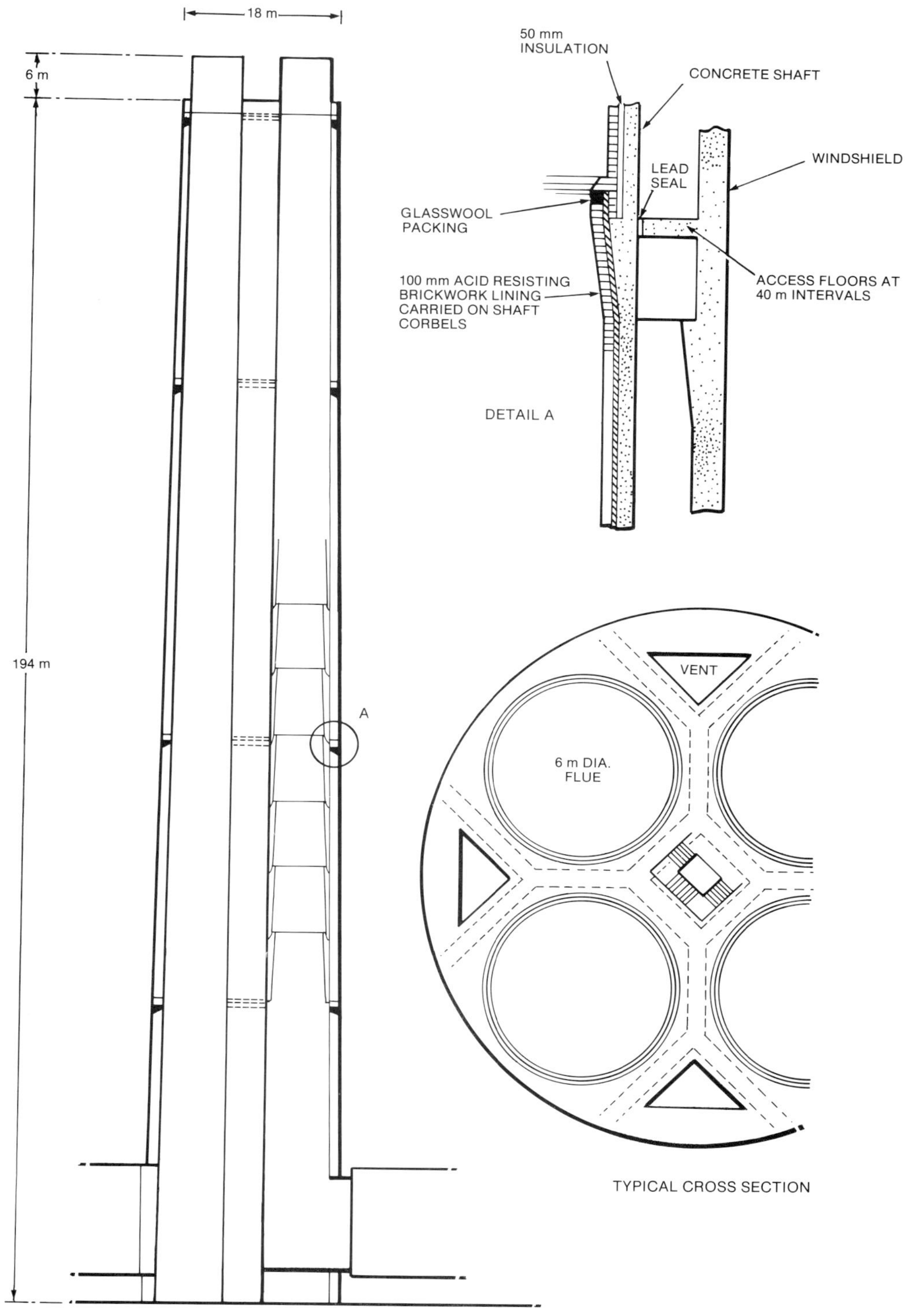

FIG. 4.21 200 m multi-flue chimney with free-standing flues

four free-standing reinforced concrete shafts with linings as already described. Floors are provided in the interspace between the chimneys and the windshield at approximately 40 m intervals for access and servicing of aircraft warning lights.

Figure 4.22 shows a reinforced concrete windshield which encloses flues formed only of lining brickwork. In this type the reinforced concrete chimney shafts are omitted, and the sections of flue brickwork are carried on a series of floors at about 10 m intervals. One feature of this type is the necessity to have deep beams supporting each load-bearing floor. In order to reduce the spans and deflections of these beams, various methods have been employed including the

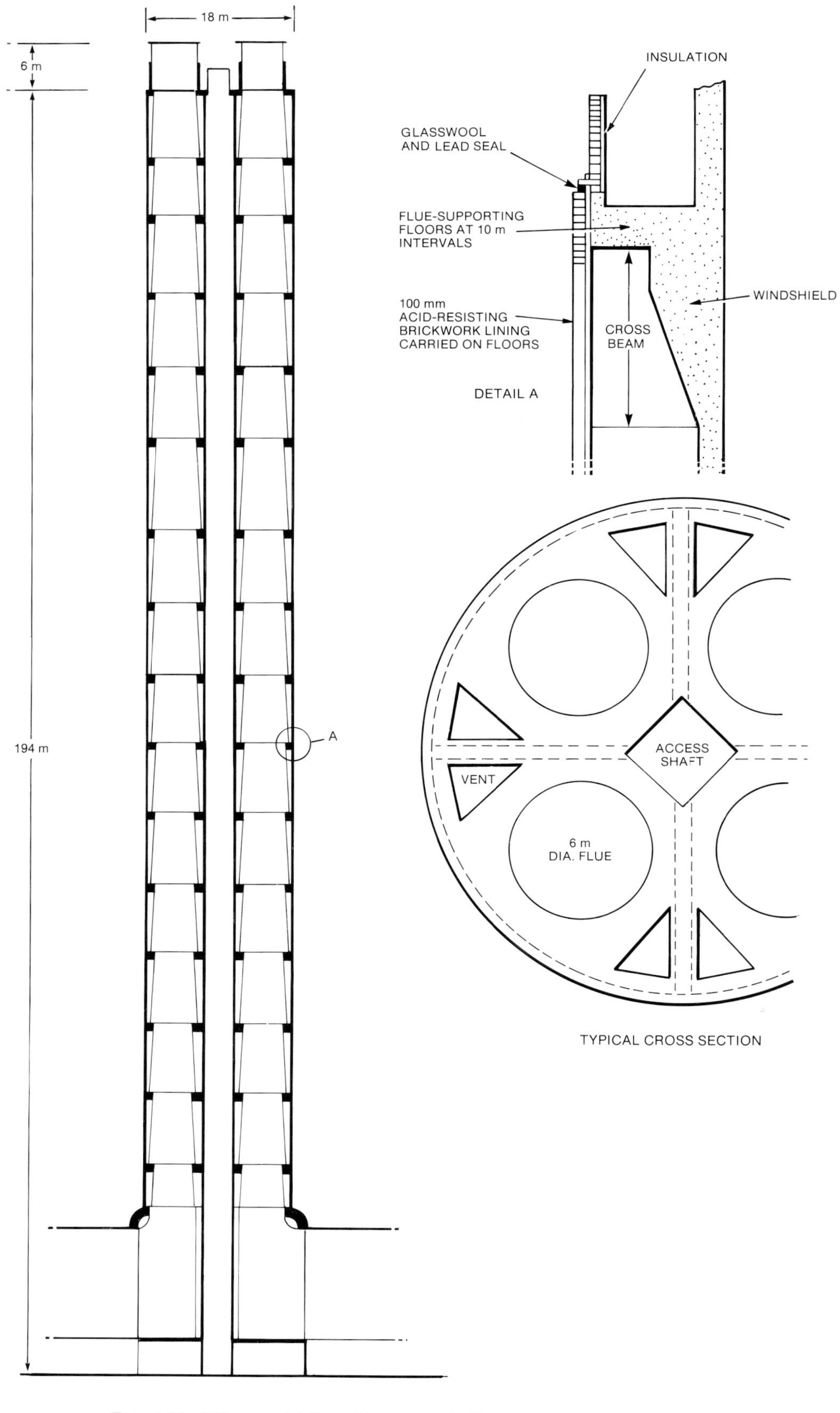

FIG. 4.22 200 m multi-flue chimney with flues supported on windshield

provision of a central column which can also serve as an access shaft, and the propping or tying of the floor beams at or near their centre points to the windshield.

9.4 Flue design

The basic parameters for flue design are the height of the flue, the temperature, the efflux velocity and the rate of emission of the flue gases. The diameter of the top of the flue will be determined from the rate of emission and the efflux velocity, the latter being kept as high as practicable to minimise downwash of the emission. The flue gases are brought from the boiler by the action of the induced draught fans through ducts as far as the base of the chimney flue. The pressure head which causes the flow of gases up the flue is the result of the difference in density between the flue gas and external atmosphere. It is good practice to maintain a slight negative pressure inside the flue to reduce gas leakage, and therefore a balance must be maintained between, on the one hand, the head available through density difference and, on the other hand, the losses caused at entry and exit and by friction in the flue. If the chimney is undersized or gas flow fluctuates excessively, a positive pressure can be created in the flue and cause gas leakage. Details at flue entry vary; where entry is from beneath it is usual to provide a *lobster-back* bend of steel plate in order to reduce friction and turbulence, in which case the flue lining commences at about 25 m above the chimney base. In the case of side entry it is usual to provide *splitters* (i.e., deflector plates) over the area of entry to reduce turbulence in the gases due to the right angle change in direction.

The multi-flue chimney of a 2000 MW station would be about 200 m high with flues about 6 m in diameter and efflux velocity about 23 m/s; for a 4000 MW station the chimney would be about 260 m high and the efflux velocity about 26 m/s.

10 Flues and ducts

10.1 Introduction

Dependent upon the boiler layout, flues and ducts can account for up to 8% of the capital cost of the boiler. Optimisation of the design and layout is essential to achieve a balance between capital cost and operating economy of the draught plant.

10.2 Flue and duct design

The layout and aspect ratios of flues and ducts are generally dependent on the space available between the boiler and the boiler support steelwork, rather than on an ideal layout requiring the least number of bends and changes of section.

Economic air and gas velocities are based on the capital cost of flues and ducts including the effect of weight on the support steelwork, set against the incremental capital and running cost of the draught plant. As the capitalised cost of power varies for each station the economic velocities will be marginally different. However it is recognised that 16 m/s velocity for gas flow and 14 m/s velocity for air flow will enable preliminary design work to proceed.

Where flue gas is carrying a heavy dust burden, particularly in the flue between the airheater gas outlet and the precipitator inlet, a higher than normal velocity is necessary to ensure that at reduced boiler load the gas velocity is not decreased to the point where the entrained dust falls out and collects in the flue. At low loads, however, fallout is inevitable. If the boiler is shut down after a period of low-load running, it is prudent to inspect the flues for excessive dust levels and to remove excessive dust before restarting the boiler. If this is not done the surface of the dust will tend to set if moisture is present so that on start-up the accumulation of dust will not be re-entrained into the gas stream.

Flues liable to dust build-up are designed to withstand a depth of dust equal to 10% of the duct height, and in a few vulnerable areas 25%. For design purposes the dry dust density is 1040 kg/m^3. Where the dust is liable to collect water from leaks and wash water, the dust density used is 1615 kg/m^3.

At the boiler design stage the correct assessment of draught losses in the flues and ducts is important as it affects the pressure rating of the fans. Overestimation results in the provision of a fan with an excess pressure characteristic. If the losses are underestimated the fan pressure margin will be eroded to meet the CMR duty.

Items which affect the flow in ducts include changes in cross-sectional area involving acceleration and deceleration of the flow. Wherever possible, changes of cross-sectional shape should be achieved with constant area to minimise losses. The biggest losses in the flues and ducts arise from the bends and attention should be paid to the following aspects of bend design:

- The flow into the bend should be even.
- The inner radius of the bend should be as large as the layout permits.
- For radius bends the mean radius of the bend should be two to four bend widths.
- Flow splitters should be provided.

The length and nature of the section downstream of a bend has a major influence on the pressure drop. The secondary flows existing after a bend will increase the pressure drop in the downstream ducts. It follows from this that when several bends are linked by only short sections of duct then the flow pattern will be so irregular as to preclude accurate calculation of the total losses.

It is essential to ensure that the flues are as gas tight as possible. The flues are under negative pressure generated by the ID fan and any leakage of air into the gas stream will have two effects:

- It will reduce the temperature of the metal round the leak, leading to corrosion.
- It will reduce the gas stream temperature, thus increasing both the density and the gas weight passing to the fan with inevitable increase in power consumption of the fan and a reduction in the draught available from the chimney.

Circular-section ducts have certain advantages and disadvantages over rectangular ducts:

- Generally no internal struts are required thus reducing the power and friction losses in the duct.
- The curved plates reduce the vibration modes of the duct walls thus reducing the noise levels.
- The sound radiating surface is reduced.
- There is a significant reduction in weight both of the duct and the supporting steelwork.
- There is a reduction of insulation costs.
- Considerable longitudinal strengthening of the circular duct is essential as the circular section does not possess the same inherent strength as the rectangular section when acting as a beam supported on trestles.
- The layout of circular ducts must be carefully checked to ensure that the configuration cannot produce vortices which will increase the boiler draught losses.
- Dampers in round ducts are more elaborate, have a reduction in free area and are more expensive to manufacture.
- Transportation and handling of round sections during manufacture and site erection are more difficult and expensive than rectangular sections.

10.3 Mechanical design aspects

The weight of the ducts, of the insulation and cladding, and of the dust and water from leaks are all significant factors in the mechanical design. The internal pressure loading, whether positive or negative, is taken as the maximum pressure which the fan can generate. A typical maximum pressure at operating temperature for FD and ID fans is 55 mbar and for primary air fans, 120 mbar. Also to be taken into account is the axial force of the gas flow acting over the cross-section of the duct. These forces must be included in designing the supporting structures, the anchor points, the interconnecting tie bars and expansion joint gimbals.

Internal tubular braces are used to reduce the section sizes of main girth steelwork. When calculating the stresses in the girth members the corners are not assumed to be completely rigid and pin-ended factors are used in the structural design. The internal tubular braces in dust-laden gas streams must be protected from erosion. It has been found that tacking heavy-weight steel angles to the upstream side of the brace preserves its structural integrity. Eroded sections of these angles are easily replaced during boiler overhauls.

Most flue temperatures do not exceed 400°C and mild steel plate to BS4360 Grade 43A is used [5]. High temperature flues such as those in economiser bypass service use 1% chrome 0.5% molybdenum steel to BS1501.620.27B [6].

Flues and ducts external to the boilerhouse must be designed to withstand wind loads and snow loads and the methods are set out in British Standard CP3 Chapter V Part 2 [7]. The snow loading includes an allowance for occasional access.

10.4 Dampers

Multiblade dampers (Fig 4.23) are used for isolation and modulation for control purposes. Although the mechanical design of the damper is the same whether for isolation or control, some modulating dampers operate by opening only one blade of the multiblade assembly when low flows are required. Separate actuators are normally required to control these blades, but one actuator working through a differential linkage can also be used.

When in good condition and properly adjusted, dampers will provide reasonable shut off, but damper blade edges and seals deteriorate in service, especially in dusty conditions, and a gas tight seal after a period of operation cannot be expected. Dampers can and do remain in one position for months at a time while the boiler is steaming. Careful design is essential to ensure that the blade spindles do not seize in the bearings, also that the blade linkage is strong enough to take the thrust from the actuator if the bearings are stiff from lack of use. Damper blade shaft bearings were a major source of trouble until new designs mounted the bearings clear of the damper frame, and provided a repackable gland which sealed the shaft as it passed through the damper frame, thus preventing hot air blowing onto the bearing housing or ambient air being drawn into the gas stream. The bearing is self-aligning with a sealed-for-life PTFE or oil-impregnated bush.

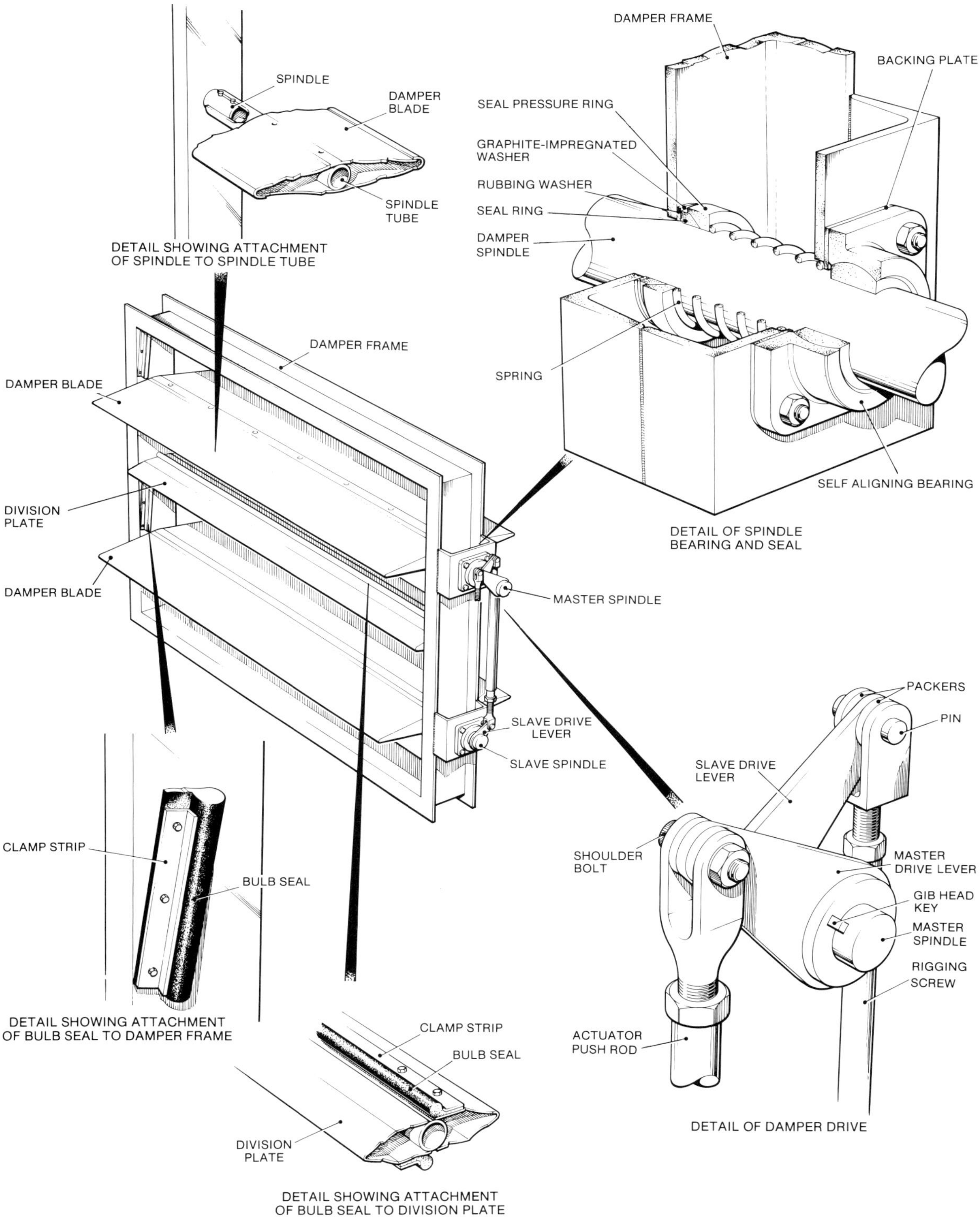

FIG. 4.23 Typical damper construction

In addition to multi-blade dampers, guillotine dampers are available with one sliding blade which seals into its frame in the duct, see Fig 4.24. Various methods of sealing are used, one type has a fan mounted on the damper frame which can feed ambient air into the seal to pressurise it. This pressurised air prevents gas seeping through the slight inaccuracies in the seal construction. The seals are made from stainless or other high alloy steels. The blade can enter the flue from the top, bottom or side, although most of those in CEGB service are top entry. The aperture in the flue through which the blade enters is sealed by a

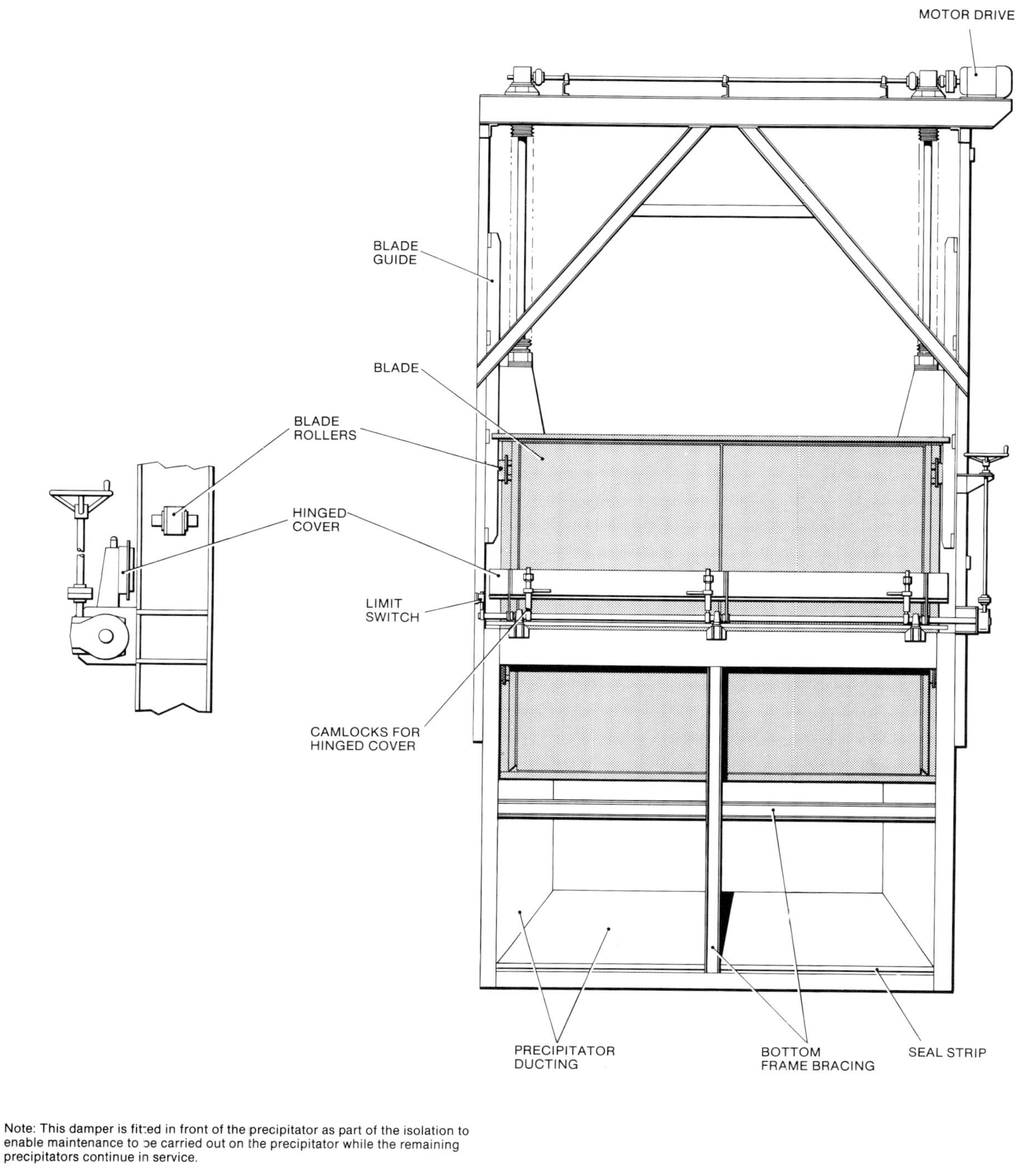

FIG. 4.24 Curtain damper for precipitator inlet isolation

hinged cover which can be opened and closed either manually or electrically. This type of damper is used as part of the system employed to isolate the inlet of each precipitator while the boiler is operated at reduced load. A conventional damper is placed upstream of the guillotine damper and the space between them is vented to atmosphere.

Isolation of primary air systems by guillotine dampers can be difficult because of the limited space available for the frame which holds the guillotine when the damper is open. To overcome this, twin-seal flap isolating dampers have been installed at Tilbury *B* for several years to isolate the primary air fans and mills so that on-load maintenance can be carried out on these items. Figure 4.25 shows the damper in a workshop jig and Fig 4.26 the seal detail. The space between the seals is vented to atmosphere to ensure that there is no pressure to cause flow past the secondary seal.

FIG. 4.25 Twin seal flap isolating damper
(see also colour photograph between pp 66 and 67)

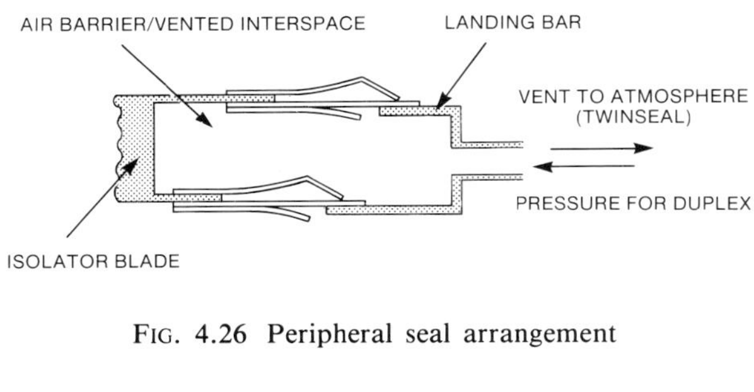

FIG. 4.26 Peripheral seal arrangement

10.5 Expansion joints

Flexible bellows are inserted at strategic positions to take up the thermal movement of flues and ducts, and to ensure that thrust from the movement of flues and ducts is not transmitted to the casings of fans, precipitators, airheaters and other plant. The anchor points must be designed to transmit all loads and thrusts to the supporting steelwork or foundations. Metallic expansion joints generate thrusts which can be calculated from the spring rate of the convolution and the movement cold to hot. Most metallic joints have one or two convolutions, but where the temperature range is high several convolutions are required.

A multi convolution joint is shown in Fig 4.27 which is designed to accommodate 25 mm axial movement. The joints are preset by stretching or compressing before erection, the preset dimension equalling the free length plus or minus half the movement to be compensated. Effective insulation of the thin metal convolutions is essential to avoid corrosion by weather and water leaks.

The fabric for non-metallic joints is woven from high quality glass fibre or ceramic threads. The woven cloth is then covered with organic polymer coatings such as fluorelastomer. The coatings vary with the temperature and the type of contaminant in the gas. Fabric joints are intended to run as cool as possible and the ambient side is not insulated. These joints

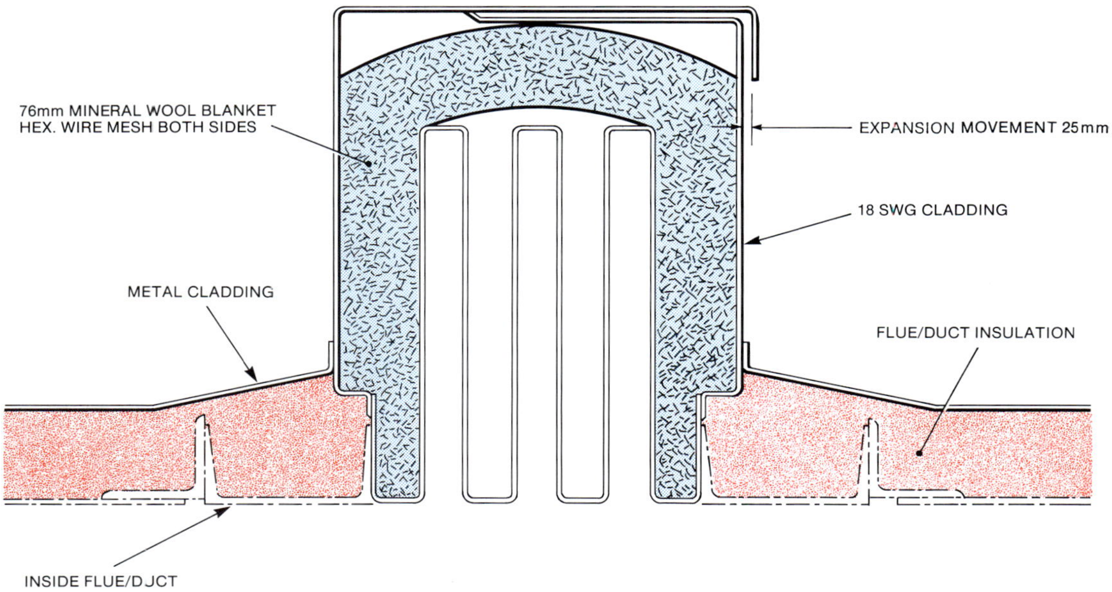

Note: The metal cladding must ensure that the insulation and bellows are fully protected from the weather and water leaks.

FIG. 4.27 Typical metallic bellows expansion joint

are used where the relative movements of ducts and plant make metallic joints impractical or expensive. Figure 4.28 shows a simple one-layer arrangement, although the seal can consist of several layers, the inner layers which are subjected to the highest temperatures are treated with graphite compounds to provide flexing and abrasion resistance. To ensure a gas tight joint a gas tight membrane of PTFE film is included in the multilayer joint.

An interesting example is the joint which connects the main flue to the chimney at Drax. The gas from two boilers passes in a common flue to the chimney inlet bend. The steel bend in the base of the chimney hangs from the chimney structure at the 38 m level, the bend expands downwards and the flue which is supported from the ground expands both upwards and longitudinally. The fabric joint thus absorbs almost 50 mm vertical shear movement in addition to the axial compression. Each expansion joint measures approximately 14 m × 9 m.

Figure 4.29 shows the location of expansion joints in the air system of a 660 MW boiler.

10.6 Future trends

For flexibility of operation, crossover ducts have always been provided on both the air and gas streams to enable part-load operation with sections of plant out of service. The high availability of fans and airheaters has resulted in little use being made of these crossovers and instead of conferring an economic advantage they have become a maintenance liability, particularly those in the gas flues which tend to fill with dust. For future plant, these crossovers will be deleted, saving the capital cost of the ducts and flues, and the associated dampers, expansion joints, supportive steelwork and foundations.

11 Thermal insulation

11.1 Introduction

Thermal insulation has three objectives, safety of personnel, plant integrity and heat conservation. The cost of insulating a new 660 MW boiler is over 3% of the total cost of the boiler and it is essential to ensure that the completed installation can withstand the arduous conditions of temperature and thermal cycling experienced in day-to-day operation. To this end the original insulation specification, the installation and the quality assurance procedures must be sound.

Economic thickness requirements for insulation will often vary widely due to such factors as fuel cost, projected plant and insulant life time, current cost accounting methods and variations in the value of the heat at different parts of the thermodynamic cycle.

11.2 Materials

Man-made mineral fibre (MMMF) is produced from molten basalt rock or slag by blowing or spinning into a fibrous mass of thin grey-brown fibres of varying

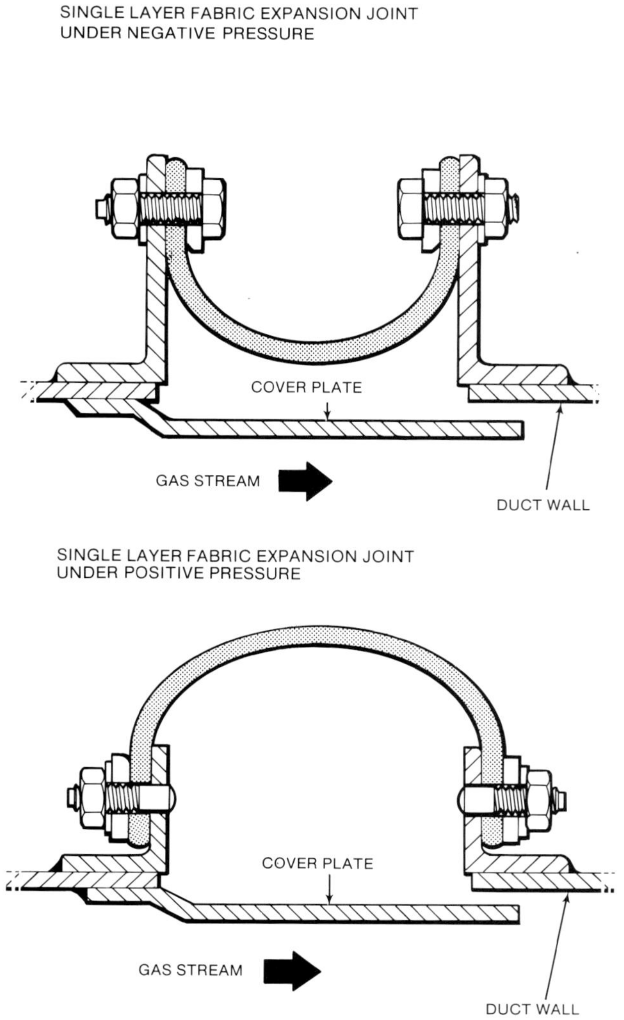

FIG. 4.28 Single layer fabric expansion joint

lengths. The formulation of rock or slag based MMMF will vary considerably, and so will their melting points and general heat stability. It is therefore essential that material is selected on the basis of its thermal stability.

MMMF produced from rock and slag is supplied in various forms such as quilted mattresses with the fibres held loosely in open weave cotton or attached to wire netting. In this form the mattress is flexible and can be used to insulate items of complex shape. Its disadvantages are that the MMMF easily compresses and will slump if not supported.

Preformed MMMF consists of fibre lightly compressed and held in a rigid predetermined shape by binders. Organic binders (usually phenolic resins) are widely used, but although the fibres can withstand temperatures in excess of 600°C, the organic binders will burn-off above 250°C. This leaves the hot face of the insulation in a flexible form and, while in some applications it is acceptable, it reduces the ability of the insulation to withstand compressive loads. This can be overcome by using material with a density of not less than 120 kg/m^3. Inorganic binders, mainly derived from bentonite clays, are now available for MMMF preformed sections. This binder significantly increases the ability of the sections to withstand compression after exposure to temperature. The thermal resistance of MMMF with inorganic binder is not quite as good as that with organic binder. Preformed shapes include pipe sections and flat slabs.

Calcium silicate is composed predominantly of reacted hydrous calcium silicate reinforced with suitable fibres. These reinforcing fibres must be asbestos-free and the finished product is coloured yellow to indicate this. The material is rigid and is supplied in preformed sections, also radiused and bevelled lagging sections. It has greater compressive strength than MMMF but is liable to crack from thermal expansion and mechanical damage.

Ceramic fibre is made from alumino silicates and has a maximum temperature range from 1000°C to 1500°C depending on the exact composition. Ceramic fibre is more expensive than MMMF and there is no justification for using it below 600°C. It is used in conjunction with refractory materials in boilers.

MMMF products based on glass are similar to rock and slag based products but have a lower melting point and 500°C should be regarded as their temperature limit. Glass is less robust than rock based fibre and should not be subjected to compressive loads.

Metallic cladding can be aluminium, sheet steel coated with hot dipped aluminium, plain galvanised steel or galvanised steel sheet coated with BSC Pvf colourcoat.

11.3 Thermal design

Heat is transmitted from furnace walls, flues and ducts by conduction through the insulation and by convection and radiation from the surface of the cladding.

The steady state heat flow can be described as conduction through the insulation followed by convection and radiation heat transfer from the surface:

$$\frac{Q}{A} = \frac{k_i}{x}(t_w - t_s) = h(t_s - t_a) + s(t_s^4 - t_a^4)$$

where
- Q = heat transferred
- A = heat transfer area
- k_i = conductivity of insulation
- x = thickness of insulation
- h = convective heat transfer coefficient
- s = radiant heat transfer coefficient
- t_w = duct or furnace wall temperature
- t_s = cladding temperature
- t_a = ambient temperature

REF	LOCATION
A1	AIR HEATER BY-PASS
A2	HOT AIR CROSSOVER
A3	AIR HEATER AIR OUTLET
A4	HOT AIR DUCT TO BURNERS
A5	SECY. AIR DUCT TO BURNERS
A6	SECY. AIR DUCT TO BURNERS
A7	PRIMARY AIR DUCT
A8	PRIMARY AIR DUCT
A9	PRIMARY AIR CROSSOVER
A10	PRIMARY AIR CROSSOVER
A11	PRIMARY AIR CROSSOVER
A12	PRIMARY AIR CROSSOVER
A13	PRIMARY AIR FAN SUCTION
A14	PRIMARY AIR FAN SUCTION
A15	PRIMARY AIR FAN SUCTION
A16	PRIMARY AIR TO MILLS
A17	PRIMARY AIR TO MILLS
A18	AIR RECIRCULATION
A19	AIR RECIRCULATION
A20	PRIMARY AIR CROSSOVER
A21	FD SUCTION
A22	FD SUCTION

A23	FD SUCTION
A24	FD DISCHARGE
A25	COLD CROSSOVER
A26	COLD CROSSOVER
A27	PRIMARY AIR FAN SUCTION
A28	TEMPERING AIR
G1	ECON. BY-PASS
G2	BOILER GAS OUTLET
G3	BOILER GAS OUTLET
G4	AIR HEATER GAS INLET

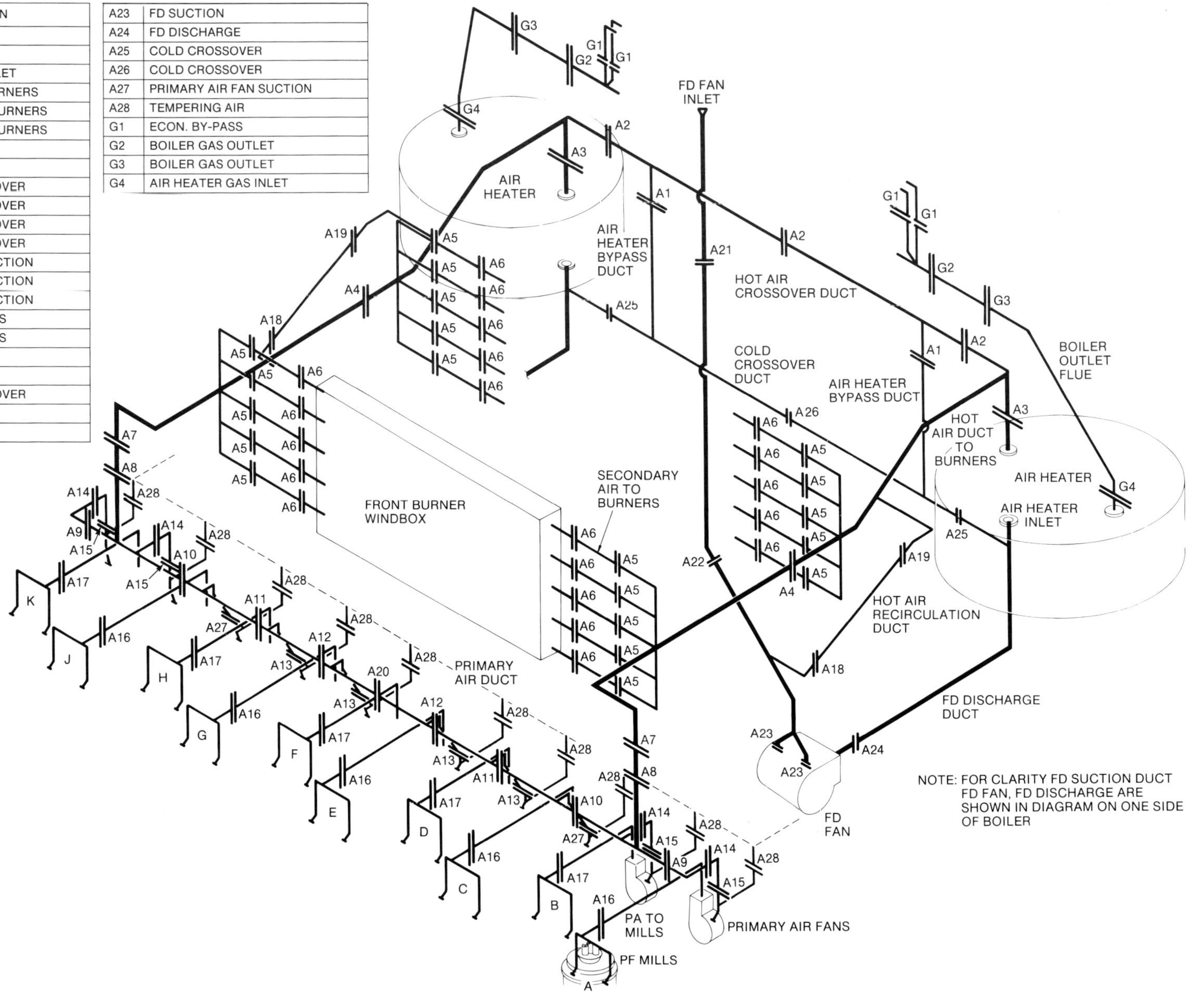

FIG. 4.29 Location of expansion joints in 660 MW coal-fired boiler air system

For design purposes, the convective and radiative components can often be combined, using a composite coefficient h_c to describe the heat flow as:

$$\frac{Q}{A} = \frac{k_i}{x}(t_w - t_s) = h_c(t_s - t_a)$$

As the coefficients of the convection and radiation components are dependent on the air flow over the surface and the radiative characteristics of the surface finish, both of which are temperature dependent, a composite coefficient can only be employed in clearly defined circumstances.

Where personnel have access to cladding, reduction of surface temperature t_s to a safe level is the main design criteria. Where surface is inaccessible, the design requirement is the combination of achievement of the best economic balance between insulation cost and heat losses and maintenance of surfaces at temperatures which avoid corrosive conditions. In many instances, safety requirements have also to be met in accordance with the requirements of the appropriate British Standard.

11.4 Installation

All recent combustion chamber and boiler walls are of finned-tube membrane construction and the fins provide a base for the studs which secure the insulation slabs and metal cladding to the boiler. Various types of stud are used, forked studs which hold the insulation slabs to the tubes, round studs with push-fix washers and screwed studs which fix the insulation but in addition provide support for the metal cladding system with which the boiler is finished. The studs are welded to the fins and spaced to give adequate support to the insulation and cladding. Figures 4.30 and 4.31 show two arrangements of boiler wall thermal insulation fixing.

The MMMF insulation slabs are impaled on the studs, the total thickness is divided into two layers to enable the butting edges of the slabs to be staggered. This avoids heat paths through the insulation which would occur if only one layer were used. After the slabs are fitted, wire mesh is applied to secure the slabs. Square expanded-metal washers are fitted to the forked studs and the tines of the studs are opened and pressed back into the insulation. Where round studs are used, push-fix washers hold the wire mesh in place.

The cladding support rails are fixed to the screwed studs by means of base plates, which are fitted with an insulating pad to reduce heat conduction and the transmission of vibration to the cladding. To prevent convection currents behind the cladding a barrier of high temperature insulation is provided at the support rails. The profiled cladding sheets are riveted to the support rails. The pattern of riveting is so arranged that differential expansion of the sheet does not cause unsightly buckling.

Parts of the deadspace roof and sides are insulated internally, the method of supporting the insulation is the same, i.e., forked studs, but the final finish is black dipped expanded metal.

Because of the varying steam temperatures in the headers and pipes in the deadspace at the top of the boiler some of the headers therein require insulation and others do not. MMMF is used in this location in the form of flexible mattresses or rigid pipe sections. The insulation is held in place with steel mesh and J-bolts, all made from stainless steel.

Zinc in contact with austenitic stainless steel pressure components at temperatures approaching the melting point of zinc can cause failure of the stainless steel by embrittlement. At high temperatures this failure can be catastrophic. Zinc should not be employed in any form in contact with stainless steel above 250°C. Embrittlement of galvanised ferritic steels by exposure to temperatures above 250°C is a phenomenon that is not widely appreciated. Galvanised steel must not be used in boiler dead spaces.

The windbox, ducts and flues are built with longitudinal and circumferential stiffeners welded externally to the plates. The external surfaces of these parts are insulated with mineral wool, slab-impaled on forked studs or round studs. The roof slabs are usually high density, i.e., 200 kg/m^3, where they are liable to be walked on. Metal cladding is usually profiled galvanised sheet inside the boiler house and aluminium profiled sheet outside. The external stiffening sections provide the basic structure for the rails on which the metal cladding is supported. It is important that the external stiffening sections are adequately insulated, see Fig 4.32.

The cladding on the top side of the flue is provided with sufficient slope to ensure that rainwater runs off.

To reduce the noise emanating from the fan casing to specified levels, lead-loaded cloth is used to cover the insulation before the metal cladding is applied, see Fig 4.33.

12 Flue gas desulphurisation

12.1 Introduction

In the UK the CEGB has recently announced that it will retrofit a sufficient number of generating units to maintain, taking one year with another, the decline in sulphur dioxide emissions which has occurred over the last decade. In addition, all new coal-fired power plants are likely to be required to be fitted with sulphur dioxide control technologies to the requirements of HMIAPI.

It is not generally appreciated that there are numerous methods of reducing emissions. Some of these methods are as follows, from which it can be seen

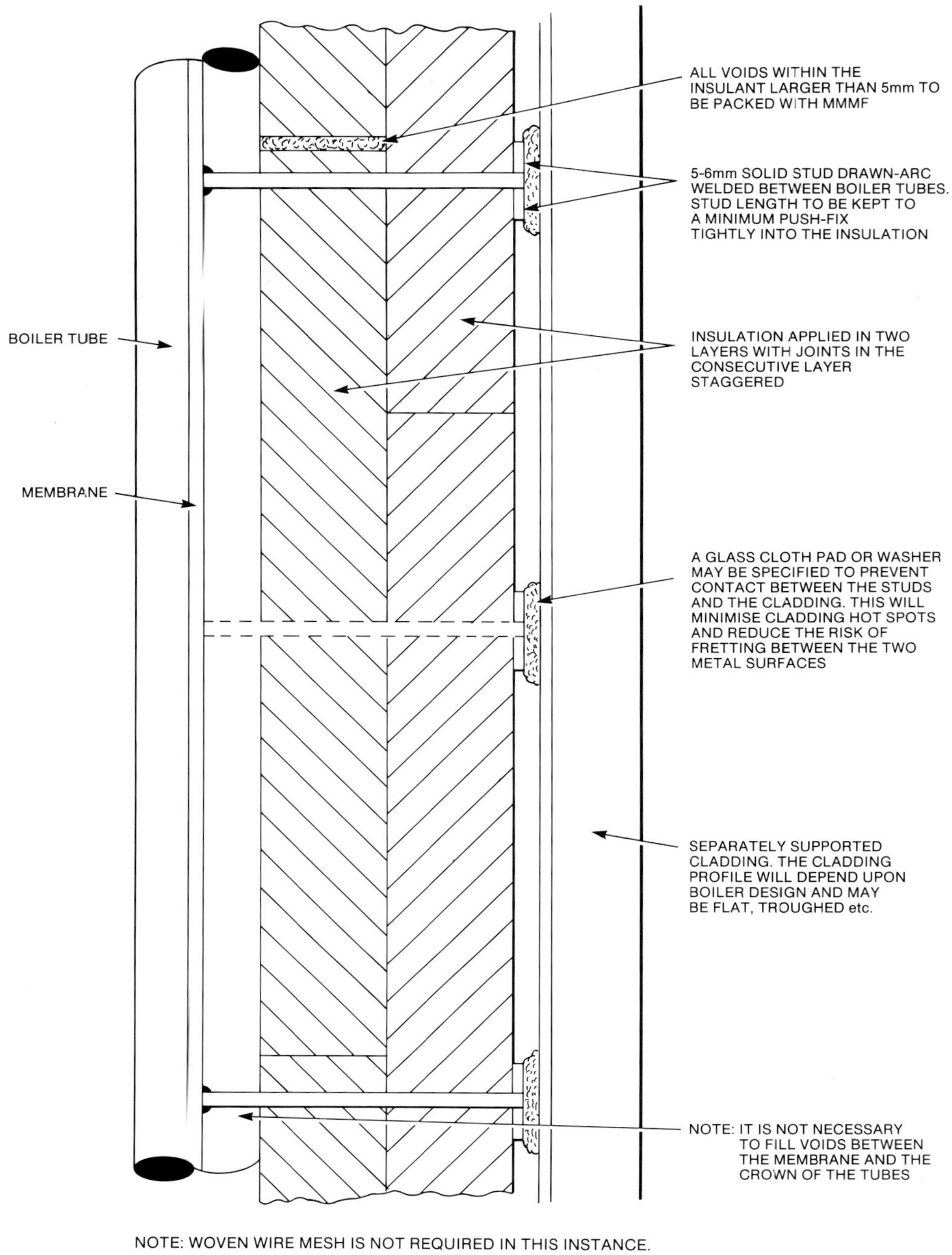

FIG. 4.30 Boiler wall thermal insulation retained by cladding

that emission reductions can be achieved without the need for the treatment of flue gases:

- Energy conservation.
- Increased combustion efficiency.
- Alternative energy sources.
- Alternative fuel sources.
- Nuclear power.
- Fuel cleaning.
- Combustion modifications.
- Flue gas treatment.

The considerable reductions that have occurred in UK sulphur dioxide emissions have mainly resulted from changes in fuel utilisation and increased efficiency of combustion. Increasing use of nuclear power, renewable energy sources and energy conservation will have important implications for the future.

The CEGB has a programme, that has now run for several years, of research and assessment to ensure

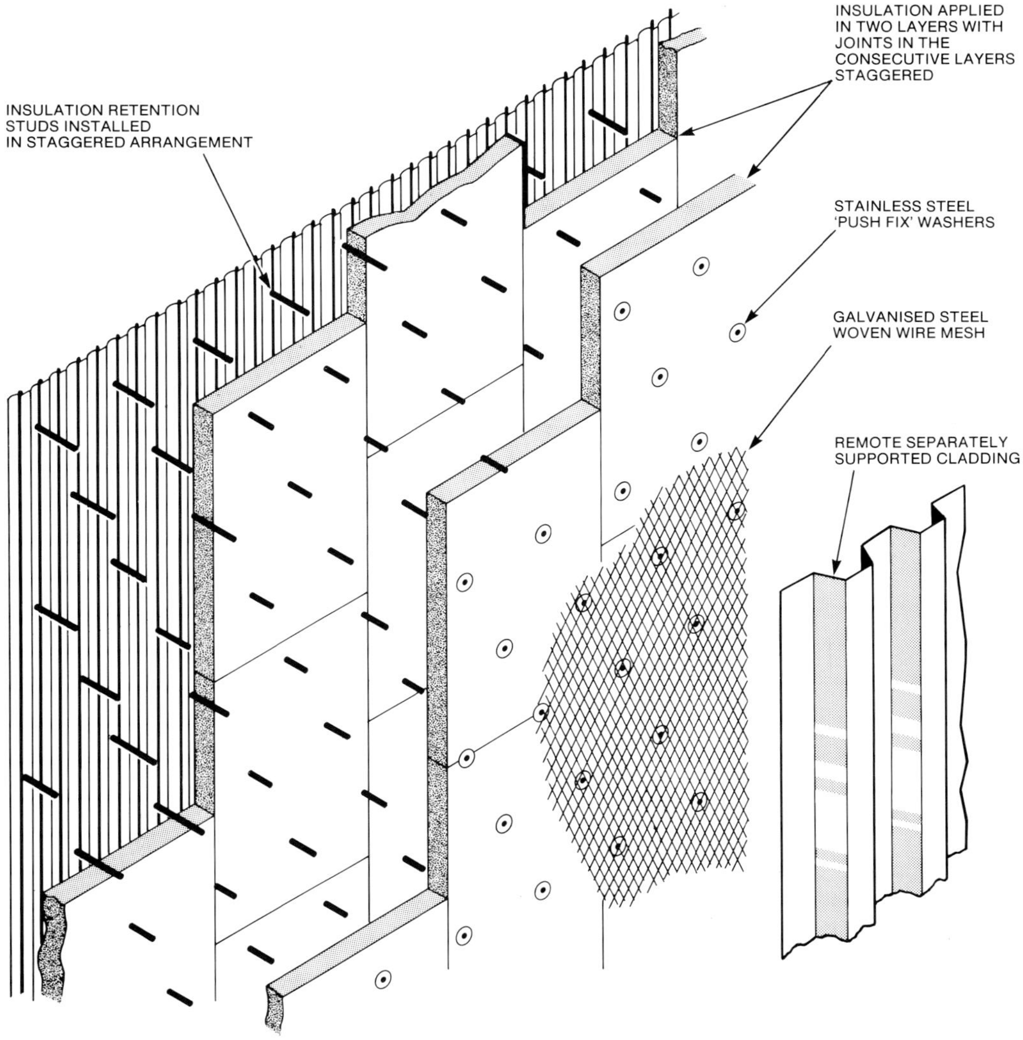

FIG. 4.31 Boiler wall thermal insulation retained by studs

that an optimised and cost effective approach is adopted towards sulphur oxides (SO_x) emission control targets once these have been set. This programme includes post-combustion removal of sulphur from flue gases, assessment of control within the boiler, precombustion removal from the fuel and advanced combustion systems such as pressurised fluidised beds and coal gasification combined cycles. It ranges in scope from detailed design and cost assessments of commercially available methods of flue gas washing to the experimental investigation of coal beneficiation techniques, some of which are scarcely beyond the conceptual stage.

There are well over one hundred different flue gas desulphurisation (FGD) processes available in various degrees of development, ranging from the laboratory scale to full scale commercial units. The main processes relative to application in the UK are the limestone-gypsum process, the regenerative process and the spray dry and seawater processes. Detailed consideration of each of these technologies for a chosen situation is required and involves optimisation of all the units and subprocesses involved. In addition careful consideration of the supply of raw material, disposal of byproducts and wastes and the interaction between capital cost, plant lifetime and energy requirements must be carefully optimised.

12.2 Wet processes

12.2.1 Scrubbing with sea/river water

The first FGD units installed in the world, at Battersea and Bankside power stations, made use of the natural alkalinity of the River Thames to absorb the sulphur oxides. The river water was cycled through the FGD plant and then, after suitable processing, discharged back to the river. The only other process in this category is that using seawater, which makes use of this principle. For a modern large power station

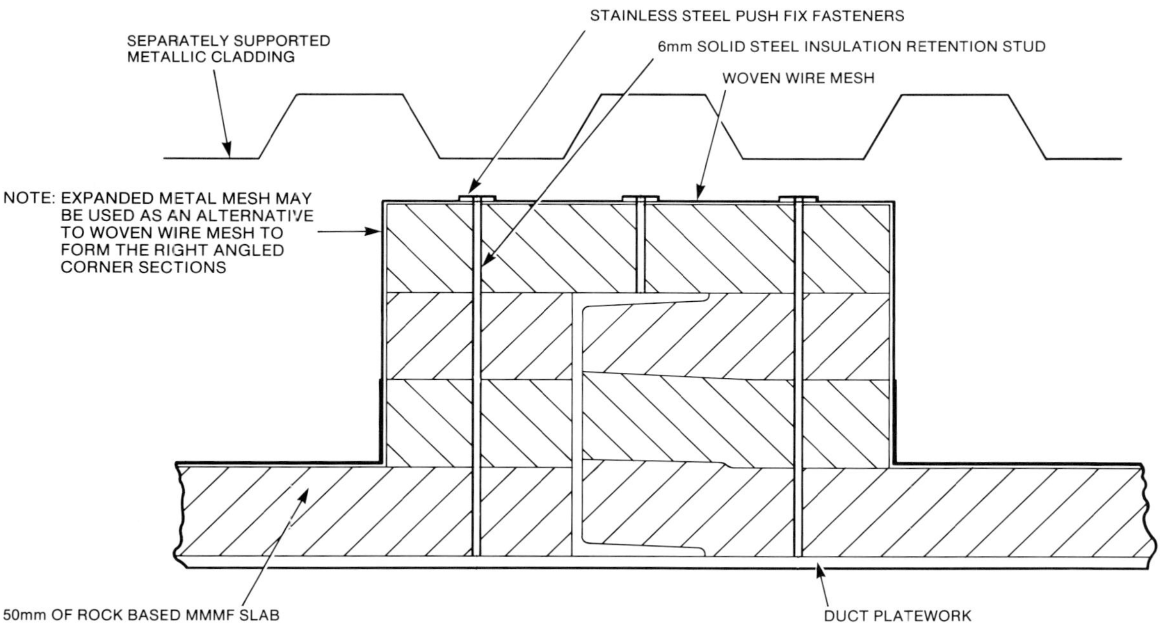

FIG. 4.32 Typical arrangement for thermal insulation of stiffening beams on flue gas ducting

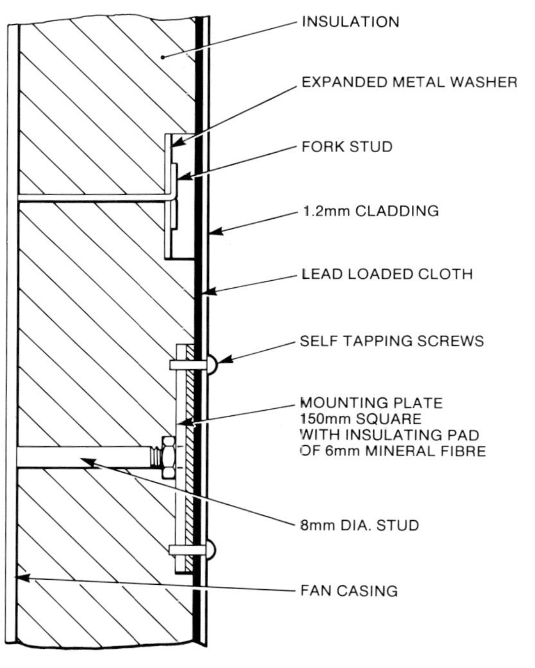

FIG. 4.33 Method of fitting acoustic insulation on fan casings

the natural alkalinity of seawater would not be sufficient and the addition of extra limestone is required. However, this process is potentially the least expensive FGD process, but is clearly site dependent and sensitive to ecological and fishing interests which may limit the nature of the disposable products and offset the potential economic advantages.

12.2.2 Throwaway processes

In these processes the alkaline absorbent liquor is recirculated in a closed loop. The reaction products are separated off and fresh absorbent is added. The only water usage is to cover losses by evaporation and that entrained with the product. In most cases a pre-cooler is used to cool the gases to avoid concentration changes in the absorber due to evaporation.

Many of the early systems used lime as the alkaline reagent because of the lower reagent utilisation in the limestone systems. However, modern limestone systems have limestone utilisations close to 100% and the marginal capital cost advantage of a lime system is heavily outweighed by the greater cost of lime (in the UK lime costs about four times as much per tonne as limestone). Thus the trend in both the USA and West Germany is towards limestone systems, whilst in Japan only one or two lime systems have ever been built.

12.2.3 Lime/limestone sludge processes

The vast majority of scrubbing systems installed in the United States use lime or limestone as the alkali to neutralise the sulphur dioxide. The resulting reaction product, a calcium sulphite/sulphate mixture is then disposed of in ponds or disused mines. This mixture is very difficult to dewater and forms a thixotropic sludge. Fixation methods, including mixing with fly ash, cement, lime, soil, etc., are often used to stabilise the sludge, but the disposal areas are generally rendered sterile and unusable for agricultural or amenity use. In addition, the cost of disposal of this sludge is high and in the UK the sludge processes are more expensive than those processes producing gypsum. There are thus strong economic and environmental reasons why lime/limestone sludge processes are not suitable for use in the UK.

12.2.4 Double alkali processes

This method avoids the use of slurries and was introduced to avoid problems of scaling in the scrubber, which can be encountered with the lime/limestone processes. The sulphur oxides are absorbed in a clear alkaline solution which is then treated with lime or limestone in a reactor outside the scrubber loop. However, since scaling problems have now been overcome in lime/limestone systems and since a calcium sulphite product is again produced, this type of process is considered unsuitable for use in the UK.

12.2.5 Gypsum processes

Since the disposal of sulphite/sulphate sludges is difficult there has been an incentive to produce a more amenable product. Gypsum is such a product which can be readily disposed of to landfill or, if produced in a suitable form, can be sold for the manufacture of plasterboard or as a cement-setting retardant.

In principle, any of the methods outlined can be made to produce gypsum by the addition of an oxidation stage in which calcium sulphite is oxidised to gypsum by air injection. Oxidation is usually carried out at a lower pH than that used for absorption.

Intentional oxidation can also be carried out in the scrubber loop. Some of these processes produce a gypsum of marketable quality and all produce a product that is suitable for landfill. Many FGD systems in Japan produce gypsum, as do most new systems being installed in West Germany. Some of the FGD process vendors in West Germany agree to take the gypsum produced as part of the construction contract.

The limestone system producing gypsum is fully commercially proven and of the throwaway systems appears to be the most suitable for UK application. Moreover, the system does not suffer the losses in energy for regenerating the sorbent.

The chemistry of the limestone-gypsum process may be described in two stages:

(a) Absorption reaction

$$CaCO_3 + SO_2 + \tfrac{1}{2}H_2O \rightarrow CaSO_3\,\tfrac{1}{2}(H_2O) + CO_2$$

(b) Oxidation reaction

$$CaCO_3\,\tfrac{1}{2}(H_2O) + \tfrac{1}{2}O_2 + 1\tfrac{1}{2}H_2O \rightarrow CaSO_4 2(H_2O)$$

In the majority of commercial systems these reactions are accomplished in two steps. One commercially available system is illustrated in Fig 4.34. In this process the flue gas, after leaving the electrostatic precipitator and induced draught fan, is boosted up in pressure by a fan to overcome the draught resistance through the FGD plant. The gas passes through a heat exchanger, usually of the regenerative type, where the gas is cooled to just above saturation temperature by the clean desulphurised flue gas stream. It enters a venturi pre-scrubber where it is washed by, and

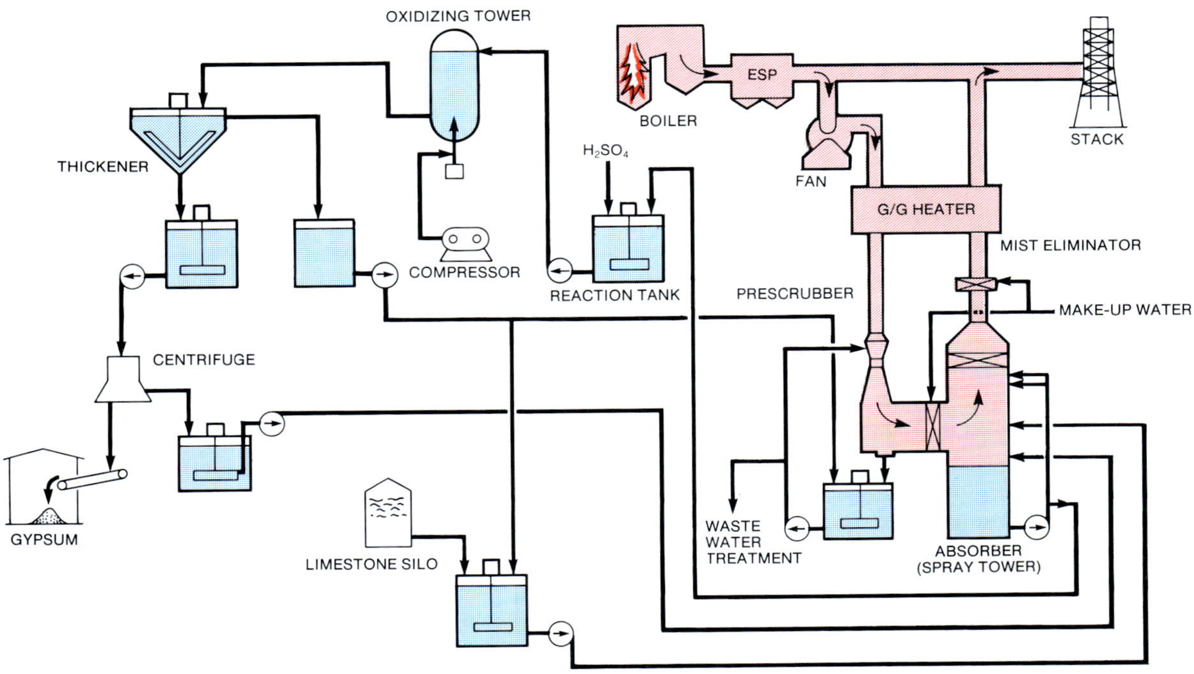

FIG. 4.34 Limestone-gypsum FGD process flow diagram

intimately mixed with, jets of water. In this unit over 95% of the halides, hydrochloric and hydrofluoric acids are absorbed and the dust removed. These are collected in the base of the pre-scrubber and removed from the system. The gas then passes into the absorber, then a series of entrainment separators, whose purpose is to remove entrained droplets of chloride-rich moisture from the flue gas.

Slurry is withdrawn from the absorber sump, pH corrected and then oxidised in a tower by air. The resultant calcium sulphate slurry is dried by centrifuging or other means and mechanically conveyed to storage.

The pre-scrubber is a vessel designed to remove the halides and dust. It is internally lined with rubber, tiles or brick to withstand the highly corrosive and erosive atmosphere. The removal of halides reduces substantially the corrosion downstream, the amount of waste water and the chloride contamination of the product; simplifies waste water treatment; precludes inhibition of the absorption reaction in the absorber. Its control is virtually independent of the rest of the process. The water is recirculated by a pump, with blowdown being made up by suitable process water. The chloride concentration is generally kept between 5000 and 20 000 PPM.

The absorber tower is lined with flake glass and stainless steel. A series of spray stages are set in the top third of the tower and the arrangement of spray nozzles ensures a complete blanketing of the gas. The sump contains the slurry which is recirculated by pumps to the sprays. Limestone slurry is fed to the sump, where it is mixed by the agitators. Slurry is withdrawn from the sump and passed to the oxidation stage. The tower is controlled independently from the pre-scrubber and oxidation stage and is regulated to respond rapidly to changes in boiler load and sulphur dioxide concentration. The stoichiometric ratio of limestone to sulphur dioxide is normally about 1.05 to 1.07 and is dependent, among other factors, on the reactivity of the limestone.

The overflow from the thickener, which contains seed gypsum, is used as absorber make-up. The seed gypsum effectively reduces the supersaturation in the absorber and thereby inhibits the formation of hard scale which is otherwise prone to form on the absorber wall.

The slurry leaving the absorber contains sulphite, sulphate and unconverted limestone. Dependent upon the limestone purity the minerals present will be mainly calcium, magnesium and aluminium. The slurry is pumped to a pH-adjustment tank where it is treated with sulphuric acid to convert the remaining limestone to calcium sulphate. It is then pumped to the oxidation tower.

This tower, which operates under pressure, has the slurry fed into the top and air injected through spinning-cone atomisers in the base. The fine bubbles of air percolate up through the tower and oxidise the sulphite to sulphate. The use of pressure accelerates the oxidation process and assists in control of the crystal size, and leads to a more compact design.

The sulphate slurry is fed to a thickener where the slurry is reduced to a water content of 80%. The thickened slurry is stored in an agitated tank and then dried by centrifuge to a water content of less than 10%. The powder is dry to touch, does not cake and can be handled by conventional means. Normally it is carried by belt conveyor to a storage shed. Typically it has a purity of 95% on a dry basis and 86% as discharged from the centrifuge. The purity depends solely on the low solubility impurities in the limestone.

12.2.6 Regenerative processes

In these processes the absorbent is chemically or thermally regenerated for re-use and a saleable product (liquefied sulphur dioxide, sulphur or sulphuric acid) is produced. Ideally, there should be no disposal problems, but in fact all the processes produce some byproducts which means that there is still a need for absorbent make-up and for byproduct disposal.

Wellman Lord process

The Wellman Lord process is based on the absorption of sulphur dioxide in a concentrated solution of sodium sulphite and its regeneration by steam stripping in an evaporator crystalliser. The process produces a pure concentrated sulphur dioxide gas stream which can be treated to produce sulphur or sulphuric acid. Some sodium sulphate is produced as a byproduct and must be disposed of. The Wellman Lord process is very popular in Japan and is operating on about 2000 MW of coal-fired power plant in the USA. It appears to be suitable for application in the UK.

Magnesium oxide process

In this process an aqueous slurry of magnesium hydroxide and magnesium sulphite is used to absorb sulphur dioxide. The resulting magnesium sulphite is then separated out, dried and calcined to release sulphur dioxide and to regenerate the reagent. The sulphur dioxide produced is suitable for conversion to either sulphuric acid or sulphur. Since the magnesium salts are separated from the absorber as a solid and solid magnesium oxide is regenerated, it is possible to separate the absorption and regeneration sites. In certain circumstances, economic savings could be made by using one regeneration complex to serve several sites. The scrubbing part of the process works well but considerable difficulties have been found in the regeneration step. If these difficulties can be economically overcome then the magnesium oxide process might compare favourably with the Wellman Lord process.

Other regenerative processes

Many other regenerative processes (e.g., the citrate and ammonium processes) have been proposed but none has been developed past the pilot plant stage. It appears unlikely that any will proceed to full scale and so these processes will not be considered further here.

12.3 Dry processes

12.3.1 Alkali injection into the combustion chamber

Not all the sulphur present in the coal ends up as sulphur dioxide in the flue gases after combustion. Some reacts with alkaline elements in the coal and the sulphur is then fixed and removed with the ash. The proportion of sulphur removed in this way depends on the amount and alkalinity of the ash present in the coal, but is generally in the UK in the range 2% to 4% of the input.

In the dry injection method additional alkali (usually limestone) is injected into the hot boiler gases where it reacts with the sulphur oxides and is then removed after reaction. Equipment for preparation of the alkali, its injection into and removal from the flue gas is required. Early trials by the US Tennessee Valley Authority on large boiler plant were unsuccessful due to the inefficient use of lime which occurred when normal PF burners were used.

Interest has, however, recently been renewed in limestone injection because of research into new types of burner for the reduction of nitrogen oxides (NO_x) emissions. These burners give lower flame temperatures and, because of the staged introduction of air, part of the flame is reducing. This leads to the possibility of increasing limestone utilisation. Even so it is very unlikely that the maximum removal efficiency achieved with realistic limestone grinds would make the system attractive. A recent review of the system undertaken by the CEGB has shown that with the twice stoichiometric rate tests on a 1 MW rig by the International Flame Research Foundation (IFRF) show that only 35% of the SO_2 is absorbed at best, whereas a limestone-gypsum process removed 90% of the SO_2 at a stoichiometry of about 1.1. Under these circumstances some two and a half stations would require fitting with this process to achieve the same absorption as one station with a 90% efficient wet process. Total operating costs have been shown to be approximately twice that of the wet system.

12.3.2 Adsorption processes

Research on adsorption processes to remove sulphur and nitrogen oxides has been conducted by Shell and Bergbau-Forschung based on copper oxide and carbon respectively. However, neither process has been fully developed to a large scale for coal-fired boilers; both have, in fact, been overtaken by technical developments in other processes.

12.3.3 Spray drying processes

During the past few years, the use of spray driers for the absorption of sulphur dioxide has attracted increasing attention. The spray drier in effect replaces the absorber, sulphur dioxide absorption taking place in the alkaline droplets created by the drier's atomiser. The alkali used can be either lime, sodium carbonate or bicarbonate. A mixture of fully reacted and unreacted absorbent is produced which is then collected in electrostatic precipitators or bag filters.

Several advantages are claimed for spray absorption among which are the minimisation of reheat, no scaling or plugging, the production of a totally dry product, and low investment. This type of system is receiving particular attention in the USA as it is claimed to be less complex and less expensive than wet scrubbers for their high calcium/low sulphur coals. However, it cannot be regarded as a commercially proven process for medium to high sulphur bituminous coals, though in the USA much work is being actively undertaken to develop the process for these coals.

The spray dry process is presently being installed in the USA on about 3500 MW of power plant which predominantly uses low sulphur coals with a highly alkaline ash content. The system is only proved on such coals though developments are taking place to prove the process on higher sulphur coals. However, the economic advantages claimed for spray drying processes will be eroded as the sulphur levels in the coals rise. In the UK, the relatively high chlorine levels in UK coals will create problems and the disposal of the products (a mixture of fly ash, calcium hydroxide, sulphite, sulphate and chloride) could be extremely difficult and therefore expensive. Unless a very high removal efficiency can be achieved, the cost of the lime absorbent will be prohibitive and will make this system unsuitable for use in the UK on all but low-load factor plant and some industrial plants.

12.4 Supply of raw materials and disposal of FGD products

A modern 2000 MW power station burns between 4.5 and 5 million tonnes of coal a year and thus the amounts of reagents required and products produced by FGD fitted to that station will be very large. Table 4.3 shows the approximate annual imports and exports in tonnes for a typical 2000 MW coal-fired power station operating on a 2% sulphur 0.4% chlorine coal at 76% load factor.

The supply of the raw materials and the disposal of the products, byproducts and wastes generated must be accomplished in an environmentally acceptable man-

TABLE 4.3
Approximate annual FGD imports/exports of a 2000 MW coal-fired power station

PROCESS		WELLMAN-LORD	LIMESTONE/ GYPSUM	LIME/ SPRAY-DRY
SO_2 removal efficiency		90%	90%	90%
IMPORTS				
Limestone	Solid	30000 t	400000 t	—
Lime	Solid	3000 t	3000 t	375000 t
Caustic soda	Liquid	40000 t	—	—
Natural gas	Gas	3×10^7 Nm^3	—	—
Sulphuric acid	Liquid	—	30000 t	—
EXPORTS				
Sulphur	Solid	75000 t	—	—
or				
Sulphuric acid	Liquid	280000 t	—	—
Gypsum	Solid	—	560000 t	—
Mixed products (including fly-ash)	Solid	—	—	1.5×10^6 t
BY-PRODUCTS				
Calcium chloride	Solid	25000 t	25000 t	—
Sodium sulphate	Solid	30000 t	—	—
WASTE WATER				
Treated water	Liquid	2×10^6 t	2×10^6 t	—
Waste-water sludge	Solid	15000 t	40000 t	—

(76% load factor, no sulphur retention in ash, coal 2% S, 0.4% Cl)

ner. For this reason it is the intention of the CEGB to market as much of the products of flue gas desulphurisation as possible. However, even for useful products such as gypsum and sulphuric acid, there are only limited markets. For example, a 2000 MW coal-fired station could produce the following FGD products as percentages of the existing UK market:

Sulphuric acid : 10–12%

Sulphur : 10–12%

Gypsum : 15–20%

For sulphuric acid it would be difficult to put the equivalent of one power station's FGD plant output into a single UK chemical works. Many sulphur burning sulphuric acid plants can only be turned down to approximately half capacity without upsetting the integrated operation of the whole chemical works. Thus complete replacement of the work's sulphuric acid demand with the FGD product cannot be undertaken without considerable plant and operational modifications. Similar implications exist for gypsum production where a single 2000 MW station would produce over 500 000 tonnes/year of marketable gypsum.

Two very important consequences arise from the consideration of the imports and exports of these FGD processes. The first is that for any substantial FGD programme it is likely that more than one type of FGD process will have to be employed. The second consequence is that the location of a particular power station may strongly influence the process to be used due to the local marketability of the product and the acceptability of any waste product. The optimisation of an overall FGD programme should be based on the requirements of all the stations in the system.

13 References

[1] BS893: 1978: Method for the measurement of the concentration of particulate material in ducts carrying gases: 1986

[2] BS2742: 1969: Notes on the use of the Ringelmann and miniature smoke charts: 1986
BS2742C: 1957 Ringelmann chart: 1986

[3] BS848: Fans for general purposes Part 1; methods of testing performance: 1980

[4] BS3921: Specification for clay bricks: 1985

[5] BS4360: Specification for weldable structural steels: 1986

[6] BS1501: Steels for fired and unfired pressure vessels; Plates Part 2; Alloy Steels; Imperial Units: 1970

[7] BS Code of Practice 3: Chapter V; Part 2; Wind Loads: 1972

SUBJECT INDEX